BROADBAND ACCESS TECHNOLOGIES

OTHER MCGRAW-HILL TELECOMMUNICATIONS BOOKS OF INTEREST

Ali *Digital Switching Systems*
Ash *Dynamic Routing in Telecommunications Networks*
Benner *Fibre Channel*
Best *Phase-Locked Loops,* Third Edition
Faynberg *Intelligent Network Standards*
Feit *TCP/IP, Second Edition*
Gallagher *Mobile Telecommunications Networking with IS-41*
Goralski *Introduction to ATM Networking*
Harte *Cellular and PCS: The Big Picture*
Harte *GSM Superphones*
Heldman *Information Telecommunications*
Heldman *Competitive Telecommunications*
Kessler *ISDN, Third Edition*
Kuruppillai *Wireless PCS*
Lachs *Fiber Optic Communications*
Lee *Mobile Cellular Telecommunications,* Second Edition
Lee *Mobile Communications Engineering,* Second Edition
Logson *Mobile Communication Satellites*
Macario *Cellular Radio, Second Edition*
Muller *Desktop Encyclopedia of Telecommunications*
Pecar *Telecommunications Factbook*
Roddy *Satellite Communications,* Second Edition
Rohde et al. *Communications Receivers,* Second Edition
Russell *Signaling System #7,* Second Edition
Russell *Telecommunications Protocols*
Simon et al. *Spread Spectrum Communications Handbook*
Smith *Cellular Design and Optimization*
Smith *Practical Cellular and PCS Design*
Tsakalakis *PCS Network Deployment*
Turin *Digital Transmission Systems*
Winch *Telecommunication Transmission Systems,* Second Edition

Broadband
Access
Technologies:
ADSL/VDSL, Cable
Modems, Fiber, LMDS

Albert Azzam
Niel Ransom

McGraw-Hill
New York San Francisco Washington, DC Auckland
Bogotá Caracas Lisbon London Madrid Mexico City
Milan Montreal New Delhi San Juan
Singapore Sydney Tokyo Toronto

McGraw-Hill

*A Division of The **McGraw·Hill** Companies*

4 5 6 7 8 9 0   DOC / DOC   0 9 8 7 6 5 4 3 2 1 0

ISBN 0-07-135060-8

The sponsoring editor for this book was Stephen Chapman and the production supervisor was Sherri Souffrance. It was set in Vendome by Pro-Image Corporation.

Printed and bound by R. R. Donnelley & Sons Company.

This book was printed on recycled, acid-free paper containing a minimum of 50% recycled de-inked fiber.

Dedication

Albert Azzam—to my wife Elizabeth, and my children JoAnna, Kathleen, Greg and Michael

M. Niel Ransom—to my mother, Ann Ransom, and my loving wife Ellen

ISDN: Integrated Service Digital Network.

① BRI (Basic Rate Interface)

$$2 B (64 kb/s \times 2) \quad \} \quad 144 kb/s$$
$$1 D (16 kb/s$$

② Primary Rate Interface (PRI)

$$23 B + 1D = 23 \times 64 + 1 \times 64 = 1536$$

in UK $30 B + 1 D = 1984 \, kb/sec$

③ mutiple PRI

④ H channel.
$$H\phi = 384 kb/s (6 B \, chnne)$$
$$H10 = 1472 kb/s (23 B)$$
$$H11 = 1536 (24 B)$$
$$H12 = 1920 (30 B) = E1 (only)$$

CONTENTS

Preface   xvii
 Contents   xix

Access Introduction   **1**

 Introduction   1
 Types of Broadband Access Networks   2
  Digital Subscriber Lines   3
  Cable Modems   4
  Optical Fiber   5
  Radio Access Networks   5
  Electric Power Lines   6
 Broadband Home Networks   7
  Paying for Broadband Access Networks   8
 Networking Protocols for Broadband Access   10

Chapter 1 **Network Evolution and Regulatory Trends**   **13**

 1.1 Introduction   13
 1.2 A Brief History of Telecommunication Competition   14
 1.3 International Deregulation   16
 1.4 Network Unbundling and Resale   16
  1.4.1 Unbundled DSL Loops   18
  1.4.2 Spectrum Unbundling   19
  1.4.3 Sub-loop Unbundling   20
  1.4.4 Technical Issues of Loop Unbundling   21
 1.5 Cable Competition   22
 1.6 FCC Cybernet Policy   22
  1.6.1 Voice Over the Internet and Regulation   23
  1.6.2 Birth of Internet Telephony   24
  1.6.3 Policy Perspectives   24
  1.6.4 Economic Model, Regulation and Chaos   25
  1.6.5 Technical Challenges of Voice over the Internet   27
  1.6.6 Future Directions for Internet Telephony   28

Chapter 2   **The Grand Unification Network**   **29**

2.1 Introduction   29

2.2 Forces Driving Network Unification   30

    2.2.1   The Technologies of Convergence   31

    2.2.2   Economies of Scope   32

    2.2.3   Customer Demand for Converged Services   33

    2.2.4   Changes in Regulatory Structure   34

2.3 The ITU Long Term Architecture Study   34

    2.3.1   Trends Toward Future Telecommunications   34

    2.3.2   Common Trends in Industrialized Countries   35

    2.3.3   Network Trends   36

    2.3.4   Network Aspects for Multimedia Services   37

    2.3.5   Overview of the LTA Business Model   38

2.4 PSTN   39

    2.4.1   Traditional PSTN   39

    2.4.2   Tomorrow's PSTN   45

    2.4.3   ATM and Broadband ISDN   48

2.5 Internet   71

    2.5.1   Internet Historic Perspective   71

    2.5.2   Internet Commercial Landscape   75

    2.5.3   Current Internet Structure   77

    2.5.4   The Next Generation Internet   79

    2.5.5   Internet2   84

2.6 Cable Networks   86

    2.6.1   A Case for Cable Network Modernization   86

    2.6.2   HFC: The Next Generation Cable Network   87

    2.6.3   HFC Access Shortfalls   89

    2.6.4   HFC Business Case   90

    2.6.5   HFC and Cable Modem Penetration   91

    2.6.6   HFC Deployment Issues   92

    2.6.7   Voice Over HFC Solutions   94

2.7 Layer 2/Layer 3 War   96

2.8 Layer 2 Switching vs. Layer 3 Switching War   97

Contents

2.8.1   Cost   99

2.8.2   End-to-End Reliability   99

2.8.3   Dumb Versus Smart Networks   100

2.8.4   Switching Techniques and Efficiency/ Scalability   101

2.8.5   And the Winner is . . .   103

Chapter 3   The Fiber Solution: ATM Passive Optical Networks   105

3.1   Fiber's Manifest Destiny   105

3.2   Impediments of Fiber Access Networks   106

3.3   Types of Fiber Access Networks   108

3.4   The Full Services Access Network Industry Group   110

3.5   ATM Passive Optical Networks   112

    3.5.1   Physical Layer Operation   114

    3.5.2   APON Bandwidth Management   117

    3.5.3   APON Ranging Protocol   118

    3.5.4   Use of Encryption on APONs   119

3.6   Prevention of Babbling ONUs   120

3.7   Trials and Deployments of APON Systems   120

Chapter 4   ADSL and VDSL—The Copper Highway   123

4.1   Introduction   123

4.2   Evolution of Capacity on Twisted Pairs   125

4.3   Twisted Pair Impairments   127

4.4   ADSL   130

    4.4.1   System Requirements Reference Model   130

    4.4.2   Performance   131

    4.4.3   Transfer Mode   131

    4.4.4   DMT Transmission   131

    4.4.5   Spectrum and Bit Allocation   132

    4.4.6   Error Correction   133

    4.4.7   Bit Rate Adaptation   133

    4.4.8   Characteristics of ADSL   134

    4.4.9   Single Carrier Transmission (RADSL)   140

    4.4.10   ADSL-Lite   141

4.5   VDSL   143

    4.5.1   System Requirements Reference Model   143

    4.5.2   Transfer Mode   144

	4.5.3 Performance	144
	4.5.4 Transmit Spectrum	145
	4.5.5 Power Consumption	145
	4.5.6 Transmission Techniques	146
4.6	Conclusions	150

Chapter 5 Hybrid Fiber Coax and Cable Modem **151**

5.1	Overview	151
5.2	Market Pull/Technology Push	152
5.3	Cable Network and Evolution to HFC	153
	5.3.1 History of the Cable Network	153
	5.3.2 Legacy Cable Network	154
	5.3.3 HFC Network	156
	5.3.4 Upstream/Downstream Cable Spectrum	157
	5.3.5 Digital Cable Network	158
	5.3.6 Cable Network Modernization Effort	159
	5.3.7 HFC Access Shortfalls	159
	5.3.8 Factors Influencing Cable Modem Operation	160
	5.3.9 Noise	160
	5.3.10 Approaches to Noise Suppression	163
5.4	Cable Modem	164
	5.4.1 ATM-Centric vs. IP-Centric Cable Modem	165
	5.4.2 Abstract Cable Modem Operation	167
	5.4.3 Cable Modem Reference Architecture	169
	5.4.4 Cable Modem Fundamental Layers	171
	5.4.5 Cable Modem Operation (Service Perspective)	185
	5.4.6 High Speed Physical Layer	187
	5.4.7 Future DOCSIS/IEEE 802.14 Milestones	189

Chapter 6 High-Speed Wireless Access **191**

6.1	Introduction	191
	6.1.1 Technology Pull, Market Push	192
	6.1.2 Organization of this Chapter	193
6.2	Satellite Constellation Fundamentals	194
	6.2.1 GEO Satellite	194
	6.2.2 MEO Satellite	195

6.2.3   LEO Satellite   196
6.3 SkyBridge   197
        6.3.1   SkyBridge History   197
        6.3.2   SkyBridge Constellation   197
        6.3.3   The SkyBridge Advantage   198
        6.3.4   SkyBridge Architecture   199
        6.3.5   SkyBridge Costs and Partners   202
6.4 Teledesic   202
        6.4.1   Teledesic History   202
        6.4.2   Teledesic Constellation   203
        6.4.3   Teledesic Architecture   204
        6.4.4   Teledesic Costs and Partners   205
6.5 Iridium   206
        6.5.1   Iridium History   206
        6.5.2   Iridium Constellation   207
        6.5.3   Iridium Architecture   207
        6.5.4   Iridium Operating Frequency   209
        6.5.5   Iridium Costs and Partners   209
6.6 LMDS   210
        6.6.1   LMDS Architecture   211
        6.6.2   LMDS Enablers   214
        6.6.3   The LMDS Business Case   214
        6.6.4   FCC Licensing   215
        6.6.5   LMDS Standardization   216
6.7 DBS   216
        6.7.1   Introduction   216
        6.7.2   DBS Architecture   217
        6.7.3   Compression Technique   218
        6.7.4   DBS Services   219
        6.7.5   FCC Regulation   219

Chapter 7   **Alternative Access Technologies: Power
            Line Carrier**   **221**

7.1 Introduction   221
7.2 Historical Perspectives   223
7.3 Power Lines as a High-speed Transmission
     Media   224
7.4 A Power Line Transmission Architecture   225
        7.4.1   Basestation Transformer   226
        7.4.2   Local Base Station   226

	7.4.3	Switching Node	227
	7.4.4	Fuse Panel	227
	7.4.5	Communication Module	227
7.5	Noise on Power Lines	227
7.6	Physical Layer Transmission	228
	7.6.1	Multifrequency Modulation	228
	7.6.2	Spread Spectrum Modulation	228
7.7	Media Access Control and the Data Link Layer	229
7.8	Applications of Powerline Data Communications in the United States	230
	7.8.1	Data over Powerline Application in Europe	231
	7.8.2	The Future of Powerline Carrier	232

Chapter 8   **Home Networking**   **235**

8.1	Introduction	235
8.2	Categories of Home Networks	237
8.3	HomeRF LAN	238
	8.3.1	HomeRF System Concept	238
	8.3.2	HomeRF Network Architecture	239
	8.3.3	Physical Plane	240
	8.3.4	Control Plane	241
	8.3.5	HomeRF Advantages and Estimated Cost	241
8.4	Powerline Ethernet	242
	8.4.1	Powerline Ethernet Applications	243
	8.4.2	Powerline Ethernet System Concept	244
	8.4.3	Physical Layer Technology Alternatives	245
	8.4.4	Control Layer Technology Alternatives	247
8.5	Home Phoneline	248
	8.5.1	Home Phoneline System Concept	249
	8.5.2	Phoneline Access Requirements	250
	8.5.3	Phoneline Network Architecture	251
	8.5.4	Physical Layer	251
	8.5.5	Phoneline: Services, Advantages/Disadvantage and Cost	253
8.6	Coax-based Home Networks	254
	8.6.1	Coax-Based Home Network Architecture	254
	8.6.2	Advantages and Disadvantages	255

| | 8.6.3 Costs | 256 |
| | 8.7 And the Winner Is . . . | 256 |

| **Chapter 9** | **Services and Advanced Applications** | **257** |

	9.1 Introduction	257
	9.1.1 Organization of this Chapter	258
	9.2 Services Application Definition	258
	9.3 Services Aspects	259
	9.3.1 Service Classifications	259
	9.3.2 ATM Service Architecture	261
	9.4 Application Domain	265
	9.4.1 Key Legacy Applications	265
	9.4.2 Ongoing Development	269
	9.4.3 Applications Under Research	275

| **Chapter 10** | **Access Performance Aspects** | **279** |

	10.1 Overview	279
	10.1.1 Basic Definition of Performance Terms	280
	10.1.2 Organization of This Chapter	281
	10.2 Network Performance Aspects	281
	10.3 Performance Requirements and Infrastructure Evolution	282
	10.3.1 How Many Nines are Enough?	283
	10.3.2 Network Performance Rationale	285
	10.3.3 Present Research Consortiums	289
	10.3.4 International Participation	290
	10.4 Likely Applications	290
	10.5 ADSL Technology Brief Overview	292
	10.5.1 ADSL: A Technology on Unconditional Twisted Pair	293
	10.6 Cable Modem Technology Overview	296
	10.6.1 Summary Description of the Cable Modem	297
	10.7 ADSL/Cable Modem Comparison	297
	10.7.1 Capacity	298
	10.7.2 Throughput	298
	10.7.3 Scalability	299
	10.7.4 Performance/Service Categories	299
	10.7.5 Security	300

10.7.6	Cost	300
10.7.7	Voice Adaptation	301
10.7.8	Reliability	301
10.7.9	Internet Application Comparison Scenario	301
10.7.10	Cable Modem Market Size	303
10.7.11	ADSL Market Size	303
10.8 The Likely Winners		305

Chapter 11   Standards — **307**

11.1 The Role of Standardization		307
11.2 Access Related Standards		308
11.2.1	ADSL/VDSL	308
11.2.2	Cable Modem/HFC	311
11.2.3	Data Over Cable Interface Specification (DOCIS) Project	313
11.2.4	IEEE 802.14	315
11.2.5	CableLabs	317
11.2.6	ATM Forum	318
11.2.7	DAVIC	319
11.2.8	Internet Related Standards	321
11.2.9	Abeline	323
11.2.10	Consortium/Organizations and Standards	323
11.2.11	XIWT	323
11.2.12	ISOC	324
11.2.13	VON	324
11.2.14	IETF	325
11.2.15	Bellcore	325
11.2.16	ECTF	326
11.2.17	IMTC	326

Chapter 12   Major Players — **329**

12.1 Introduction		329
12.2 RBOCs		329
12.2.1	SBC Communications	329
12.2.2	Ameritech	330
12.2.3	Bell Atlantic Corporation	331
12.2.4	BellSouth	331
12.2.5	US West	332

12.3	MSOs	333
	12.3.1 TCI Group	333
	12.3.2 Cablevision	334
	12.3.3 Comcast Corporation	334
	12.3.4 Continental Cablevision	334
	12.3.5 Cox Communications, Inc.	335
	12.3.6 Time Warner	335
12.4	Internet Service Providers	335
	12.4.1 UUNET	337
	12.4.2 AOL	338
	12.4.3 BBN	338
	12.4.4 PSINet	339
12.5	Competitive Local Exchange Carriers	339
	12.5.1 Time Warner Telecommunications	339
	12.5.2 Frontier	340
	12.5.3 Allegiance	340
	12.5.4 Level 3	341
	12.5.5 NextLink	341
	12.5.6 McLeodUSA	342
Terminology		**343**
Acronyms		**361**
Bibliography		**365**
Index		**369**

PREFACE

The information age is upon us. Multimedia applications are quickly becoming an integral part of our lives not only in the workplace but also at home. Integrated voice, video and data has moved from clever technical concepts to market reality. Bandwidth hungry applications have already outpaced the capabilities of traditional dialup access networks, and entrepreneurs are busily developing even more sophisticated applications that will push the limits of what even recently has been considered high-performance networks. Delays in accessing complex web pages, audio files or video clips have given the Internet the derisive nickname, "World Wide Wait." Yet despite these performance limitations, the Internet continues its explosive growth revealing a strong market demand for multimedia information services. Today's users are becoming very sophisticated about network performance, and these users are demanding, and have shown a willingness to pay, for higher performance network services.

One can classify today's networks, by and large, as service specific. Telecommunication networks were designed and deployed to handle voice traffic. Their switching and transmission facilities were optimized to handle voice traffic efficiently. Conversely, Cable TV networks were optimized for one-way broadcast of analog video entertainment signals. The Internet network was initially developed and optimized for data transport—bursty traffic without real-time constraints. Service specific networks were well-matched to the environment of the past where user communication needs were segmented, where different technologies were cost-effective for different traffic types, and where regulation forced separation of communication industries. All of these factors have undergone dramatic change.

Business users are now demanding to set up video conferences with shared whiteboards, video clips and computer generated slides—all with the ease of a telephone call. Residential users, now comfortable with the interactive multimedia experience of CD-ROMs, are demanding this same experience in communication services, from interactive television to network games with real-time voice and video communications between players. Network operators whose networks are constrained to one traffic type will find themselves increasingly locked out of high revenue multimedia services.

Service specific networks cannot achieve economies of scope, that is, economies gained by carrying multiple traffic types on the same network. In the past, the information format of the various traffic types put very different demands on the underlying networks: 3 kHz analog for voice, 6 MHz analog for entertainment video, and bursty bit streams for data. Now all traffic types are being carried in the common format of digital bit streams differing only in their bit rates, burstiness, delay tolerance and error rate tolerance. These differences can be accommodated in emerging switching systems and routers allowing cost-effective networks to be built capable of carrying all of today's traffic types and likely future traffic types as well.

Until recently, legal and regulatory constraints kept differing traffic types on their own networks. Telephone companies were forbidden from providing CATV traffic in the same areas they provided voice traffic. The FCC's Computer Inquiry II ruling forced telephone companies to provide advanced data service from structurally separated subsidiaries. The result was that even if these economies of scope existed in the past, network operators could not take advantage of them.

Rivalry between the data, telco, and CATV operators can be traced back even before 1982 when the MFJ, overseen by U.S Federal Judge Harold H. Greene, broke the Bell System into AT&T and seven Regional Holding Companies (RHCs) with their regulated Bell Operating Companies. Since then, both the RBOCs and the CATV operators have lobbied the U.S. Congress to allow them into new markets (while maintaining own their turf to the extent possible). That, of course, changed with the passing of the Telecommunication Act of 1996. The principal objective of Congress in passing this Act was to create a competitive environment in which traditional and emerging communications providers would build networks of interconnected networks supporting new innovative multimedia services on a widespread basis. With few exceptions, the Local Exchange Carriers can now provide video programming and CATV operators can market voice telephony over cable. In its role as implementer of the Act, the FCC is issuing regulations encouraging new Competitive Local Exchange Carriers to build alternative access networks.

The Internet is the wild card that is playing a pivotal yet confusing role in this regulatory arena. Congress and the FCC have thus far kept the Internet free of regulation and support fees. However, the Internet is playing havoc among regulators and network operators as it begins to carry traditionally regulated services such as telephone services.

The result of these market, technology and regulatory changes is that network operators are scrambling to build new multimedia networks. The most challenging portion of this construction is the access network. Emerging access technologies hold the promise of providing individual users with high-speed and flexible network access at low cost. However, traditional network operators understand that while none of their service specific networks can support integrated interactive services today, it is not economically feasible to consider full access network replacement. Some means are needed for reusing their existing coax and twisted pair facilities. New competitive carriers do not face this constraint, but they are faced with the daunting problem of installing an access network after the streets are paved and the bushes and trees are planted.

A host of broadband access technologies are emerging to meet this challenge. ADSL and cable modems allow telephone CATV companies to provide high-speed packet data over their twisted pair and coax networks. A number of technologies are emerging to help new operators overbuild existing networks such as Local Multipoint Distribution Service, power line carrier, Direct Broadcast satellites, and Low Earth Orbit satellites. New fiber access technologies such as ATM Passive Optical Networks promise the ability to build economical access networks with almost unlimited capacity. The choice of a particular access technology will depend on the demography, business model, competitive conditions and a host of other factors. The choice of access technologies may differ from city to city and town to town.

In light of this, network operators must deal with ever more complex issues when planing the deployment and evolution of their access networks. Indeed, network *revolution* may be more characteristic of the future industry as network operators who endeavor to provide multimedia services meet stiff competition while remaining viable in the market place.

The goal of this book is to describe technical aspects of the principal broadband access technologies available or becoming available in the next few years. The relative advantages of these technologies are discussed. Various market and field trials of these technologies that have taken place are described.

Contents

The book is composed of 12 chapters and an Introduction. The first two chapters provide a background of the technical, economic and regulatory forces that will guide the deployment of broadband access networks. The subsequent five chapters describe the major alternative technologies under development for broadband access networks. The remaining chapters of the book deal with issues surrounding broadband access network deployment such as applications, performance requirements and standards.

The **Introduction** considers legacy networks and the forces that are driving their owners to consider making the heavy investments necessary to upgrade these networks to support interactive broadband services.

Chapter 1 describes network evolution and migration strategy in the regulatory environment of the Telecommunications Act of 1996. This chapter focuses on the how well the Act is working specifically with regard to unbundling. Focus is given to the difficult issue of loop unbundling for Digital Subscriber Line services. Voice-Over IP (VOIP) is highlighted along with its effect on arbitrage international calling. Congressional actions and FCC rulings in regulating this industry will have a profound impact on how each network will evolve to accommodate the emerging interactive multimedia services.

Chapter 2 describes the concept of a grand unification network architecture, one that encompasses all services in a common network protocol. It discusses the forces that are driving networks toward this goal. The networks which the major operators have in place today are described as well as plans to evolve these networks to this unified network goal. The chapter discusses the relative strengths of ATM vs. IP to provide this network unification.

Chapter 3 focuses on emerging fiber-based broadband access networks. The challenges of deploying a Fiber-to-the-Home access network are discussed. The chapter describes technologies under development which address these challenges, in particular ATM Passive Optical Networks (APON).

Chapter 4 gives a general description of the ADSL and VDSL. It highlights the copper highway and the surprising capacity of the twisted pair. It discusses the System Reference model, performance, DMT transmission, spectrum, bit allocation and error correction. The chapter also describes the new ADSL-lite transmission technique.

Chapter 5 gives a brief history of the cable network and the modernization plan to HFC (Hybrid Fiber Coax) needed for bi-directional communication. The chapter also describes in detail the high-speed cable modem, the physical layer modulation techniques, the MAC (Medium Access Control) Layer needed to accommodate bi-directional traffic in a shared medium environment with a notion of quality of service. The new advanced physical layer modulation techniques for the downstream and upstream channels is described. Both the IP-centric modem, developed by DOCSIS, and the ATM-centric cable modem, developed by IEEE 802.14, are described.

Chapter 6 describes emerging high-speed wireless access technologies. The chapter will focus on landline cellular technologies such as MMDS and LEOS (Low Earth Orbiting Satellite) systems such as SkyBridge and Iridium.

Chapter 7 describes unconventional ways companies are addressing the broadband access market. It focuses on the role the electric utility companies may have in this market through use of their electric power line infrastructure. The chapter describes the challenges of sending high-speed data over power lines and describes technologies that have been developed to make this possible.

Chapter 8 describes broadband premises networks that are emerging as natural complements to broadband access networks. The various types of premises networks are described according to the underlying physical media upon which they are based: phone wires, CATV coax, wireless and powerline carrier.

Chapter 9 describes the emerging network services and advanced applications now under active research by the Internet2 research community as well as the White House sponsored (NGI) Next Generation Internet Consortium. Other advanced applications are highlighted such as WWW interactive, electronics commerce, residential and business video conferencing, and interactive distant learning.

Chapter 10 analyses the performance aspects of the various access technologies, particularly as they apply to voice-over IP and other time critical applications. Moore's law will also be described in the context of network performance. An objective comparison of the various access solutions is given.

Chapter 11 describes the various standard bodies who are involved and contributing to the development and standardization of all the access technologies described in this book. The status of the specifications and controversial issues surrounding these standards is described. A guide is provided on how to obtain specifications and recommendations from the various organizations such as the ADSL Forum and ATM Forum.

Chapter 12 concludes by providing a survey of the major network operators who will be deploying broadband access networks. They include the Regional Bell Operating Companies, the CATV Multiple System Operators, the Internet Service Providers, and the Competitive Local Exchange Carriers.

ACKNOWLEDGMENTS

I wish to thank my colleagues who directly contributed to this edition of the book: Nada Golmie of NIST, for her contribution on MAC performance analysis in chapter 5, and JoAnna Azzam for her contributions on Internet history and other BOC materials researched for the book.

Access Introduction

Introduction

The world's access networks are about to undergo a broadband overhaul. The telephone and CATV networks of the past cannot meet the networking needs of the future. This is seen in the major broadband deployments underway by existing network operators and by their upstart challengers.

What forces have come together to force what may become one of the most expense network upgrades in communications history? Probably the primary force has been technological advances. Simply put, it is possible to do things economically today that was not possible at any price a few years ago. Computers with over 200 MIPS, 64 M byte of RAM, a 4 G byte hard drive, CD ROM drive, and 56 kb/s modem are available for less than $1000. Digital signal processors are compressing everything from voice, images and video into lower rate bit streams than were ever thought feasible. And this same digital signal processing technology is allowing lowly telephone wires to carry 100 Mb/s or more within walls and several Mb/s over the wires to one's house.

Radio technology has become efficient, compact and cheap. The result of these technological advances is that it is possible to build access networks capable of supporting interactive broadband services and to do so for little more than the cost of old fashioned telephone and CATV access networks.

The second major force has been the Internet which finally solved the chicken and the egg problem. For years communication companies have attempted to build videotext and teletext systems giving users online access to news, information, shopping and entertainment. The problem was that once these systems were built, there was not enough good material online to hold the users' interest. Meanwhile, potential content providers looked at the small number of users on these pilot systems and declared them insufficient to spend time and money creating online content. At last both the content and the users are there. Content providers are fighting with each other for audience attention and are doing so by spicing up their sites with animations, video clips and video. Users find this material compelling but are frustrated with the limited speed of access. As well, they have indicated that they are willing to pay for it when it becomes available.

The third major force is competition. The legal and regulatory restrictions that had kept the telephone and cable companies away from each other's throats are now history. And those are not the only ones want to be king of the access mountain. Several low earth orbit satellite companies believe they can bring economical broadband access to businesses. The FCC has released 1 GHz of spectrum around 28 GHz into the market that upstart companies will be using to offer Local Multipoint Distribution Services to these same businesses. Even the electric utilities are considering joining the fray with telecommunication services over power lines.

Since it is easier to hold onto a customer than gain a new one, each of these companies are showing a willingness to invest to gain an early market lead.

Types of Broadband Access Networks

To meet the needs of these existing or aspiring network operators, a number of broadband access technologies are either under development, in trials or in some cases in active deployment. The result may well be multiple competing broadband bit pipes brought into the home, strong competition and consumer choice. These networking technologies are the subject of the subsequent chapters.

Digital Subscriber Lines

The same wires carrying telephony into the home can be used to carry high-speed data. This is illustrated in Fig. 1. By stripping off this data in the central office before the line reaches the telephone switching system, two benefits are achieved.

First, the amount of data that can be sent is greatly increased as it is the voice switch and not the telephone line that primarily limits how much data can be sent. Secondly the telephone switching systems and trunks are spared those long holding time data calls.

There are a variety of types of Digital Subscriber Line (DSL) techniques to choose from. Since most users download rather than upload data, two types—Asymmetrical Digital Subscriber Line (ADSL) and Very High Speed Digital Subscriber Line (VDSL)—provide higher speeds downstream than upstream. Depending upon the loop length, DSL systems can provide speeds from 128 kb/s up to 52 Mb/s.

Early DSL trials concentrated on switched digital video. There were a number of reasons these were not successful: video coders at the time were expensive; the 1.5 Mb/s video in these trials was not very good; only one TV set per home could be supported; and the cost of the system, including video storage systems, video switches, etc. was high relative to the alternative of the $3 movie rental at the video store.

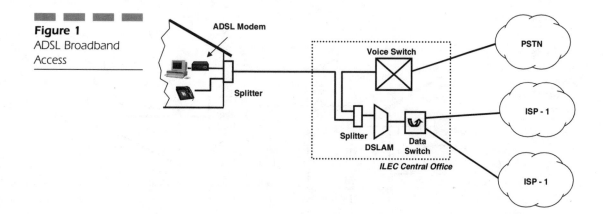

Figure 1
ADSL Broadband
Access

The current emphasis of ADSL deployment is on high-speed Internet access and there the technology seems to have found a ready and eager market. Every telephone carrier in North America and Europe and elsewhere is in at least some stage of DSL deployment.

Cable Modems

The CATV systems deployed today as shown in Fig. 2 have a tremendous bandwidth capacity, but also have one glaring problem: most only provide downstream transmission. In the late 1980's, cable companies began experimenting with upstream transmission. At first these were little more than telemetry channels, useful for such things as audience polling and purchasing pay-per-view movies. Now modern cable modems over Hybrid Fiber Coax networks can provide downstream rates over 40 Mb/s, although this is shared over a large number of subscribers.

Once this structure is in place, the cable company can offer more than just high-speed Internet access. Voice over IP (VoIP) technology can be used to offer telephony services that the cable company could market itself or could sell in partnership with a Competitive Local Exchange Carrier (CLEC). In acquiring TCI and partnering with Time Warner, AT&T intends to offer local telephone service using this scheme.

The upstream channel can be used for interactive television such as being able to immediately purchase what is being shown on a commercial.

Over 300,000 cable modems have already be installed in the U.S., putting this technology far ahead of other means of providing high-speed Internet access.

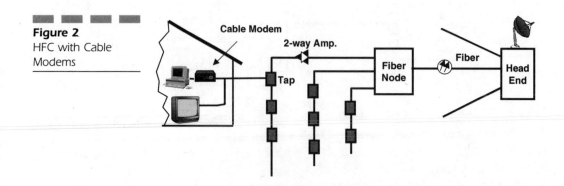

Figure 2
HFC with Cable Modems

Optical Fiber

The dream of the telecommunications companies remains having an optical fiber running to every customer's premises. This dream is largely realized for office buildings in downtown areas. Here SONET point-to-point and optical rings snake through the office towers bringing telephony, data, video conferencing, and other services today, and these are able to be easily upgraded to offer even greater amounts of bandwidth in the future. The issue the telcom companies have struggled with is the cost to extend fiber to small and medium size businesses and eventually even to residences.

A promising technology for solving this cost problem is the ATM Passive Optical Network (APON) as shown in Fig. 3. Rather than run the optical fiber all the way from the customer premises to the central office, a passive optical splitter is installed allowing the fiber to the central office to be shared by several homes or businesses. Since this splitter is passive, it requires no powering and should be of very low maintenance. This approach also shares the laser transmitter and optical receiver in the central office, further reducing costs.

NTT has announced a deployment of APON in Japan in 1999 in a fiber to the small business arrangement. BellSouth is conducting a trial of APON in a fiber to the home arrangement. Trials and deployment of APON by other network providers are expected to follow.

Radio Access Networks

The fastest deployment of broadband access network would be one based on wireless technology as illustrated in Fig. 4. Worldwide, cellular telephone net-

Figure 3
ATM Passive Optical
Network

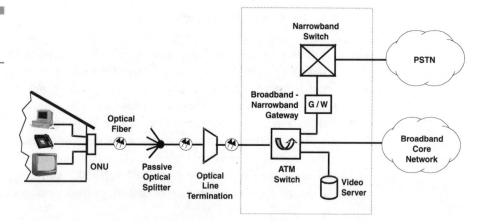

Figure 4
Wireless Broadband
Access

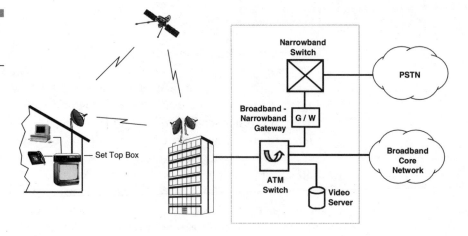

works and wireless local loop systems have been installed in many countries, but these are narrowband systems. To create a more competitive local access market, the FCC and equivalent agencies in several other countries have made sufficient spectrum available to offer broadband wireless access networks.

Broadband low earth orbit satellite systems such as Teledesic and Sky-Bridge are in the final planning stage, and will bring tens of Mb/s to end users. Recently, the FCC auctioned one GHz of bandwidth around 28 GHz to be used for Local Multipoint Distribution Service (LMDS). Both of these will offer alternatives to telco and cable last mile facilities. These technologies are most likely to be used for business customers. It is to be seen whether they will ever become economical enough for residential applications.

Teligent and WinStar began deploying LMDS networks in 1998; several others including NextLink will deploy systems starting in 1999.

Electric Power Lines

The electric utilities have been quietly deploying the largest non-telco fiber network in the world. Though built for their own internal communications needs, these utilities have been eyeing the telecommunications market as an attractive avenue for business expansion. What they bring to the table is another way to bring information to and from home: by powerline carrier.

Though they present an extremely noisy environment, electric power lines are capable of passing high bit rate telecommunication services. They can connect with an in-home powerline network creating a powerful architecture illustrated in Fig. 5. Any device that plugs into a wall socket could have access

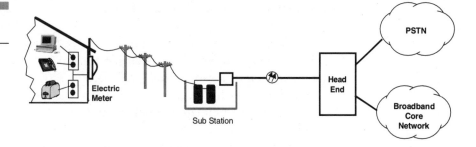

Figure 5
Powerline Carrier

to broadband networking. This architecture integrates smoothly with home automation systems, allowing control of electric appliances not only within the home but also remotely through the Internet.

A number of trials of this technology have taken place in the U.K. In the United States, Duke Power has been the most aggressive in deploying this technology.

Broadband Home Networks

With broadband access networks on the cusp of significant deployment, a number of companies are looking to the market potential of broadband home networks. In businesses, Local Area Networks face an easier environment: wiring is professionally installed and maintained, is of high quality, and there is sufficient quantity of it to dedicate to the LAN. This is not the case in the home. Although some modern homes are wired with multi-pair cable in a home-run architecture, this is by far the exception rather than the rule. Numerous home networking products are being announced that provide broadband networking while requiring no more wiring that what the customer already has.

New products are emerging capable of establishing a home LAN using the same twisted pair that is also providing telephone service. One Mb/s networking is available today with 10 Mb/s systems soon to be on the market.

Other products establish LANs on the CATV coax running around the house. The advantage of these systems is that they can utilize unused or undesired channels on the CATV coax for local video communications within the home, such as security cameras that can be monitored from any TV set. These systems also provide high-speed data networking and integration between the TV and the computer.

Powerline carrier home networking products are becoming available that go far beyond turning on and off lights. These systems can provide networking up to 10 Mb/s between any capable device plugged into a wall outlet.

Wireless LAN products are available for the home capable of providing untethered home networking at 1 Mb/s or more. Interfaces to these networks are available in PCMCIA form factors for as little as $150.

Finally, for the ultimate in home networking, prototypes are being assemble of plastic plexiglass fibers for the home. Unlike silica fiber that requires difficult and expensive mechanical or fusion splices, these plastic fibers can be simply glued together. Typical bit rates are up to 100 Mb/s. The question is whether plastic fibers are arriving too late and have already been surpassed by copper pairs pushing the 1 Gb/s envelope.

Paying for Broadband Access Networks

Where will the revenue come from which will pay for these advanced broadband access networks?

Given the uncertainty of revenues from future interactive broadband services, it would be easier to justify deployment of an advanced broadband network if it could be cost-justified for existing services alone. This was the approach some of the Bell Operating Companies attempted in the 1980's. In the wars between the phone companies and the cable companies, the phone companies probably fired the first shot. In the mid 1980's, the Bell Operating Companies began to explore the potential of deploying fiber to residences. Their motivation was two-fold. One motivation (and the one emphasized to the Public Service Commissions) was that the life span of a fiber cable is much longer than copper cables which are subject to water corrosion. The stronger motivation, however, was the potential of offering broadband services. The most obvious and near-term market was TV video distribution. Although the phone companies were forbidden at the time from offering CATV services, the BOCs recognized this restriction would be removed one day. In the meantime, they needed to get the fiber in the ground. To that end, they argued before the PSCs that shorter depreciation intervals were needed for copper loops. Accounting rules for outside plant sometimes required copper loops to be depreciated over 30 years. The BOCs argued for new depreciation intervals, claiming the true economic life of copper plant is now closer to 10 years. Under the rate-of-return regulation that the BOCs were under at the time, this extra depreciation expense would have resulted in extra revenues used to pay for the fiber deployment.

The cable companies objected that the BOCs were attempting to build a fiber network to get into the CATV business and were doing so on the backs of telephony ratepayers. This debate was cut short by a change of regulatory structure from rate-of-return regulation to price cap regulation where the Public Service Commissions regulated prices, not profits. With the extra cost of a fiber network coming out of profits, the BOCs were unwilling to invest unless a clear profit potential could be shown.

Broadband access networks will have to be paid for by the revenues resulting from future broadband services. But whose revenues? The most straight-forward approach would be to finance these network from subscription fees charged to users of these networks. Market studies have shown that heavy Internet users accustomed to paying $19.95 a month for unlimited narrowband access are willing to pay $40 per month for high-speed Internet access. This extra $20 a month has to pay for the broadband access network and the much heavier demand placed on the Internet backbone. Assuming this is equally split, the $10 for access must pay for operations expenses and capital recovery. Assuming $5 for monthly operations, the remaining $5 monthly translates into about $200 of capital investment. This is probably half of what is needed to deploy a broadband access network. Potentially, the rest could come from additional services made available over this same infrastructure. For example, upgrading a CATV network to support cable modems opens the door for the cable company to offer telephony services over IP. Upgrading the network to full Hybrid Fiber Coax (HFC) creates an infrastructure supporting hundreds of digital video channels, which is important as the cable companies respond to competition from Direct Broadcast Satellites such as DirecTV.

Probably the greatest potential source of revenue to pay for broadband access networks is electronic commerce. Already electronic commerce on the Internet is a multi-billion dollar business. It has the potential to grow to over $100 billion during the next five years. Broadband could hold the key to this growth. A customer would be more willing to buy a dress over the Internet if they could see a high-quality video of someone modeling the dress. In some areas, order fulfillment could take place over the Internet. For example, the emergence of low-cost CD ROM recorders may create a business opportunity for selling CD's over the Internet. The music would be downloaded into the customer's recorder along with a graphics file for printing the label. Similar fulfillment could be used for software sales. With software applications now shipping on multiple CDs, it is not practical to do such distribution over dial-up modem links.

The issue is how to flow some of the electronic commerce revenue back to the access network operator. The cable companies are currently in a better position to do so. Under current regulations, the cable companies can restrict their cable modem service to their own embedded ISP.[1] @Home may be the only Internet service available to you over your cable modem. The cable companies can then sell links to their home page or offer shopping channels in ways similar to their cable shopping channels. By contrast, the BOCs are required to allow any ISP to utilize their ADSL access network and in fact must provide their own ISP service from a separate subsidiary. As a consequence, the approach taken by many BOCs is to sell broadband access not to the users but to the ISPs. A consumer would buy broadband Internet service from an ISP who would imbed in its price the cost of ADSL access, taking into account potential electronic commerce revenues.

Networking Protocols for Broadband Access

True multimedia networking requires not only that all services be carried on the same physical medium, but also that all services be managed by the same networking protocol. For the past decade, the telecommunication carriers were sure they knew the protocol that would ultimately bring about this unification: Asynchronous Transfer Mode (ATM). Indeed, ATM was designed right from the start with this goal in mind. Terminals and switching systems could handle ATM streams of hundreds of Mb/s enabling ATM to handle services all the way to HDTV and beyond. Quality of Service (QoS) was designed to ensure that the telecommunication carriers could offer services with the high quality befitting their reputations. An end-to-end ATM network was the ultimate goal.

Now, even the telecommunication carriers are not so sure. For many years, data networking products were based on proprietary standards. IBM had its SNA, and each of its competitors had their own proprietary equivalent. Users ended up getting locked into one of these systems which was, of course, the goal of the manufacturers. A few manufacturers began to offer products based on TCP/IP—the Internet's networking protocol. No one claimed that TCP/IP was the best nor certainly the most full featured networking proto-

---

[1] America Online has asked the FCC to force the cable companies to provide non-discriminatory access to their cable modem service under rules similar to those under which the telephone companies operate. Thus far the FCC has been reluctant to do so, fearing that would slow down deployment of cable modem service.

col, but it was an open standard. As the popularity of the Internet grew and as competitive products based on this compatible TCP/IP protocol flooded the market, the proprietary solutions of the computer companies began to fade into oblivion.

From its ARPANET days, the Internet has been a laboratory for networking research. Over the past decade, graduate students and corporate researchers have experimented with ways to pass speech, music, video and other types of information over the Internet. At first, these were little more than academic exercises, but as the Internet grew, a number of these technologies began to seep into web browsers and specialty applications. Also, the quality of these applications began to improve.

Thus ATM has a worthy challenger for the title of foundational multimedia protocol of the future. Some would claim that this is not an either-or situation. They would point to IP over ATM protocols and the fact that most large-scale IP networks are using ATM in their core to suggest that these protocols are really complementary. They would also note that where QoS is needed, IP simply cannot do the job today.

Whether IP eventually eliminates ATM is difficult to predict, although few today would be willing to bet against anything associated with the Internet.

Network Evolution and Regulatory Trends

1.1 Introduction

The rapid transformation of the telecommunications industry is a reflection of today's exploding market for Global Connectivity. In 1995, worldwide telecommunications was a $600 billion industry, in the year 2000, it is projected to expand to more than a trillion dollars. Though not the largest industry, it is certainly the one of the fastest growing.

This chapter discusses recent structural changes in the communication industry affecting broadband access. It examines the effects of changing government regulation and deregulation on the market and the rapidly developing competition in the telephone, cable, and the Internet marketplace. All this is setting the stage for intense competition in broadband access.

1.2 A Brief History of Telecommunication Competition

Only few decades ago, telecommunications was virtually synonymous with Plain Old Telephone Service (POTS). Technology consisted primarily of copper wires and electromechanical switches. Other telecommunications services of this era included telex, telegraphy and facsimile. Television and radio services were considered unrelated to telecommunications. Now through digitalization and the resulting technological convergence, telecommunications connotes the transfer of all forms of information: digital data, voice, image, video, and sound. Although this industry has been heavily regulated, market forces nonetheless conspired to spur the development and application of a dizzying array of new technologies.

The accelerated growth rate of the communications industry with the recent relaxation of regulation illustrates a fundamental market principle that consumers are better served in an open system of free markets rather than through government regulation. The explosive growth of the Internet is an excellent example of this.

Technological advancements in telecommunications helped spark the break up of the Bell System. At one time the high cost of transmission equipment fed the view that telecommunications was a "natural monopoly" to be run by a single utility with regulated profits. However, by the late 1970's, low cost microwave radio transmission facilities created an environment where it became practical to have multiple, competing long distance carriers. Hence, on January 1, 1984, the telecommunications landscape in the United States (and soon the world) was forever changed when AT&T settled an antitrust suit with the Justice Department by agreeing to break itself up into a long distance company (AT&T) and seven regional holding companies (RHCs) also called RBOCs. These RBOCs or "Baby Bells" were Ameritech, Bell Atlantic, BellSouth, Nynex, Pacific Telesis, SBC Communications, and US West (Fig. 1.1). This arrangement lasted barely a decade followed by a rash of recent mergers: Bell Atlantic merging with NYNEX (and seeking to merge with GTE) and SBC merging with Pacific Telesis and SNET (and seeking to merge with Ameritech).

Before divestiture, the FCC had begun to let other companies compete with AT&T for long distance telephone service, first for private line facilities and later for switched services. With divestiture, the number of these long-distance competitors greatly increased. Nonetheless, local telecommunications service was still considered a "natural monopoly."

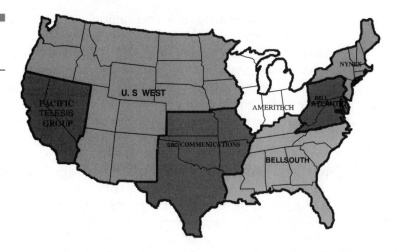

Figure 1.1
The RBOCs Landscape at Divestiture

Under the rules of the Modified Final Judgement (MFJ), the RBOCs could provide service only within Local Access and Transport Areas (LATAs)—not long distance service between LATAs; they could not manufacture telecommunications equipment or provide "information services" such as electronic Yellow Pages. It is interesting that while AT&T retained its manufacturing operation (Western Electric) and its research and development subsidiary (Bell Labs) it later chose on its own to divest itself of that portion of its business.

The RBOCs soon chafed under these restrictions and repeatedly asked the Federal District Court overseeing the MFJ to loosen these rules. The restriction on information services was lifted in late 1991 and soon thereafter permission was granted the RBOCs to offer long distance service to their wireless customers. Failing to obtain the relief on wireline long distance services they sought in court, the RBOCs turned to Congress with the result of the passage of the Telecommunications Act of 1996. The result may not have been what they had hoped.

The Act (referred herein as TA '96) forced the RBOCs to take specific proactive steps to open up their markets to local competition and granted them long distance relief only after a 14-point checklist has been implemented. Responding to this, a number of Competitive Local Exchange Carriers (CLECs) have sprung up to compete with the RBOCs for local service. The various long distance companies (or Interexchange Carriers—IECs) are responding to the impending competition with the RBOCs by attempting to compete for local service themselves. A particularly interesting example is the merger between AT&T and TCI. AT&T has announced its intention to provide local telephone service over TCI's CATV network and has also formed a joint venture with Time Warner to provide local telephone service over its CATV network. The result is that AT&T will be able to offer a complete range of local, long distance, cellular, CATV, and Internet access services—a broader service portfolio than AT&T had before divestiture. Currently AT&T offers local service in 45 states.

Replacing the long distance telephone monopoly with competition has had some interesting effects. The average cost of long distance service has dropped dramatically, from about $ 0.50 per interstate minute in 1982, to less then $0.20 in 1994, and to $0.10 in 1999. However, drastic downsizing and reorganizations were also the result. Nonetheless, as a whole, employment in the telecommunications industry has remained high. Thus the break-up of AT&T did not destroy jobs, but relocated them.

Since the Telecommunications Act of 1996 became law, over 300 new carriers have signed interconnection agreements to provide local service in 450 U.S. cities.

1.3 International Deregulation

While telecommunications deregulation started in the United States, it is spreading quickly around the world. On February 8, 1997, more than 60 countries that are members of the World Trade Organization endorsed a landmark agreement in Geneva to open up their telecommunications markets to all rivals. Countries agreed to end protections for satellite and telephone monopolies, in most cases, by the year 2000. More than half of the countries agreed to adopt legally binding regulatory principles that force traditional telephone monopolies to let new rivals connect with their networks at "cost based" prices. This will dramatically change the lives of consumers in countries where monopolies control the telecommunications industry and provide poor service at inflated prices.

The United States refused to participate in this negotiation in 1998 because the country was holding out for more concessions and a freer trade agreement which it eventually got.

1.4 Network Unbundling and Resale

At this point imposing equal regulations on ILECs and CLECs and then declaring the local telecommunications market competitive can be likened to starting a baseball game with one team already having 10 runs. The ILECs have the huge advantage of possessing a local access network that they built during the period of regulated monopoly. Recognizing this, Congress wrote into TA '96 special rules easing the barriers to entry faced by the CLECs.

First Congress allowed the CLECs to enter the local telecommunications market by reselling ILEC services. The ILECs would be required to wholesale these services at a price that reflected the costs they would avoid by having the CLEC market their services. This was, in fact, how long distance competition was achieved. Before MCI and Sprint could complete a nationwide network, they could still offer nationwide long distance service by reselling AT&T's service in areas where they had not yet built out their networks. Even today, there are large numbers of long distance companies that own no facilities but simply resell services from the major long distance carriers.

In practice, few CLECs have found local service resale an attractive business opportunity. The ILECs were required to discount their services according to avoided costs. The ILEC, in fact, spend little on marketing basic telephone services. They argued therefore that few expenses were avoided and thus the discount should be small. The CLECs recognized they would have to offer significant price discounts and spend heavily on advertising to gain market share. They required therefore deep discounts to have a viable resale business. The bottom line is that, at this point, local service resale is not working.

A second mechanism put in place by Congress to spur local competition is network unbundling. The ILECs are required to sell "unbundled elements" of their network to the CLECs for them to use as components in building their own networks.

The CLECs are finding unbundled elements a more attractive way to break into the local telecommunications business. Some portions of a local telecommunications network are relatively easy to put in place with few entry barriers. For example, switching is not much of a barrier. Getting into switching can be done by leasing a building and installing a switching system. Buildings and switching systems are readily available. In fact the switching systems vendors, facing a slow market, are quite willing to finance the switching system.

The major barrier to entry the CLECs face is in the access network. Installing an access network requires working through a dizzying array of local permits and ordinances as well as digging up streets and yards. As a result, the CLECs have concluded that their best option, at least outside downtown metro areas, is to lease unbundled access from the ILECs.

The FCC's scheme for pricing unbundled elements is rather strange. Instead of requiring the ILECs to price these elements according to their costs plus a fair profit (traditional rate-of-return regulation), the FCC mandated a scheme called the Total Economic Long Run Incremental Cost (TELRIC). Basically, it requires the ILECs to price unbundled elements according to what it would have cost them had they used the best equipment available today. Understandably, the ILECs cried foul. In essence, they stated that they were being asked to subsidize their competitors. Thus far the courts have sided with the ILECs. The 8th Circuit Court upheld the FCC's authority to determine what must be unbundled, but ruled that pricing is a matter for the state regulators.

As to what must be unbundled, several RBOCs petitioned the FCC in 1998 with regard to advanced telecommunication services. They noted while there may be logic in requiring them to unbundle the networks they installed before the playing field was leveled, the new network equipment they install now should not have to be unbundled. Specifically they argued that the ADSL network they were about to install should not have to be unbundled. The FCC was unmoved by their arguments. It noted that there was nothing in TA '96 that differentiated old networks from new networks.

The ILECs are thus in the unenviable position of having to take risks to deploy new technologies like ADSL or APON. If the technology is a flop, then they lose their investment. If it is a success, then their competitors can use these networks at bargain prices.

Indeed, having an existing network infrastructure has a down side. New carriers like Quest or Level 3 without legacy networks and can take advantage of emerging technologies like wavelength-division multiplexing and high-speed synchronous links. They can engineer their networks to take full advantage of IP technology. Even though the ILECs can deploy these same technologies, they have the added challenge of integrating these technologies with existing equipment from multiple vendors.

1.4.1 Unbundled DSL Loops

Note that most CLECs who lease a copper pair from the ILEC are doing so not to offer POTS service, but in order to deploy DSL over it. A CLEC can lease an unbundle pair for about $20 per month, invest $500 for HDSL equipment for the two ends, then sell the equivalent T1 line for $1000 per month. There are other expenses to consider such as leasing collocation space in the building to house the HDSL equipment; nonetheless, the economics is compelling.

Hence the loops the CLECs wish to lease from the ILECs are not just any loops, but DSL capable loops. The problem is that there is no clear definition as to what is a DSL capable loop. Much depends upon which DSL technology is to be deployed. A loop may work fine for IDSL, but not work for ADSL. However, there are loops that would not work with any type of DSL. Those are loops with loading coils and loops serviced from a Digital Loop Carrier (DLC) system. Beyond that, the loop length, bridge taps, gauge changes, noise, adjacent interferers, etc. that can be tolerated vary from one type of DSL to another and from one equipment manufacturer to another.

At this point, the FCC seems satisfied for the ILEC to respond to a CLEC's request for a DSL capable loop to a customer by noting whether a DLC is present, removing any loading coils, and giving whatever information the ILEC has on the loop makeup. It is up to the CLEC to determine whether the loop meets their needs.

1.4.2 Spectrum Unbundling

Most ILECs will provide ADSL service by using splitters and ADSL modems to place high-speed data in spectrum above the voice spectrum. Indeed, a major feature of ADSL is its ability to support voice and high-speed data on the same loop. In this way, the expense of a second loop is avoided.

A CLEC wanting to compete with the ILEC could lease an unbundled loop and similarly place voice and ADSL data services on this loop. However, some CLECs only provide data services and would not have use for the voice portion of the spectrum. Hence the cost of the loop must be borne by the data service alone. This could put the CLEC at a cost disadvantage relative to the ILEC.

For this reason, the question has been raised as to whether the CLEC should have the option of leasing only the spectrum above the voice band on a loop. This might be called spectrum unbundling. In implementation, the ILEC would install splitters at each end of the loop and pass the upper spectrum to the CLEC at each end of the loop. This would be particularly important when all of the existing loops to the customer are already being used for voice services.

The ILECs have opposed spectrum unbundling, and it has not been demanded by the CLECs. The CLECs have been satisfied with the concept that if you lease the loop, you get the whole loop. The data CLECs believe that the underlying voice channel is a valuable resource that they could conceivably resell to a voice CLEC.

Thus far the FCC has discussed, but has not mandated spectrum unbundling.

1.4.3 Sub-loop Unbundling

The majority of copper loops extend all the way from the customer's premises to the central office. However, a significant and growing percentage do not. In some areas as many as 30 percent of loops terminate on remote Digital Loop Carrier (DLC) systems. From the DLC, traffic is carried to the central office over T1 carrier or over optical fiber. To offer ADSL service to customers serviced by DLCs, an ILEC has two choices. First, some DLCs accommodate ADSL line cards and will pass the high-speed data over the optical fiber to the central office. The second alternative is to install a "mini-DSLAM" at the DLC site. The cabinets housing the DLC have just enough spare room to slide in a pizza-sized mini-DLSAM.

At issue is how CLECs will be able to offer DSL service to customers served by DLCs. A number of alternatives are under discussion at the FCC. Where there are copper pairs from the DLC to the central office (as would be the case when DLCs are installed for feeder pair relief), then the ILEC could be required to wire around the DLC so as to create a continuous loop from the customer's premises to the central office. Another alternative under active discussion at the FCC is called "sub-loop unbundling." Instead of leasing the whole loop, the CLEC could lease only the portion of the loop from the DLC site to the customer's premises. This immediately raises two issues: (1) where does the CLEC install their DSL equipment; and (2) how does the CLEC get the traffic from the DLC site to the central office. Typically, there would not be enough space for the CLEC to install their own mini-DSLAM in the cabinet (much less two or three competing CLECs). The CLEC could install a concrete pad and cabinet next the one owned by the ILEC. This would be an expensive proposition and subject to local building permits. Most people do not like to see DLC cabinets in their neighborhood much less two or three standing right next to each other.

A solution under discussion by the FCC might be called "logical collocation." Under this approach, whenever an ILEC installs ADSL at a DLC site, they must be willing to resell access to this ADSL equipment from CLEC equipment in their central office. In essence, the ILEC's mini-DSLAM or DLC would multi-home; that is, instead of a single network interface the mini-DSLAM or DLC would present multiple network links, one to the ILEC and others to various CLECs. There are, however, a number of problems with this approach:

■ ADSL equipment available today does not provide this multi-homing capability.

■ The CLEC would be dependent on the ILEC as to when ADSL would be available at each DLC site.

- The only type of DSL service available would be that deployed by the ILEC and supported by the manufacturer of DLC or mini-DSLAM deployed by the ILEC.
- The CLEC would be dependent upon the ILEC for operation and maintenance of the ADSL equipment at the DLC.

At this point, the FCC has not ordered sub-loop unbundling or logical collocation.

1.4.4 Technical Issues of Loop Unbundling

Unbundling local loops raises a number of difficult technical, economic, administrative, and business issues. The FCC is hoping that these can be solved as the industry struggles to implement its unbundling order.

Since the architecture of the telephone network assumed that loops were terminated on switching systems, loop testing equipment gain access to the loop via the switching system. Hence with an unbundled pair, the ILEC cannot test the loop for preventative maintenance and must respond to trouble complaints by dispatching someone to the central office to attach test equipment.

As the CLECs begin to deploy various types of DSL on unbundled pairs, there becomes a real danger that these systems will interfere with each other. Hence loop unbundling raises a number of important questions:

- How can one limit interference between the many possible technologies that may appear on the loop?
- Are CLECs required to tell the ILEC the type of DSL service they are operating or intend to operate on each unbundled loop?
- Who will establish the rules for spectrum compatibility?
- Who is to arbitrate what is allowed to coexist in a particulars binder group?
- Who is certify equipment intended for loop application?
- Who is to develop the test specifications that apply to certification?

All of these questions are contentious in nature and will be difficult to resolve.

Little experience is available on how the various flavors of ADSL, T1, HDSL VF and VDSL may live together in the same binder groups. Early evidence shows that VDSL and T1 may never coexist in the same distribution cable, much less in the same binder group. Does this mean that these services must be segregated by a ILEC because a CLEC wants to introduce VDSL? Who will pay for this extra effort? The FCC is trying to bring together the industry to come up with a solution that accommodates the needs of all sides in this controversy.

1.5 Cable Competition

The Telecommunications Act of 1996 recognized that the cable industry faces increased competition from alternative video providers, including wireless Multi-Channel/Multi-Point Distribution System (MMDS) cable, Direct Broadcast Satellites (DBS), telcos, and broadcasters. In order for the cable companies to raise the necessary capital to compete with these powerful competitors, Congress deregulated cable rates under certain terms and conditions.

This law also created greater flexibility for acquisitions, mergers and trading between and among Internet Service providers, telcos and MSOs. Within two weeks after the law took effect, U S West made a $10.8 billion bid for Continental Cablevision. A month later, SBC Communications announced its $16.7 billion deal to buy Pacific Telesis, which would form the first union between two Baby Bells. GTE announced its acquisition of UUNet, and so on. In order to survive the stiff competition resulting from the move to a freer market, many telecommunications companies have decided in favor of acquisitions, mergers, joint ventures, and trades.

1.6 FCC Cybernet Policy

The FCC has staunchly avoided regulating the Internet. Nonetheless, the FCC faces a dilemma. What happens when the services it traditionally regulates like telephony migrates onto the Internet? This was brought to the fore when America's Carriers Telecommunications Association (ACTA) filed a petition before the FCC in March in 1996 to ban Internet telephony and asked that it be regulated. One of ACTA's grievances was that it was not fair for Internet telephony not to be regulated while long distance resellers competing in the same market were regulated.

The FCC is looking for help to develop new competition-based policy models. In doing so, the FCC cyber policy seems to be driven by eight P's:

1. Pricing—inexpensive services through competition

2. Protection—against disaster and hackers

3. Privacy—bottom-up model controlled by consumer

4. Piracy—a show stopper, must protect intellectual property value for creators

5. Programming—widest variety through competition

6. Pornography—blanket restrictions are unconstitutional; need filtering tools for consumers

7. Policing—how to prevent crimes and track criminals

8. Paranoia—fear of danger on the Internet

The Internet has grown faster than any other communication means in history and its growth was largely due to the government staying away from regulating it. President Clinton, Vice President Gore, and Secretary Daley have spoken openly about the need to let new telecom communications grow in a private sector without interference from federal agencies.

1.6.1 Voice Over the Internet and Regulation

The momentum of Internet telephony services is unmistakable. Leading analysts are predicting that 50 percent of the worldwide voice traffic will be carried over the Internet after the turn of the century. Some predict that Internet voice revenue will exceed $60 billion in five years. These impressive numbers are luring entrepreneurs to offer telephone service using data networks as a backbone. The new so-called next generation carriers are basing their entire networks on IP.

Only three years ago Internet telephony was more like a hobby than a commercial enterprise. It was certainly of no great interest to the large telephone companies. Today, however, over 60 companies offer telephone services over IP networks. Moreover, traditional providers such as AT&T, GTE, Telecom Corporation of New Zealand, Daycomm of South Korea, KDD, France Telecom, Dutch Telecom , Telecom Finland, and Deutsch Telecom have all recognized the market reality and have started Internet telephony trials. Even in Japan where traditionally competition is restricted, statistics suggest that by the year 2000, Internet telephony will account for about 25 percent of telephone use in Japan.

1.6.2 Birth of Internet Telephony

Although IP telephony on PCs started earlier, the real growth in IP telephony starting in 1995 when VocalTec introduced the world's first IP telephony gateway to the telephone network. At last one could use the most common, mobile, low-cost, easy-to-use terminal to make Internet voice calls: the phone. Although lack of standards plagued early IP telephony efforts, in 1998 the IP Forum was founded by VocalTec, Cisco, Microsoft, Dialogic, U.S. Robotics, and others. It is now part of the International Multimedia Teleconferencing Consortium. This group meets on a regular basis to agree on interoperability.

The obvious next area of focus is network access. Computer users can gain access to Internet telephony via their data connection. For telephone users, a gateway is needed between the PSTN and the Internet. Fourteen companies have announced gateway products. Among the more visible are Ascend, Camelot's Digiphone, Quarterdeck's Webtalk, Internet Telephone Company's Webphone, Tribal Voice's PowWow, Electronic Magic's Netphone, Telescape's Intercom and CyberScience's Cyberphone.

There are about 65 consortiums that have been formed and working together to provide voice over IP services.

1997 was the year of the Internet telephony trials. Since then, commercial announcements have come about from AT&T, MCI, Sprint, Cable & Wireless, and BT. Everyone seems to have an Internet telephony strategy.

1.6.3 Policy Perspectives

Internet telephony will have a profound impact on regulation and the health of the voice network in the United States. As FCC Chairman Hundt observed, Washington can't make the Internet succeed, but it can be an obstacle to its success through unwise action and unwise inaction.

The FCC is going to have to deal with new issues in uncharted territory. The international sphere is where we are likely to see the greatest growth potential for Internet telephony.

Some of the more specific regulatory issues facing the FCC are these:

- When to start treating software companies as common carriers. After studying the issue, the NTIA administrator concluded that this is a bad idea and would be unworkable.
- Concern from the industry that the Internet will collapse because there will be too much voice traffic. The FCC feels that this is something that market forces can solve. The Internet service providers can

offer more capacity or start charging different amounts for differentiated services. Technically, the FCC believes that voice applications should use the UDP protocol instead of TCP which backs off gracefully in the face of congestion. TCP-based applications are acting like good citizens in an Internet environment, while UDP applications are not.

■ Concern that increased Internet traffic resulting from voice telephony would degrade the public switch telephone network. The FCC does not see evidence for this except in isolated areas such as Northern California where network congestion has started to occur in the late afternoon due to school kids logging onto the Internet. ADSL and cable modems are good solutions to this problem.

■ Regulatory parity concerns. The FCC is evaluating what would happen if Internet telephony became a real substitute for POTS service. In that case rather than regulate Internet telephony, the FCC could instead use the forbearance authority it has under TA '96 to lessen regulation on traditional telephony networks.

■ Regulatory arbitrage and mechanisms for avoiding access charges. The FCC recognizes the current regulatory structure is going to come under a lot of pressure due to Internet telephony.

■ What will be the impact on universal service?

In summary, voice is becoming an application and not a network optimized service. The traditional top-down regulatory model will not fit the Internet model. Premature government involvement can easily stifle the hundreds of millions of dollars of ongoing infrastructure investment. We have seen many times in telecommunications where people did not make investments for fear of regulation.

1.6.4 Economic Model, Regulation and Chaos

There is no doubt that Internet telephony is growing rapidly, but why? What are the drivers that are causing this phenomenon? A major driver is price. It is simply cheaper to place a call over the Internet than over the telephone network. Let's examine why this should be so.

A major factor in the price differential is that using the Internet avoids access and settlement charges. This is especially true for international calls. The convention known as the International Accounting Rate or settlement system requires that the originating carrier pay the terminating carrier a negotiated settlement rate to terminate the call. Conceptually, it is similar to the negotiated charging settlement of the long distance carriers in the United States who pays local carriers to complete calls.

Internet telephony has the pleasant side effect of putting pressure on accounting rates and on the excessive profits of international telephony providers. International settlement rates are absurdly inflated, far higher than U.S. domestic charges. U.S. carriers pay much more of these costly settlements than any other carriers internationally because Americans originate more calls than they terminate. In the United States, a domestic long distance call averages about 13 cents a minute. An average international call from the United States to overseas is priced at about 88 cents a minute. The underlying costs do not justify to this price difference.

Internet telephony will push the price that consumers pay for international calls much nearer to the actual cost, closer to 13 cents than 88 cents. The FCC, therefore, welcomes Internet telephony and is hoping to use it to get these settlement rates down over time, particularly as the last WTO basic telecom agreement becomes implemented.

Unfortunately, not all foreign regulators feel this way. Some European countries have already banned Internet telephony. For example, the Czech Republic, Hungary and Portugal have put in a total ban on this service. Other countries are now contemplating treatment of Internet voice applications under the same rules as traditional long distance service. As a result, ISPs have complained that foreign governments are imposing burdensome licensing restrictions as a condition of operation. For example, the telecommunications authority of Singapore has announced that it will require ISPs to have a basic telecom license as if they wanted to provide telephone-to-telephone service. The FCC, however, is lobbying their foreign counterparts in European capitals to change their mindset toward a market-oriented and non-regulatory approach to Internet telephony.

If the same good is sold at different prices with different people, arbitrage or other mechanisms will ultimately arise to eliminate these differences through recontracting or trading between people. Is Internet telephony simply arbitrage, an artifact of regulation? Does it exist just because the telecommunications network is regulated? Would Internet telephony survive under competitive conditions? The answer, in our view, is that Internet telephony and voice over IP will flourish even without these regulatory differentials.

There are projections that within five years the United States will have 10 times more capacity in IP networks than in telephone networks. There are a number of start-up companies, aggressively funded by venture capitalists, developing terabit routers. There are start-up carriers in the United States who are building out national networks with 100 to 200 fiber pair cables with each fiber supporting OC 192 on each of 16 to 64 wavelengths using dense wavelength division multiplexing.

Fax over the Internet is an ideal application for IP telephony. The latency, a half a second or so, does not affect the quality of a fax transmission. There is an enormous amount of money being spent to send long distance faxes. UU-Net and others have already begun to offer a fax-over-IP service as an extension of their ISP services.

We will see competition at every level. Companies will create new applications for the way that we communicate. We see this phenomena on the Internet today. And as a result, we'll see some companies compete on cost, others on reliability, or quality, on features and so on.

There are over 5,000 ISPs in the United States. The average ISP employs 16 people. This has become an incredible industry. Despite analysts' predictions that by 1997 only 6 ISPs would survive consolidation and shakeout, this has obviously not happened.

1.6.5 Technical Challenges of Voice over the Internet

The Internet faces a number of challenges that it must meet to be a good network for voice communications:

- Real time—Voice communications are very sensitive to delays in the connection. It is interesting that person-to-person video, while needing more bandwidth, is less sensitive to delay variation than is voice. Internet telephony puts significant demands on the Internet for low end-to-end delays.

- Network congestion—On a heterogeneous multimedia network like the Internet, different applications place very different demands on the network and place different values on having those demands met. Someone downloading a video clip must not interfere with a sensitive million dollar transaction.

- Quality of service—This is critical for service acceptance. IP telephony voice quality has improved significantly with improvements in voice coding. Current IP Telephony implementations generally have delays exceeding 250 msec which makes them sound much like calls over satellite links.

■ Reliability/availability—FCC and other foreign regulators should work with industry, and with industry taking the lead, to proactively ensure the reliability of the Internet. One possible option would be to create a global version of the U.S. Network Reliability Council.

■ Directory services—In the PSTN, if you know a person's telephone number, you can complete a call. On the Internet, a person's IP address typically changes each time that person connects to the Internet. Directory services which can keep up with the relationship of user names and current IP address is an unresolved issue for IP telephony.

1.6.6 Future Directions for Internet Telephony

Internet telephony is not about cheaper voice but is about new ways to communicate. It is as different from the telephone today as telephone was from telegraph a few years back.

We are not talking about all or nothing. We are talking about future uses which will include both the packet switching world and the circuit switching world. Both are going to be used to provide integrated services, and the user is going to benefit by receiving the services he or she wants at the lowest possible price.

Most businesses have a PBX and a LAN. The classic view of telecommunications is computers on the Internet, and phones on the PSTN. The new view of telecommunications is what is called the "soft PBX." That is a computer connected to the Internet, but that is also connected to the telephone system.

Telecommunication companies including MCI, AT&T, Alcatel, Lucent, Nortel, Motorola and others are making significant investments in new technology. Today, we have a voice network that's struggling to carry data. In the future, we are going to have a data network that can easily carry voice plus other vertical and value-added services. Voice is becoming an *application,* not a distinct network.

2

The Grand Unification Network

2.1 Introduction

A sea of change is taking place in the world's communication infrastructure. The result of this will profoundly affect the technologies upon which future networks will be built, the services they make available to the world's populations and the structure of the telecommunications and other industries. In important ways it will affect how we work, educate ourselves, receive health care, shop, and are entertained.

For nearly 100 years, the communications industry grew through the addition of new types of networks, each overlaid on existing networks, to meet the unique needs of new classes of service. The Public Switched Telephone Network (PSTN) was developed and optimized to provide two-way voice communications between individuals. It was deployed separate and apart from the then-existing telegraph networks that were providing low-speed text transmission between relay stations. Later, when a need arose for the wired delivery of television signals to homes, a separate overlaid cable network was built, optimized for the one-way delivery of video signals where each home received the same content. Later, when the Internet arose from its ARPANET genesis, a specialized packet data network was deployed, with high-speed routers interconnected with broadband packet pipes. Each network was built around and cost-optimized for the specialized requirements of the targeted service. More profoundly, not only were these networks themselves distinct, the companies that deployed them were distinct, the companies that manufactured the network equipment were distinct, the legal and regulatory governing each were distinct, and even the industry infrastructure supporting these networks were distinct.

The regularity over the past 100 years of the specialty network approach makes the change in direction of the industry that much more startling. Today, in all parts of the network, operators are making sweeping changes to their networks with the goal of being able to carry all traffic types. Telephone companies are deploying fiber loops and Digital Subscriber Line technology and ATM switching with the goal of adding high-speed Internet access and interactive multimedia to their service offerings. Cable companies are adding fiber and two-way transmission plus cable modems to their cable networks so that they can compete for Internet and voice customers. And new companies are springing up building networks around Internet technology with the goal of offering all services–voice, data, imaging and video–over the same network. All these companies are making the claim that once their networks are in place, not only will they be able to support all of today's services, but by being able to seamlessly combine service types, they will be able to offer powerful, interactive multimedia services.

This chapter will examine the forces that are driving network convergence. It will describe the Long Term Architecture study being carried out by the ITU as part of the Global Information Infrastructure study. It will then examine, in turn, the PSTN, Internet and cable industries to understand the current structure of their networks and to discover the fundamental changes they are making to these networks to compete in the future converged industry. The chapter then concludes by examining the technologies that are wrestling over the right to carry multimedia services: ATM and IP.

2.2 Forces Driving Network Unification

Why is network convergence happening? What is fundamentally different about the industry today than what it was during the first 100 years of the communications industry? The forces enabling and driving network convergence are technology, economies of scope, customer demand and changes in regulatory structure.

2.2.1 The Technologies of Convergence

The networks upon which today's information infrastructure is built are fundamentally different because basic characteristics of the services each supports are fundamentally different. Voice telephony originates as a switched, two-way, point-to-point, continuous 3 kHz analog signal with severe delays restrictions. TV distribution is 50 or more point-to-multipoint, one-way, non-switched, continuous 6 MHz analog signals with no delay restrictions. Internet data is bursty, two-way but usually asymmetrical, connectionless, digital information with modest delay restrictions. Historically, a network that is cost-effective for one of these types of traffic will not be cost-effective for the other.

Technology advancements have suddenly put all traffic types on an equal footing. The first of these changes is digitization. Advances in digital signal processing allows all forms of information to be cost-effectively converted to digital bit streams. Telephony networks now carry voice internally as 64 kb/s μ-law (or A-law in Europe) bit streams. New Direct Broadcast Satellites and MMDS networks distribute entertainment video as 6 Mb/s MPEG-2 signals. As all forms of information become converted into "bits," the potential to cost-effectively carry them in the same network becomes realistic.

The second technology enabler for convergence is optical fiber communications. All networks, irrespective of the services supported, now make heavy use of optical fiber communications. The cost of a fiber cable is now low enough that it is cost effective to install a fiber cable where only a few megabits of capacity is needed. Once this fiber is in place, there is enormous bandwidth capacity. Hence one can install a fiber cable to provide one type of service, then have the available capacity to carry other types of services as well.

A third enabling technology is high-speed, low-cost Digital Signal Processing (DSP). DSP allows digitally encoded video and voice signals to be compressed into much lower speed data streams. DSP allows bit streams to be squeezed into relatively narrow analog channels. The combination of these capabilities allows an analog signal to be converted into a compressed digital bit stream that is then transmitted in an analog channel one-tenth the size of the original analog signal. This efficiency is resulting in CATV systems providing hundreds of channels and digital cellular systems quickly replacing analog cellular. DSP allows the echo in voice connections to be cancelled out, thereby relaxing the end-to-end delay requirements enabling voice to be carried over packet switching networks. DSP is enabling conventional copper phone lines running from the central office or running through walls of buildings to carry high-speed data. It is even allowing power lines and home electrical wires in the wall to be used as high-speed data transmission lines.

Even with all forms of information converted to bits and even with low-cost optical fiber transmission, the widely differing switching demands of the different information types might require different networks. Conventional circuit switching systems is optimized to carry information streams of a specific bit rate. While one can aggregate multiple streams together to form higher rate streams, this is usually not cost-effective. And circuit switches are not cost-effective at all for bursty data traffic. Hence the final enabling technology for convergence is broadband packet switching. With packet switching, all types of information are placed in packets, labeled, then routed through the packet network according to the address in the label. Higher bandwidth services simply result in more packets being sent. Up to its total capacity, a packet switch is not sensitive to the bandwidth of the connection, its symmetry, or its burstiness. However, the cost per bit of early packet switches was high, and thus these switches were only cost-justified for highly bursty sources—those bursting at high bit rates but with a low average bit rate. Those packet switches would make very expensive voice switches, and carrying entertainment video was not even possible. All this has changed. Modern broadband packet switches such as gigabit routers and the promised terabit routers have enormous capacities and a much lower cost per bit than even circuit switches. As a result circuit emulation can be done over a packet switch at a lower cost than through a voice switch: packet switches beating circuit switches at their own game.

2.2.2 Economies of Scope

With the above advancements in technology, there is now a belief among the network providers that the incremental cost of supporting an additional service class on an existing network is less than the cost of a separate overlay network for that service. For instance, cable companies believe that while the phone network may carry voice more cost-effectively than a cable network built just for voice, they can still add voice incrementally to a cable network (already paid for video) for less than the cost of the voice network. Similarly, phone companies believe they can install Fiber-to-the-Curb systems to serve voice traffic, then add video entertainment services for less than the cost of the CATV network. Internet providers believe that once companies and individuals pay for high-speed access to the Internet, the incremental cost of adding voice services over the Internet is trivial.

These effects are called economies of scope. It purports that it is cheaper to build one network, enhanced to carry voice, video and data traffic, than to build three separate networks, each optimized for a particular traffic type. It suggests that any company trying to compete with a network carrying only one traffic type will not survive. As long as technological or regulatory boundaries stood between these networks and industries, peace could prevail. Once these boundaries fall, the member companies have little choice but to attack each other's market.

2.2.3 Customer Demand for Converged Services

Ultimately networks will converge if that is what is needed to meet user needs. Until recently users seemed satisfied by having their voice, video entertainment, Internet access and cellular services provided by differently companies through different networks. Indeed, their information transport needs tended to be stratified along those service categories. This appears to be changing. Users have now become interested in mobile data. Devices that allow one to surf the Web via a wireless Personal Data Assistant (e.g., Palm Pilot or Windows CE device) are attracting considerable market attention. Voice and video applications are beginning to be added to web sites as these sites use every trick in the book to gain audience. Numerous interactive video services are about to be brought to the market. Sports channels, competing for market share, are experimenting with technology allowing users to select the camera angle in football games. There is a huge potential for targeted advertising where different commercials are shown to different segments of the viewing audience according to their interest profile. And when users see something on a commercial that interests them, then with a click of the remote they can be taken to an interactive screen to get more information on the product or place an order. Technology may at last make the video phone a consumer hit.

The bottom line is that networks tightly optimized for specific services of yesterday will not meet the varied a rapidly changing user demands of tomorrow. As a result, network operators recognize that to be successful they must generalize their networks, not only to carry multiple service types of today, but in order to make them more adaptable to the killer services of tomorrow.

Even if users were not demanding multimedia services, there would still be a compelling consumer demand forcing network convergence: surveys and marketing trials have shown that when given the choice, consumers simply prefer getting all of their information transport services from one company on one bill.

2.2.4 Changes in Regulatory Structure

Many have claimed that the recent rush to convergence is a result of legal and regulatory changes. It is true that regulatory restrictions had prevented the cable and telephone companies from directly competing, and that these restrictions were lifted by TA '96. However, laws come about through a political process driven by the demands of the voters. It would be more correct to say that these legal and regulatory changes came about as a result of the convergence forces. Once technology advancements removed the underlying technical differences in these networks, creating the potential for significant economies of scope, and once users started demanding multimedia services that blended the services of these separate networks, then changes to the laws and regulations were inevitable.

2.3 The ITU Long Term Architecture Study

The telecommunication Long Term Architecture (LTA) vision is the focus of the special study being conducted by International Telecommunications Union (ITU). That study was inspired by the GII (Global Information Infrastructure) initiative promoted by U.S. Vice President Al Gore. The goal of the study is to understand the future migration of the telecommunication industry. It focused on social attitudes of people in industrialized nations and draws conclusions on the type of services and applications they will be demanding. This study is useful in that it provides insights into the capabilities broadband access networks must provide for network operators to compete successfully.

2.3.1 Trends Toward Future Telecommunications

The LTA recognized a number of indisputable important trends driving the industry. The most notable are the following:

- Convergence of the separate segments of the communication industry, in particular telecommunications and broadcasting, due to integration of the common services and applications.

- An increased emphasis on standards rather than proprietary solutions thus facilitating interoperability and creating an open competitive market.

- A shift from a national to a more global approach concerning services and regulation.

- A shift from monopoly supply to open competition in both services and equipment.

- End users are demanding operational simplicity in the applications.

- Falling costs of digital processing is leading to cheap information appliances combining the functions of computers, televisions, and smart telephones.

- As monopolies crumble and industry converges, telecommunications is gradually developing the wholesale/retail characteristic seen in other industries.

These factors are creating an environment where packaged applications obtainable from a single source become an economic necessity. Nonetheless, the costs for upgrading existing communications infrastructures to provide advanced value-added services are immense. For new comers to make the required investments, regulations, global trade policies, and the commercial potential have all to be right.

2.3.2 Common Trends in Industrialized Countries

Customer demands are a strong function of the social environment in which the customer lives. Since every country has its unique lifestyle, one would expect disperate trends within this global community. Nonetheless, some trends seem to be shared by the different societies of industrialized countries. These trends help predict customer demand in the telecommunications market. The common trends are these:

Individualization. People throughout the industrialized world have become more focused on their individual rights, demands, wishes, etc. The consequence of this individualization is an increasing demand for mobility. Telecommunications is expected to play a major role in satisfying this need.

The Internet has demonstrated that global communication can substitute personal mobility and can minimize differences between societies and widen the scope of the individual in the global context.

Health. People are putting more emphasis on their health, so much so that they are willing to accept restrictions to their personal lifestyles in return for better fitness. Numerous trials of telemedicine have shown the significant role telecommunications can play in healthcare.

Personal Security. People are becoming increasingly concerned over their personal security, including the protection of information about themselves. This can sometimes place conflicting demands on the design of telecommunication services such as legal wiretaps of encrypted communications.

Environment. People are concerned over the harm current lifestyles are inflicting on the environment. This is leading to individual and governmental demands for telecommuting.

Longevity. Life expectancy has increased due to medical advancements particularly in the industrialized countries. This shift in demographics towards a graying population will change buying habits. Shop at home services may become more popular.

Trustworthiness. People are perplexed over the increased complexity of the products they purchase. People rely upon the seller of the products for its quality or harmlessness.

Work at Home. Global markets will dramatically change attitudes towards work. The cozy office job will be replaced by a virtual office with less dependencies on the company and more emphasis on entrepreneurs. Work will be performed anywhere convenient.

From the above social trends, one can derive a list of the future demands in the telecommunication market.

2.3.3 Network Trends

Telecommunications networks have become increasingly complex. The network evolved from POTS lines to the introduction of ISDN and now broadband capabilities are being introduced. Several different mobile networks have been deployed to meet different market needs for mobility: GSM being a good example. The distribution of video entertainment may soon be added. Recently deployed satellite systems will be used for personal communications services. Private and public networks are becoming indistinguishable with the rapid the introduction of Virtual Private Networks.

In the multimedia arena, the most obvious trend is the rapid acceptance of the Internet. Its differences from traditional telecommunications networks are significant, from tarriffing, service quality, architecture, to technology.

Some operators are bringing about a convergence between fixed and mobile networks with Intelligent Network-based services expected to play a major role in providing mobility in fixed networks. Customers will access their personal profiles via a terminal using a Smart Card, their personal profiles defining their identity and the range of services they are entitled to use, as well as details of their billing records

2.3.4 Network Aspects for Multimedia Services

A list of services is being considered in the GII study as representative of the type of demands future networks will be required to meet. Many of these will require significant changes to today's telecommunications networks.

- Interactive speech
- Real time image transfer
- Electronic mail
- Multimedia document retrieval
- Video on demand
- Interactive video services
- Computer supported services
- Broadcast TV/Radio/data contribution
- Broadcast TV/Radio/data distribution
- Real time multipoint retrieval

These multimedia services will be supported by not one telecommunications system, but by a set of systems and sub-systems working together. Therefore the approach taken by the LTA study team was to define a framework within

which the different elements can co-exist and interwork. The network architectures for multimedia communication systems should meet the following
objectives:

- Be able to cope with the diversity of services to be supported.
- Enable service provision by a large diversity of network operators and
  service providers.
- Allow the same application to be transparently accessible from differ-
  ent terminal equipment.
- Offer all services in different environments.

2.3.5 Overview of the LTA Business Model

The LTA proposed a business model of the future telecommunications industry. This business model is composed of six domains. Figure 2.1 illustrates
this business model showing the following domains:

Consumer.   A consumer may be an individual, a household with multiple
end users, or a small or large business with multiple end users.

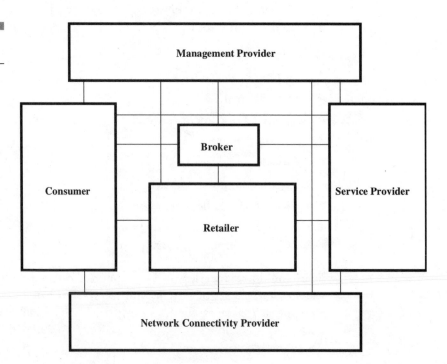

Figure 2.1
LTA Business Model

Broker. A broker provides location information that the business domains use to locate other entities for the purpose of establishing communications.

Retailer. A retailer provides the contact point for the consumer to arrange for, and the contact point for the Service Provider to offer, communication services.

Service Provider. A service provider supplies any of a variety of services that employ communication services as the delivery mechanism. Note that a content provider is a sub-class of Service Provider that supplies information content to another Service Provider.

Network Connectivity Provider. A network connectivity provider supplies the communication services that transports information that may either be control plane information or user plane information. The service control information can be transported by either means depending on the business application.

Management Provider. A management provider administers one or more domains in whole or in part and provides functions needed for administrative functions and off-line billing.

Note that nothing in this model precludes any combination of these six domains in single business entity.

2.4 PSTN

In this section, the Public Switched Telephone Network (PSTN) will be described both in its current incarnation and by the ways telecommunication carriers are attempting to evolve meet the consumer's appetite in the future competitive broadband network environment.

2.4.1 Traditional PSTN

This section gives a general overview of the traditional PSTN. From its inception, the PSTN was developed to handle voice service, although over the years, a variety of data communication capabilities have been added.

Figure 2.2 depicts a simplified but typical configuration of a voice network traditionally deployed by U.S. Incumbent Local Exchange Carriers (ILECs).

Figure 2.2
Traditional STM
Networks

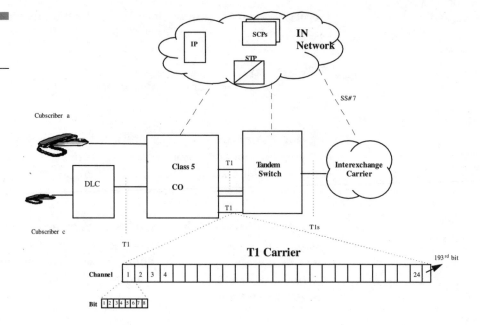

The network hierarchy, in simplified terms, can be classified into three network elements:

1. Access layer

2. Transport layer

3. Switching layer

2.4.1.1 Access Layer. The access network is the portion of the network closest to subscribers. A Digital Loop Carrier is normally installed in neighborhood areas providing access to customers via copper loop (a point-to-point twisted pair of copper wires). This pair of wires, identified as Tip and Ring, is used to perform several functions:

■ Carrying the analog voice signals (or digital signals for ISDN lines).

■ Providing power to end user device (20 milli-amp DC).

■ Providing an indicated for requests for service (off-hook conditions).

■ Transmitting ringing voltage (20 Hz, 88 V. RMS).

■ Sending coin collect signals for public pay telephones (by reverse tip and ring polarity)

2.4.1.2 Access Systems. Digital Loop Carrier (DLC) systems are multi-plexing nodes that connect a number of subscribers (typically 96 to 2000 subscribers) to a class 5 switch. The class 5 switch is where all switching functions are performed and where the services are provided. The rationale for deploying DLC is quite simple. It is much less expensive to connect, for example, 96 subscribers, to a nearby DLC box and carry the information on four T1 lines (requiring eight copper pairs plus two additional for spare) to the Central Office (CO), than to provide 96 copper pairs all the way to the central office. Further, by minimizing the length of the subscriber's loop, higher speed Digital Subscriber Line services can be offered such as ADSL and VDSL.

Early DLCs operated in what is called the "universal" mode. Here the subscriber's line is re-created in the central office for termination on the switch. In this arrangement, a customer served via a DLC appears identical to the switch as one serviced by a dedicated copper pair all the way to the subscriber's home.

Newer DLCs make use of special purpose switch interfaces defined by Bellcore TR-08 and GR-303 (V5.1 and V5.2 in Europe). In TR-08, the T1 lines from the DLC are terminated directly on the central office switch. By eliminating the conversion equipment in the central office, this arrangement provides cost savings and improves reliability. Also, the elimination of the extra digital-to-analog conversion in the connection is needed for customers' 56 kb/s modems to operate properly. TR-08 can be operated in two modes. Mode-1 is non-concentrating: each customer's loop is dedicated to a channel on one of the T1 lines to the central office. In mode-2, two-to-one concentration is supported by the central office assigning an off-hook line to one of 48 channels on two active T1 lines to the central office (a third T1 line is kept as spare). With 96 onto 48 concentration, blocking is extremely rare. A fraction of the framing bit in the T1 line (the 193rd bit of each frame) is robbed for use as a messaging system between the DLC and CO to assign one of the 24 channels. In the vast majority of DLC deployments in the United States, DLCs operate with TR-08 mode-1.

GR-303 is a more recent DLC interface (such systems called an NGDLC—next generation DLC). GR-303 supports DLCs much larger than 96 lines, supports fiber interfaces to the central office and supports switching. GR-303 based DLCs can support ISDN. In GR-303, a 64 kb/s channel is dedicated for signaling.

It is important to note that with DLC, the CO switching system has the sole responsibility of handling the call, at least as far as services and call features are concerned. If all the T1 lines connecting to the central office were to be cut, then service to the subscribers served by the DLC would be disrupted.

The presence of DLCs in the loop makes loop unbundling problematic. A CLEC wanting access to a loop for termination on the CLEC's switching system or to offer DSL services would be blocked. One solution under consideration is called "sub-loop unbundling." Here the CLEC would be given access to the customer's distribution pair at the DLC site. For NGDLCs, an alternative approach has been called "virtual collation." Here the NGDLC would present multiple GR-303 interfaces, one to terminate on the ILEC's switching system and another to terminate on switching systems of requesting CLECs. The administration of the NGDLC would remain the responsibility of the ILEC. The FCC issued a Notice of Proposed Rulemaking (NPRM), FCC 98-188, to decide which of any of these approaches would be required for DSL services.

An alternative to a DLC is to remote a switching module of the central office switching system. Most digital switching systems support this type of remoting. Most such remote switching modules support survivability, that is, if the connection to the CO is cut or if the CO switching system fails, the small remote switch can still handle local calls.

2.4.1.3 Transport Layer. Thirty years ago, central offices were interconnected with individual copper pairs, or with analog microwave radio systems, or with analog RF over coax (L-carrier systems). Today, digital transmission systems have replaced analog, both between central offices, and in the feeder plant between the central office and the digital loop carrier systems. Digital transmission systems are also used to directly connect to large business customers.

The most popular transmission system in the loop plant today is the T1 line (E1 line in Europe). T1 is a digital carrier that operates at 1.544 Mb/s (2.048 Mb/s for European E1 systems) over two twisted pairs (for transmit and receive). The bit stream is divided into 193 bit frames. The 193rd bit in each frame is used for synchronization, framing of the T1 line, as well as for signaling (using the alternate frame approach). A voice connection will be assigned one of the 24 channels in each frame. Each time slot contains 8 bits and is used to carry a voice sample every 125 μsec (8 kHz sampling). This 8 kHz timing is maintained throughout the network. In a rather awkward design called robbed-bit signaling, the least significant bit of each channel of every 6th frame is used for signaling. The European E1 system sends signaling in a separate channel and bit robbing has been eliminated in newer incarnations of T1 systems. Nonetheless, because of the presence of bit robbing in the network, most data communication systems in North America making use of the PSTN send data at only 56 kb/s.

Digital repeaters placed every mile in the loop can allow T1 lines to be operated over 50 miles. Digital repeaters are equivalent to the amplifiers deployed in the cable network. The digital repeaters are powered by DC current carried by the T1 copper wires. Alarm logic and signals are monitored by the host CO verifying the repeaters operating conditions. If a repeater fails, the CO will reroute the traffic using other T1 trunks that are available in the network.

A more recent technology called High-speed Digital Subscriber Line (HDSL) carries the payload of a T1 line over two loop pairs without digital repeaters. A new standard called HSDL-2 can do the same over a single pair of copper wires.

Optical fiber transmission systems, first developed for inter-office trunk facilities, has recently made its way into the access facilities. SONET systems (SDH in Europe) operating at OC-3 (155.52 Mb/s) and at OC-12 (622.08 Mb/s) are used to connect NGDLC systems to the central office or in some cases to connect large business customers to the central office. SONET systems are often deployed in "self-healing" rings wherein traffic is picked up and dropped off at nodes around the rings and the system remains operational even if the ring is cut.

Crossconnect systems are used to manage traffic on digital transmission systems. For example, a 1/0 crossconnect system terminates T1 lines; it takes time slots on one T1 line and flexibly places them on other T1 lines. This would be used, for instance, to "groom" traffic on a T1 line from a customer to send some of this traffic on a T1 lines to the switching system and send other traffic on a trunk facility to a different office. A 3/1 crossconnect does the same thing but manages DS1 payload capacity within a DS3 or SONET facility.

2.4.1.4 Switching Layer. The class 5 switching system is directly connected to the end user. Typical, class 5 systems can handle up to 100,000 subscribers. Most custom calling features such as call waiting, call forwarding, caller ID, etc. are provided by the class 5 switching system.

The line cards in the switch are periodically scanned for off-hook conditions to handle a subscriber call. Off-hook is detected by DC current flow due to the closure of the tip and ring loop. Based on the dialing information received from the subscriber, the switching system performs number translations and routes the call to the appropriate destination. The processors in the class 5 switch perform the control functions, while the switch fabric is used to carry the 8 bit voice samples through the core. If the called party is not in the class 5 local area (a long distance call), then the call is dispatched up in the hierarchy to a class 4 or tandem switch.

Because of the large number of customers served, most service effecting elements in modern digital switching systems are duplicated to ensure reliability. This includes the call processors, the switching fabric and the power buses. An exception is the line cards which are usually simplex as their failure typically affects only 4 to 8 customers.

The various call processor in the switching system usually operate in an active/standby mode. Beside protecting against hardware failures, this can also provide protection during software upgrades. When software is loaded into a switch, it is first loaded into the standby processor. If a problem occurs when attempting to make this processor active, the processor with old software copy will be switched back into place.

AC power is used indirectly to power the central office. DC batteries (−48 volt) are used to power the central office equipment. DC-to-DC converters within the equipment itself provide lower voltages for electronic circuits. Commercial AC power is used for charging the −48 volt batteries. When AC power fails, the batteries must have enough energy to operate for 8 hours. Diesel powered AC generators provide backup for longer outages.

Intelligent Networks. In the 1980's, a new way of implementing network features was introduced called Intelligent Networks (IN) and later Advanced Intelligent Networks (AIN). With IN, instead of implementing all feature logic in switching systems, switch features could be implemented an external computer call a Service Control Point (SCP). As the switching system would process a call, its software would hit various "trigger points" where the call process would stop and the switch would consult the SCP to determine what to do next. These SCPs typically provide a programming interface for adding features by the network operator. There are several advantages to this approach:

- A network operator who uses a variety of brands of switching systems could ensure that the same feature is available (and operates identically) to all customers on their network.

- A network operator could program features or purchase them from third parties instead of having to rely on the switching system manufacturer.

- Centralized databases could be used to manage resources across multiple switching systems. For instance, calls to line groups, such as catalog sales operators, could be managed across multiple central offices.

- Services could be developed quickly reducing time to market.

An example of an AIN service is the Area Number Calling feature offered by BellSouth. In this service, a business with many locations (such as a pizza parlor chain) is given a single number. Calls to that number are directed to the location closest to the customer.

IN and AIN are built on a specialized signaling network called Signaling System 7 (SS7). SS7 replaced the in-band Multi-Frequency (MF) signaling of the past with a highly reliable packet switching network. This greatly increased call set-up speed and reduced fraud (some thieves would use "blue boxes" to simulate MF signaling to many free calls). The ITU is currently working on upgrading the protocols and call models to accommodate broadband services in the IN environment.

2.4.2 Tomorrow's PSTN

As noted earlier, today's network, be it telephone, cable, or the Internet, was designed and deployed to support specific applications. Subscribers today are much more sophisticated and are now demanding integrated broadband services for interactive multimedia applications. The telecommunication carriers recognize this and are defining next-generation PSTN architectures that will enable them to compete for multimedia services with a network more flexible and more cost-effective than the one deployed today.

Service integration is not new to the telecommunication carriers. The telecommunications industry envisioned such trend in the 1970's and supported the effort, in ITU and ANSI, to standardize narrowband Integrated Service Digital Network (ISDN). ISDN did not fully succeed, especially in North America, because of several factors:

- Progress in modem technology kept raising the bit rates for analog modems. Even tying the two B channels together provided little more than a factor of two over what is possible with analog modems.
- ISDN was not ubiquitous.
- End user equipment vendors support was cool at best.
- User friendly interactive applications that took advantage of ISDN were virtually nonexistent.
- The public was not socially or psychologically ready.

Since then, deregulation, social attitude, technology, and political leadership played a major role in defining the interactive market. Today it appears that high speed interactive services are what consumers want and, more importantly, are willing to pay for it. As a result, the telecommunication carriers are willing to give network integrated services another try.

2.4.2.1 Multimedia Communication Systems. Based on the LTA described in section 2.3 and the services that are being considered as part of the GII study, a future multimedia network model is emerging. Network consists of four conceptual domains as illustrated in Fig. 2.3:

1. Terminal equipment
2. Access networks
3. Core transport networks
4. Application services

The division of Multimedia communication systems into four domains accommodates different types of multimedia services networks and different operator roles. The following describes the four domains along with how current systems and can evolve towards a multimedia communications systems.

2.4.2.2 Terminal Equipment Domain. The advanced multimedia services envisioned will require advanced terminals capable of interfaces with the broadband access networks that will be put in place and will be capable

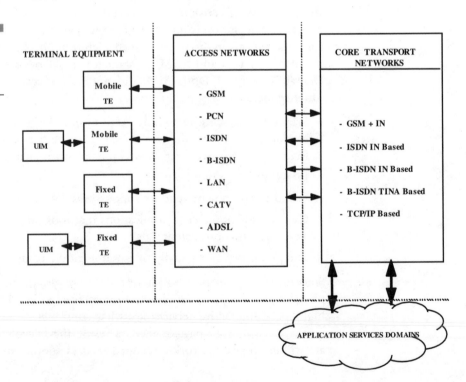

Figure 2.3
Multimedia Communication Systems:
Conceptual Model

of interacting with high-level services in the network. These will consist of single-mode terminals that access a single type of access network and multi-mode terminals that will be able to access several different types of access networks. For example, a multimedia PC might have both an ADSL interface and a cable modem interface. The User Identify Module (UIM) would be used to identify the user and to support personal mobility related services (and possibly, additional functions) in the different multimedia domains. For instance, inserting a smart card into a terminal would identify you to the network, whereupon the network directly calls to you and customizes the services and parameters according to how you like to have these set.

2.4.2.3 Access Network Domain. The access network of the future will include current systems, evolution of those systems, and new components offering capabilities beyond those of the present systems. The access network is used to gain access to the application services domain via a core transport network. The use of several different access networks in parallel will allow for a gradual introduction of high bit-rate multimedia application services in different areas where market demand and economics justify.

A number of new access technologies are under development for the PSTN that is the subject of the following chapters including ADSL, VDSL, APON, Broadband ISDN and LEO satellites. It is expected that at least one new radio access network component will be needed in order to support mobile applications of high-bit-rate multimedia communication services.

2.4.2.4 Core Transport Network Domain. The future public broadband core network as envisioned in the ITU is Broadband ISDN (B-ISDN) (see Fig. 2.3). The B-ISDN network would be based on Asynchronous Transfer Mode (ATM). The role of the core transport network is to interconnect different access networks and to provide access to the application service domain. The core transport network provides various network services and features such as mobility management and service logic for the intelligent routing of calls.

Although B-ISDN is based on ATM, other alternatives of core transport network based on an IP structure are under consideration by ILEC and CLEC customers. The relative advantages of these approaches are discussed at the end of this chapter. The likely future of ILECs core transport network would consist of a combination of these and legacy systems. New operators are more likely to go directly to networks based on IP and ATM. The relative success of these approaches will be determined in the marketplace.

2.4.2.5 Application Services Domain. Application services in the future multimedia network are likely to evolve to a client-server model. Intelligence in the user's terminal will interact with server systems that can be located anywhere in the network. This greatly expands beyond today's AIN concept. The user will be able to gain these services from any point in the access network subject to the capabilities of the terminal and capabilities of that part of the access network to which the user has roamed. The application services are either transparent to the access and core transport networks, or they may make use of network specific logic, e.g. location information. The only limitations would come from the capabilities of the networks and terminals used and subscription conditions.

Existing application services have typically been developed to be used in different fixed environments. It is expected that users would wish to access these services also when mobile with as few limitations as possible and at a price acceptable to the user. An example of a transparent application service is the World Wide Web (WWW). This could be provided from any point via multimedia communication systems, but the bit rate (for instance) will be dependent on the specific local access and core transport networks. As an example, the possible bit rates for access the WWW via a GSM access network may be limited to somewhere between 9600 bit/s and 100 kb/s (the highest rate would require an extension to the current GSM standard).

2.4.3 ATM and Broadband ISDN

The centerpiece of the telecommunication carrier's vision of the next generation access network is broadband ISDN. In this section, the concepts behind broadband ISDN and ATM and its underlying technology, will be described. Enough details will be included to better understand the rationale of why ATM may play the major in future telecommunication networks.

2.4.3.1 History of ATM. ATM (Asynchronous Transfer Mode) was first introduced in ITU (previously CCITT) in 1984. In 1987, ITU selected ATM as the switching technique for broadband ISDN. Finally in 1990, ITU released a set of recommendations on B-ISDN, using the ATM techniques. ANSI, ITU and the ATM Forum are still developing and completing aspects of broadband ISDN standards for both public and private networks.

ATM is a fast packet switching technology transferring digital information in fixed-length cells. Cells are constantly convey the information in an asynchronous manner occupying no identified position in time (hence asynchronous). This is unlike the circuit switch technique in which the information is transmitted in regular time slots. ATM is also differentiated from the packet switching techniques, such as X.25 or TCP/IP, in that ATM operates at very high speed and is connection oriented, with fixed cell size and with no re-transmission at the network layer. The concept of quality of services is strongly embedded in the ATM protocols.

The ATM cell shown in Fig. 2.4 can deliver voice, data, and video services. The cell size is 53 bytes and is divided into a 5-byte header and a 48-byte payload or information field. The header contains address information identifying up to 2^{28} logical address, service class information, management information, congestion information, error protection/correction. The 48 byte may contain signaling, management, or user information such as digitized speech, images etc.

2.4.3.2 The ATM Advantage. Before describing the advantages of ATM and why ITU selected ATM as the target solution for B-ISDN, it would be worthwhile comparing it with existing circuit switching technologies, namely Synchronous Transfer Mode (STM). This comparison should be viewed in the context of integrating services of voice, data, and video.

Today's STM switches in the telecommunications network worldwide were developed and deployed in the 1970's and 1980's. The switch was optimized to handle 64 kb/s circuit connections. The network is synchronous (time

Figure 2.4
ATM Cell

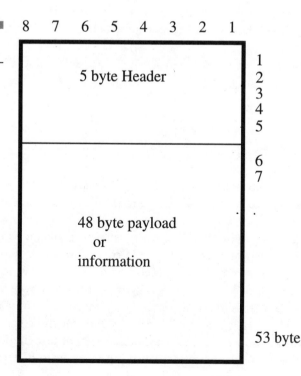

dependent); the time relationship between user was continuously maintained to ensure that voice samples were collected according to a centralized network clock and played out at the far end according to this same clock. Each voice channel was assigned a symmetrical and fixed slot every 125 μsec and delivered to its peer every 125 μsec (traditional sampling rate of 8 kHz). This time slot remains allocated to the connection throughout the life of the call, even during the silent periods when no one is speaking.

STM is poorly matched for data traffic. Data traffic is bursty in nature and often highly asymmetrical. Using a symmetrical fix time slot in an STM network is obviously not an efficient way of communicating data. A data user may need to burst a megabit of bandwidth in one direction for a short period of time with little traffic between bursts. Often the network is wasting its resources waiting on user input. Packet networks solve this problem by sending data in labeled packets. These packets are then sent in a store-and-forward basis around the network. While good for data traffic, typical packet networks are poorly matched for voice or video. First, the delays in typical packet networks can be highly variable. For synchronous traffic like speech, this can cause voice samples to arrive after they were supposed to be played out, resulting in gaps in speech and annoying blobs on a TV screen. Further, typical packet networks cannot operate at the hundreds of Mb/s speeds envisioned for B-ISDN. With its fixed length cells and simple header processing, ATM was seen as a good compromise between STM and conventional packet switching.

There are several reasons that ATM was selected as the target solution for B-ISDN:

1. Service integration
2. Service transparency
3. Multiplexing gain

Service Integration. Figure 2.5 represents the typical multi access lines arrangement to homes today, a connection for the telephone, another one for Internet access, and a third connection to the cable TV. B-ISDN and ATM promise to replace all of that with a single broadband connection. ATM transfers information in cells that could contain voice, data, or video. Hence with a single physical line over which one can deliver a variety of services to the user and all of the these services can be provided by a single network.

B-ISDN envisioned a fiber run to the home operating at OC-3 (155.52 Mb/s) or at OC-12 (622.08 Mb/s) rates. While still the target, many network operators are now considering an interim step of bringing the fiber to with 1000 m to 3000 m of the home, then using some form of Digital Subscriber Line

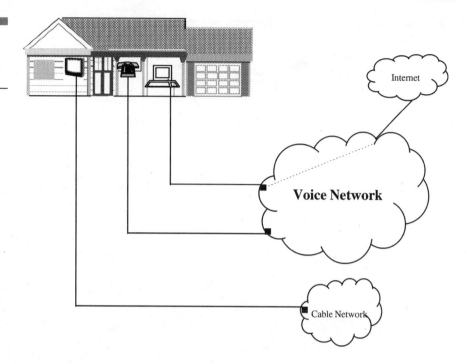

Figure 2.5
Typical Multiple Access Lines Arrangement Today

(DSL) over a copper twisted pair for the final connection to the home. The ATM forum has defined that interface to be a UNI_{FTTC} (User Network Interface for Fiber-to-the-Curb) (see Fig. 2.6). This UNI_{FTTC} represents the model of an ATM broadband network in the telco environment.

Service Transparency.   A fundamental advantage ATM brings when integrating services is service transparency. In conventional networks, the network operator must provide the hardware interface at the central office for each type of service offered. With ATM it will be possible to provide additional services to a user simply by negotiating the needed bandwidth and quality of service (QOS) with the network. No additional hardware will be needed at the central office to provide this new service. For example, a user wishes to add a video camera to his home network. The concept is that the user will be able to buy video cameras with standard generic interfaces much like it is done today when buying a telephone with RJ-11 connector and simply connect it to the home network. To go online, the terminal device (video camera is this case) negotiates with the network the bandwidth needed and the sort of quality picture the user is willing to pay when communicating with the outside world.

Figure 2.6
Typical UNI$_{FTTC}$ Con-
nection to a Home

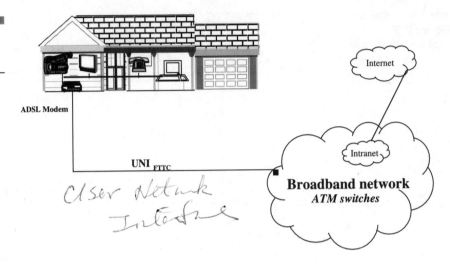

Figure 2.6
Typical UNI$_{FTTC}$ Con-
nection to a Home

Multiplexing Gains. A major benefit of ATM over circuit switch techniques is that it can take advantage of the burstiness of user information. This can be in Fig. 2.7 below.

Assume a scenario in which a user is connected to the Internet through the traditional PC modem. The line, representing 64 kb/s capacity through the network will be tied up for the duration of the Internet connection. In this simplified model, if one assumes that the connection is instead ATM, the multiplexing nature of ATM and its independence on time domain will allow the user to communicate with the Internet while talking on the phone

Figure 2.7
ATM Statistical Multi-
plexing Gain

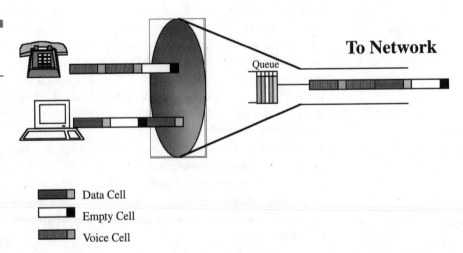

using the same physical line. ATM cells can deliver voice samples and during the silent period, data could then be transferred transparently.

Data cells simply stay in queue and are transferred in a multiplexed manner. Management of the queue is a function of the quality of service needed for the connection. Voice cells are time critical and will have priority when extracted from the queue, while data cells are more immune to delay but unlike voice, are less immune to cell loss.

2.4.3.3 ATM Interfaces. The interfaces as defined for ATM-based transport networks are the following shown in Fig. 2.8.

2.4.3.3.1 Public UNI. The public UNI connects ATM users directly to the public network. This interface was developed by ANSI and ITU standards and fully conforms to the B-ISDN reference model.

2.4.3.3.2 Private UNI (P-UNI). The ATM Forum developed the concept and specifications of the private UNI. This is the interface connecting an ATM user to a corporate enterprise network. Unlike the public UNI, its is optimized to operate in a campus environment with limited physical reach.

2.4.3.3.3 B-ICI. The B-ICI interface was developed by ANSI and completed by the ATM Forum. It is the interface between two different public network providers or carriers. The specification also includes service specific and man-

Figure 2.8
Broadband Network Architecture and Interfaces

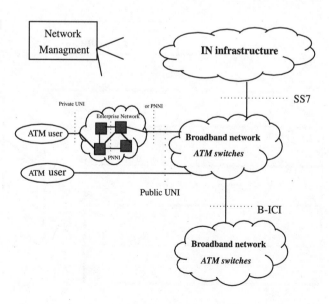

agement functions needed when the inter-carrier crosses the LATA boundaries.

2.4.3.3.4 P-NNI. P-NNI interface was developed by the ATM Forum. It is the interface between private ATM switches. The PNNI contains two protocols: one for distributing topology and routing based on the link-state routing technique, and the other for signaling to establish connections across private networks.

2.4.3.4 B-ISDN Protocol Reference Model (PRM). The protocol reference model is the blue print one needs to understand the fundamentals of a protocol. The ATM PRM, as shown in Fig. 2.9, was first developed in ITU.

The PRM contains three planes:

1. User plane: mainly used to transport user information

2. Control plane: used for signaling

3. Management plane: maintains the network and operational functions.

B-ISDN lower layers: The lower layers of the Protocol Reference Model follow the OSI discipline and regiment in all respects. Each layer functions independently of the others and is designed to perform services that are needed to the layer above or below. The three layers are:

Figure 2.9
B-ISDN Protocol Reference Model (PRM)

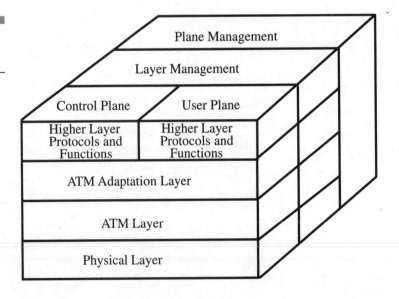

[handwritten annotation: Physical Medium Dependant / Transmission Convergence sublayer]

1. Physical layer
2. ATM layer
3. ATM adaptation layer

2.4.3.4.1 Physical Layer. The physical layer, contains two sublayers: The Physical Medium Dependent (PMD) sublayer and the Transmission Convergence (TC) sublayer. The PMD sublayer addresses aspects which are dependent on that particular transmission medium. The functions of this sublayer are:

- Correct transmission and reception of bits
- Insure bit timing construction/reconstruction
- Line coding at the transmitter and receiver

At the TCC sublayer, the bits are assembled to fit into the frame of that particular physical media. ITU defined this sublayer in I.442. The functions performed by the TC are:

- HEC (Header Error Check) is a one byte code generated using a specific polynomial. It is generated by the transmitter, and inspected at the receiver to check the sanity of the received ATM cell.
- Cell delineation. HEC is also responsible for identifying cell boundary. A number of consecutive and correct HEC received, marks the boundary of the cells that is in a frame.
- Cell scrambling/descrambling are performed at the receiver and transmitter side. Scrambling avoids continuous non-variable bit patterns and improved the efficiency of the cell delineation algorithm. Scrambling also prevents any malicious operation on the user information.
- ATM Cell Mapping is performed in order to align by row, the byte structure of every cell with the byte structure of the SONET STS-3c payload capacity (Synchronous Payload Envelope). This is specific function when SONET is used as the physical layer.
- The insertion of idle cells is performed at the transmitter to accommodate the rate of the transmission system. If the physical rate is at DS3 and the data transmitted on that link is less, then idle/unassigned cells are inserted to maintain the DS3 clock rate.

2.4.3.4.2 Physical Link. Both the PMD and TC are the building blocks needed to assemble the physical layer. There are many interfaces defined for the public and private interfaces. These specifications were defined in ANSI

(T1S1 and T1E1), and ITU. A partial list is as shown below. More interfaces are still being specified in the ATM Forum and T1E1 to address specific markets for residential broadband or to further reduce component cost. The list now includes:

- SONET 155.52 (SM and MM fiber)
- DS3 (PLCP and Direct mapping)
- ADSL
- DS1/E1
- E3/E4

2.4.3.4.3 ATM Layer. The ATM layer is independent of the physical layer. That means designers will not have to develop different ATM modules for different physical interfaces described above.

ATM cell structure. The ATM cell contains 5 bytes header and 48 bytes payload as shown in Fig. 2.10. Two ATM header cell structures were defined by ITU:

1. ATM header for the UNI

2. ATM header for the NNI (Network Network Interface)

With the exception of the GFC field, the two structures are functionally similar and therefore only the UNI header will be briefly described. The GFC 4 bits are added to the VPI field for the NNI

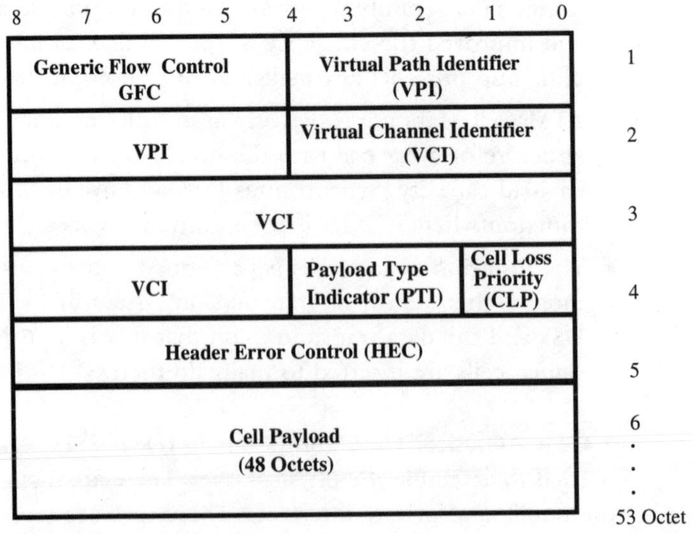

Figure 2.10
ATM Header Structure for the UNI

The structure of the ATM cell at the UNI, shown in Fig. 2.10, contains the following fields:

Generic Flow Control (GFC): This field has local significance only and can be used to provide standardized local functions (e.g. flow control) on the customer site. The value encoded in the GFC is not carried end-to-end and will be overwritten by the ATM switches. In the uncontrolled mode, the GFC will be set to a zero value.

Virtual Path/Virtual Channel (VPI/VCI) Identifier: The number of routing bits in the VPI (8 bits) and VCI (16 bits) subfields are used for routing. The VPI/VCI value is allocated by the network during call negotiation between the user and the network.

Preassigned VPI/VCI values: ITU defined a set of pre-assigned VPI/VCI values for specific use. These values are unique and cannot be used for any other function. Table 2.1 shows the values and associated functions.

Meta-signaling cells are used by the meta-signaling protocol when establishing and or releasing signaling connections.

TABLE 2.1

Predefined VPI/VCI
Header Value

USE	Value[1,2,3,4]			
	Octet 1	Octet 2	Octet 3	Octet 4
Unassigned cell indication	00000000	00000000	00000000	0000xxx0
Meta-signalling (default)[5,7]	00000000	00000000	00000000	00010a0c
Meta-signalling[6,7]	0000yyyy	yyyy0000	00000000	00010a0c
General broadcast signalling[5]	00000000	00000000	00000000	00100aac
General broadcast signalling[6]	0000yyyy	yyyy0000	00000000	00100aac
Point-to-point signalling (default)[5]	00000000	00000000	00000000	01010aac
Point-to-point signalling[6]	0000yyyy	yyyy0000	00000000	01010aac
Invalid pattern	xxxx0000	00000000	00000000	0000xxx1
Segment OAM F4 flow cell[7]	0000aaaa	aaaa0000	00000000	00110a0a
End-to-End OAM F4 flow cell[7]	0000aaaa	aaaa0000	00000000	01000a0a

1: "a" indicates that the bit is available for use by the appropriate ATM layer function
2: "x" indicates "don't care" bits
3: "y" indicates any VPI value other than 00000000
4: "c" indicates that the originating signalling entity shall set the CLP bit to 0. The network may change the value of the CLP bit
5: Reserved for user signalling with the local exchange
6: Reserved for signalling with other signalling entities (e.g., other users or remote networks)
7: The transmitting ATM entity shall set bit 2 of octet 4 to zero. The receiving ATM entity shall ignore bit 2 of octet 4.

General broadcast signaling cells are used by the ATM network to broadcast signaling information independent of service profiles.

Segment OAM F4 flow cells have the same VPI value as the user-data cell transported by the VPC, but are identified by two unique pre-assigned virtual connection.

Payload Type (PT): The PT is a 3-bit field and used to indicate whether the cell contains user information or layer management, network congestion state or network resource management. The detailed coding of this 3-bit field is as shown in Table 2.2.

Code points 0 to 3 are used to indicate user cells information. Bit 2 indicates the congestion condition of the connection referred to it as EFCI (Explicit Forward Congestion Indicator).

Code points 4 and 5 are used for VC level management functions. Code point 4 is used for identifying OAM cells, while the PTI value of 5 is used for identifying end-to-end OAM cells.

Cell Loss Priority (CLP): This one 1-bit field, in the ATM cell header, allows the user or the network to indicate the explicit loss priority of the cell. The user may want to set priority on ATM cells that could be discarded if the network is congested.

Header Error Control (HEC): The HEC field is used by the physical layer for detection/correction of bit errors in the cell header. HEC is also used to perform delineation of cell boundary in the physical layer.

When a framing error occurs, the HEC sequence will invariably be corrupted. The state machine goes into a hunt state looking for a correct HEC. It will slide the window by one bit and recalculates HEC until a correct one is found. The calculation is HEC multiplied by 8 and divided by the poly-

TABLE 2.2	**PTI Coding**	**Description**
PTI Coding	000	User data cell, congestion not experienced, SDU-type = 0
	001	User data cell, congestion not experienced, SDU-type = 1
	010	User data cell, congestion experienced, SDU-type = 0
	011	User data cell, congestion experienced, SDU-type = 1
	100	Segment OAM F5 flow related cell
	101	End-to-end OAM F5 flow related cell
	110	Reserved for future traffic control and resource management
	111	Reserved for future functions

nomial $X^8 + X^2 + X + 1$. Once the correct HEC is found, the state machine goes into a Pre-Sync state. In the pre-synch state, the correct HEC must be derived "n" numbers of times before proceeding into a synch state (to insure validity). Otherwise it goes back to hunt state. In the synch state "m," incorrect consecutive HEC will bring the system into hunt state. The numbers for "n" and "m" are defined by ITU to be 6 and 7 for the SONET based physical interface.

2.4.3.5 ATM Adaptation Layer. The AAL is the layer above the ATM layer. Again this layer in independent of the layers above or below it. The function of AAL is to enhance the service provided by the ATM layer and support the specific adaptation of services above it. The upper layers provide control, management and user services. AAL has two sublayers as shown in Fig. 2.11.

1. Segmentation and Reassemble (SAR) Sublayer

2. Convergence Sublayer (CS)

2.4.3.5.1 Segmentation and Re-assembly (SAR). The SAR sublayer on the transmit side will segment the information from a higher layer into a size to fit the ATM payload (48 bytes, without overhead). On the receiving end, the SAR will assemble the data into packets for delivery to the higher layer.

2.4.3.5.2 Convergence Sublayer (CS). The CS sublayer performs several functions. It is service dependent. Services performed over the ATM layer are grouped into four classes. The four classes are:

1. Class A

2. Class B

3. Class C

4. Class D

Three basic parameters are used to classify each of the service. The parameters are: time relation between source and destination; bit rate; and connection mode.

Figure 2.11
SAR and CS: Sublayers of AAL

Segmentation And Re-assembly (SAR) Convergence Sublayer (CS)	AAL Adaptation Layer

Time Relation Between Source and Destination: Voice and video services usually require time relationship between end users. Data services, on the other hand, needs no time relation. Data can wait in buffers before dispatching it to destination.

Bit Rate: This parameter differentiates between a constant bit rate and variable bit rate.

Connection Mode: The connection parameter can either be a connection-oriented connectionless service.

ITU in recommendation I.362 defined the 4 classes of services and mapped with these associated parameters as shown in Fig. 2.12.

Class A Service.   Class A service was classified as a connection requiring: time relation between end users; connection oriented; and at Constant Bit Rate (CBR). This service is sometimes referred to it as Circuit Emulation (CE). Transporting a DS1 between end users (PBX trunk) is a typical example of the service. MPEG-2 video may also use Class A service.

Class B.   Class B service is classified as a connection requiring: time relation between source and destination; operation in a connection oriented mode; and has a Variable Bit Rate (VBR).

Class C.   Class C service is classified as a connection requiring: no time relation between end users; operation in connection oriented mode; and has a variable bit rate. Typical service over Class C is the transfer of medium or large files.

Class D.   Class D service is classified as a connection requiring: no time relation between end users; operation in connectionless oriented mode; and has a variable bit rate. SMDS (Switched Multimegabit Data Services) is one typical application of this service.

Figure 2.12
Adaptation Service
Classes

	Class A	Class B	Class C	Class D
Timing Relation	Required		Not Required	
Bit Rate	Constant	Variable		
Connection Mode	Connection Oriented			Connectionless

There are three AALs that had been well defined in ANSI, ITU and the ATM Forum. Another AAL-CU (AAL—Composite User) is in its final stage of definition. Recently ITU gave it the name AAL2. The Three AALs that are well defined are: AAL1, AAL2, AAL3/4, and AAL5. AAL1, AAL5 and AAL-CU (AAL-2) will play a major role in the ADSL and IEEE 802.14 cable modem development and therefore brief description of these AALs are as shown below.

2.4.3.5.3 AAL1. AAL1 is defined under Class A service. It is connection oriented, requiring time relation between the source and destination and constant bit rate.

2.4.3.5.3 AAL5. The data equipment manufacture first started AAL5 development in ANSI (T1S1). It was then called <u>SEAL (Simple and Efficient AAL)</u>. The ATM forum and ITU completed this AAL5 specification. The objective of AAL5 was to reduce the overhead in the SAR/CS. The AAL5 other objective was to devise a better error detection mechanism for data services. ITU selected AAL5 for transporting signaling information for the control plane.

2.4.3.5.4 AAL-CU (AAL-2). ITU and ANSI (T1S1) have long discussed voice over ATM. Voice samples carried over the ATM payload is very inefficient since voice samples are very short. Filling up the ATM cell with voice samples will cause echo due to the delay encountered in transmission and cell reassembly. A partially filled cell was another approach discussed, but did not go far because of its obvious inefficiency of transporting voice. The Composite User (CU) method is a way of multiplexing several users voice samples into minicells and embodied them into a single ATM cell payload. In summary, the two fundamental benefits of AAL-CU are:

1. Efficient bandwidth utilization: Multiplexing users over a single ATM VC allows for efficiency in bandwidth.
2. Low-delay: Multiplex of short packets (mini cell) in the payload of a single ATM cell can be re-assembled with less much less delay.

This ends the brief descriptions of the foundation layers (low layers). Next, the higher layers of B-ISDN are briefly described.

2.4.3.6 B-ISDN Higher Layers. The higher layers, as mentioned in the protocol reference model are:

- User plane
- Control plane
- Management plane

2.4.3.6.1 User Plane. The user plane contains user information and service-specific protocols that ride on top of ATM. IP over ATM or Frame relay over ATM are a few examples in which such service is delivered end-to-end. The associated service-protocol is encapsulated using the ATM data unit and transported via the appropriate AAL. In most cases, these services have their own protocol but are preempted by the B-ISDN protocol layers.

2.4.3.6.2 Control Plane or Signaling. This control plane provides signaling message transport and call control/connection control capabilities.

Signaling Definition. Signaling is a mean of dynamically establishing, clearing and or modifying a call or a connection. A call may have several connections. Permanent connection between users is practiced today. The operator simply establishes the call (via a network management element) assigning VPI/VCI connection for the requested user(s). Tearing down that connection to manually clear it will also require operator intervention. This is obviously a clumsy and an expensive task if the call duration is short. Signaling, therefore, is required when the average holding time of a call does not exceeds a certain period of time. Four to six minutes is the average holding time of a voice call today.

ITU released several recommendations dealing with signaling protocol, procedures, and interworking. Recommendations Q.2931 for user signaling and Q.2761 for NNI (B-ISUP) signaling were based on a remnant of the signaling recommendations developed for Narrow Band ISDN (NB-ISDN). Both of these recommendations are the foundation needed to develop signaling for broadband private and public networks. Unlike NB-ISDN, broadband added capabilities included features such as: modifying a live connection e.g., change bandwidth; modify the connection's QOS; Interwork with Narrow band signaling; and multimedia signaling capabilities, etc.

Both the ITU Q.2931 and Q.2761 use out-of-band signaling, meaning that the signaling messages have an independent communication channel. VPI = 0 and VCI = 5 virtual channel was assigned exclusively. This VPI/VCI = 5 is used for signaling communication between the user and the network.

Figure 2.13 illustrates the set of the Q series recommendations (mapped into a broadband network) developed or under development by ITU to fully implement signaling in the network. It is beyond the scope of this book to describe each one of the Q series recommendations, but the authors will briefly describe a few that will have immediate impact on ADSL and HFC/cable modem development.

Signaling at the UNI. UNI 4.0, released in 1996 from the ATM Forum, is aligned with ITU signaling recommendation. Below are some important features:

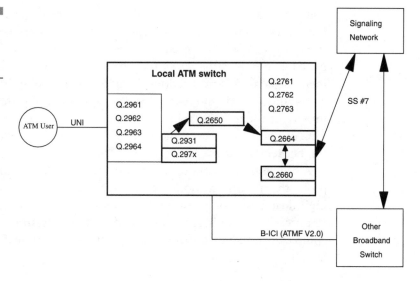

Figure 2.13
Q Series Recommendation Mapped into Broadband Network

- Point-to-point Calls
- Point-to-multipoint Calls
- Signaling of Individual QoS Parameters
- ABR Signaling for Point-to-point Calls
- Switched Virtual Path (VP) service
- Proxy Signaling
- Frame Discard
- Traffic Parameter Negotiation
- Calling Line Identification Presentation (CLIP)
- Calling Line Identification Restriction (CLIR)
- Connected Line Identification Presentation (COLP)
- Connected Line Identification Restriction (COLR)
- Subaddressing (SUB)
- User-user Signaling (UUS)

2.4.3.7 Signaling in Multimedia. More demanding features such as multimedia are important aspects for residential broadband. ITU has extended the call model of an ATM call such that a call can include multiple connections or zero connections. Hence, a call between two users becomes an association that describes a set of connections between them. This concept was extended to three or more users to become a multiparty/multi-connection call. There can be multiple calls between two users. By extending the call between two users to support multiple connections, multimedia applications can be supported more effectively. Each media type of a multimedia communication application has its own bandwidth and QOS requirement. Separate VC connections became feasible and satisfied the multimedia communication requirements. The concept of a call with a multiparty is very important for residential broadband service, because many applications running on the set-top box could require communication with various servers in the headend or at the network.

A call without a connection is also useful. When making a call, a user can probe the network without actually setting up the connection, thereby wasting valuable network resources. If a connection is established first, network resources could be held up temporarily, only to learn that the call cannot be completed because the end user is busy or its terminal capabilities do not match.

2.4.3.7.1 Connection Types. ITU defined five types of connections covering all possible aspects needed for connecting users specially for residential broadband. It is beyond the scope of this book to describe the details of these connections and their associated performance requirements. A general and brief definition is as shown below. The five connection types are:

Type 1 connection is a point-to-point configuration with bi-directional and asymmetric communication.

Type 2 connection is point-to-multipoint unidirectional configuration with unidirectional communication and with a Root and Leaf control. Type 2 connection is required to support multicast services e.g., broadcast TV. The leaf-initiated join mechanism allows the viewer to join a particular TV channel.

Type 3 connection is a multipoint-to-point unidirectional configuration with unidirectional communication, and with leaf and single root control. Typical application for a Type 3 connection is advertisement insertion. This requires merging of multiple (video clips) from different servers

Type 4 connection is a multipoint-to-multipoint bi-directional configuration with bi-directional and full party control. Type 4 connection is important for supporting group communication, such as off campus education, multi-user games and multiparty video calls.

Type 5 connection is a point-to-multipoint bi-directional connection with bi-directional communication, and root and leaf control.

2.4.3.8 Management Plane. The management plane is composed of two layers:

1. Plane management

2. Layer management

2.4.3.8.1 Plane Management. Plane management is responsible for overall functions of the system end-to-end. Its main role is to report, extract and coordinate all management information between all the planes shown in the Broadband Reference Model.

2.4.3.8.2 Layer Management. Layer management performs management related functions such as performance, operation, administration, resource management and parameters associated with each of the user planes. ITU recommendation I.610 defined basic OA&M principles that provide the status, testing performance, monitoring etc. The basic functions of OA&M defined by ITU are:

- performance monitoring
- defect and failure detection
- system protection
- performance information
- fault localization

2.4.3.9 Other Aspect of B-ISDN.

In this section, the following will be described:

1. User-network interface

2. Traffic control

3. Service categories

4. Routing

2.4.3.9.1 User-Network Interface. ITU recommendation I.413 defined the user-network interface for B-ISDN as shown in Fig. 2.14. The reference configuration was generic and covers all aspects of the user interface configurations and connectivity with various private to pubic operator switches.

Figure 2.14
B-ISDN Reference Configuration of Customer Equipment

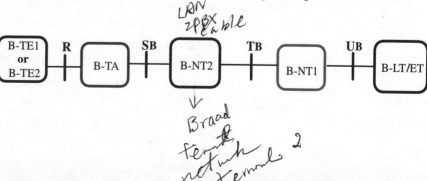

There are four reference points: R, S, T and U. Definition and functions of each are:

B-TE1 and B-TE2 are broadband terminal equipment 1 and 2, which terminates all protocols from the low and higher layers.

B-TA is a terminal adapter when needed.

B-NT2 is the broadband network termination 2. It could be a LAN, PBX, smart cable modem i.e., set top box. It performs signaling protocol, adaptation layer handling, policing, cell multiplexing, traffic control, etc.

B-NT1 is the broadband network terminal 1. It performs line termination and OAM functions.

B-LT is broadband line termination at the end office.

Possible variation of this configuration is the star and shared-medium configurations. Functionally, however, they all behave in a similar manner.

2.4.3.9.2 Traffic Control.   Traffic control is the central nervous system that is responsible for the overall operation of the broadband network. The primary role of the parameters and procedures used for traffic control and congestion control is to protect the network and the end-system in order to achieve network performance objectives. The network must operate under the worst case condition and must also match today's network reliability and availability which the user had enjoyed for decades. To do less will undermine the deployment of broadband in general and ATM in particular.

The previous chapters, described how a user can have all these Multimedia ATM features with associated QOS on demand and so on. These features are unique and extremely valuable in making multimedia work the way it is envisioned. The industry has little practical experience in this area of complicated traffic control couple with maintaining a QOS. The issues that are partially addressed are:

- How to deal with all these traffic variables
- Keeping the network sane under fault condition and avoid congestion collapse
- Protecting the network by maintaining the performance objectives of cell loss, CDV, and cell transfer delay
- Maintaining objective performance in presence of unpredictable statistical traffic fluctuation (Mother's Day syndrome)
- How best to engineer the network in such an environment, while optimizing network resources to achieve practical network efficiency.
- In summary, traffic management objectives must exhibit:
    - Robustness

■ Flexibility

■ Simplicity (e.g., must be operator friendly to avoid network congestion collapse, or total traffic outage)

ANSI, ITU (I.371/I.311 recommendation) addressed most of these issues. They are described below. The critical topics covered briefly are:

■ Connection Admission Control (CAC)

■ Usage Parameter Control/Network Parameter Control (UPC/NPC)

■ Priority Control (PC)

■ Traffic Shaping (TS)

■ Network Resource Management (NRM)

■ ABR Traffic Control

Figure 2.15 illustrates the positioning of these funcitons in the network.

Connection Admission Control (CAC). CAC contains a set of actions performed by the network during initial call set-up with an ATM end-user. CAC accepts the requested connection if sufficient resources are available in the whole network to accommodate such QOS and bandwidth requests. The call is accepted if and only if the system can maintain, and preserve the QOS of all other connections already established. If resources are not available, the network will simply deny the connection or may re-negotiate the parameters with the end user.

Figure 2.15
Traffic Control Network Functions

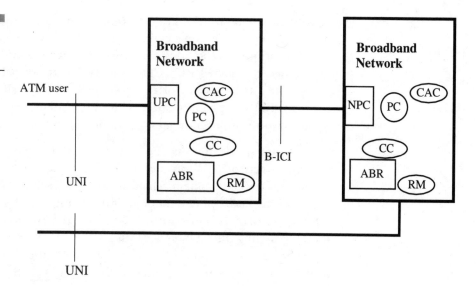

CAC traffic contract may change during the life of the call. An ATM user may want to add, delete, increase or decrease the QOS or bandwidth that was originally negotiated during the initial call set-up time. In this case, CAC performs the same functions as described above, as if a new call is established. CAC is operator implementation dependent. An operator may even choose to control his network resources to first accommodate his best paying customers.

Usage Parameter Control (UPC). In an ATM environment, unlike in a Synchronous Transfer Mode, the ATM user may at any time opt to send information up to his physical access rate. Therefore, throughput must be policed at the UNI interface by the Usage Parameter Control (UPC) function in the network. This ensures that the contract originally negotiated during setup phase with the user is respected and fair to the other connections. The set of actions taken by UPC are:

1. monitoring

2. enforcing

In the monitor mode, the UPC validates routing of the VPI/VCI connections that are being used, and monitors every connection and validates its traffic characteristics and volume per the negotiated contract. As the "traffic cop," the UPC enforces the negotiated traffic contract of any and all misbehaving users (or due to equipment failure).

The Network Parameter Control (NPC) at the B-ICI (interface between two public broadband switches) functions in the same manner.

Priority Control (PC). The priority control is a feature the user may elect to use by setting priority on his traffic. He may choose to set the CLP bit to "1" (low priority traffic) so the network, if congested, will discard them first as opposed to the cell with CLP = 0. This network behavior insures that performance is preserved for those of high priority cells in case of traffic congestion of equipment failures. This also guarantees a priority service for the user who inadvertently submits traffic beyond his negotiated contract.

Traffic Shaping. Traffic shaping partially alters the characteristic the traffic and the effects of CDV on the peak cell rate of the ATM connection. An example of traffic shaping mechanisms is to throttle and therefore limit burstiness of the peak cell rate traffic so it may comply with the negotiated peak cell rate contract. This mechanism also allows the user to prioritize his bursty traffic by setting the CLP bit accordingly.

At the ATM switch, traffic shaping alters the traffic characteristics cell flow so it can reduce the network jitter or Cell Delay Variation (CDV). It is done by suitably spacing cells in time.

Network Resource Management. The main function of the network resource management is to allow an operator to segregate his traffic characteristic based on the differing QOS. This mechanism simplifies CAC and reserves capacity in virtual path connections separated from other traffic. This mechanism may reduce operational control cost vs. increased capacity cost.

2.4.3.9.3 Service Categories. The architecture for services provided at the ATM layer consists of the following five service categories:

1. CBR   Constant Bit Rate
2. rt-VBR   Real-Time Variable Bit Rate
3. nrt-VBR   Non-Real-Time Variable Bit Rate
4. UBR   Unspecified Bit Rate
5. ABR   Available Bit Rate

This section defines the ATM service categories using the following QOS parameters:

- Peak-to-peak Cell Delay Variation (CDV)
- Maximum Cell Transfer Delay (Max CTD)
- Mean Cell Transfer Delay (Mean CTD)
- Cell Loss Ratio (CLR)

Constant Bit Rate (CBR) Service Category Definition. The CBR service category is intended for real-time applications, i.e. those requiring tightly constrained delay and delay variation, as would be appropriate for voice and video applications.

Real-Time Variable Bit Rate (rt-VBR) Service Category Definition. The real time VBR service category is intended for real-time applications, i.e. those requiring tightly constrained delay and delay variation, as would be appropriate for voice and video applications.

Non-Real-Time (nrt-VBR) Service Category Definition. The non-real time VBR service category is intended for non-real time applications which have bursty traffic characteristics and which can be characterized in terms of a PCR, SCR, and MBS.

Unspecified Bit Rate (UBR) Service Category Definition. The Unspecified Bit Rate (UBR) service category is intended for non-real-time applications, i.e. those not requiring tightly constrained delay and delay variation. Examples of such applications are traditional computer communications applications, such as file transfer and email. UBR sources are expected to be bursty. UBR service supports a high degree of statistical multiplexing among sources.

UBR service does not specify traffic related service guarantees. Specifically, UBR does not include the notion of a per-connection negotiated bandwidth.

Available Bit Rate (ABR) Service Category Definition. ABR is an ATM layer service category for which the limiting ATM layer transfer characteristics provided by the network may change subsequent to connection establishment. ABR service was described in the previous section.

2.4.3.9.4 Routing. ATM is basically connection oriented. The ATM header values are assigned to each end of a connection for the duration of the call. The header identifier is translated within a switch from one port to another. Signaling and user information are carried on separate virtual channels. There are two types of connections that are possible: virtual channel connections (VCC); and virtual path connections (VPC). A VPC can be considered as an aggregate of VCCs. When cell switching is performed, it must be done based on the VPC, then on the VCC. The 5 byte ATM header of an incoming port is translated by the ATM layer and switched to an output port based on the assigned VPI/VCI value. Figure 2.16 illustrates how switching is performed in ATM. In general, all ATM switches behave in the same manner. Input ports are ports$_1$ to ports$_m$ transport ATM cells based on their header information to output$_1$ to output$_m$ Headers of input ports are translated using the translation table to the appropriate output port having a new header value. For example, an ATM header "c" on input ports$_1$ becomes (per translation table in Fig. 2.16) ATM header "w" output$_2$.

Figure 2.16
ATM Switching
Principle

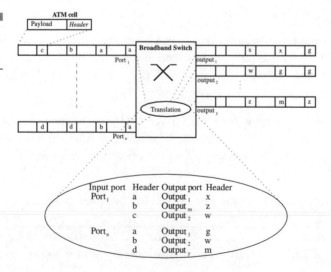

In this way ATM cells are switched based on their respective VPI/ VCI values. These VPI/VCI values have only local significance. The table is created during call setup when the switch assigns these VPI/VCI values to the calling party. The table is maintained throughout the life of the call and disassembled (removed VPI/VCI association from the table) when the call is terminated.

VCC = Virtual Channel Connection
VPC = Virtual Path Connection

2.5 Internet

No one person had the vision, knowledge, skills, and resources to create what we see in the Internet today. Instead, it was a spontaneous ordering process involving numerous individual contributors. The names listed in this summary (below) represent only a portion of everyone involved in the creation of the Internet. There are many sources of information on this subject with slight variations in the names and dates of those involved. Therefore, it is entirely possible that something within this historical account is inaccurate, or more likely, not the best choice of reference.

2.5.1 Internet Historic Perspective

On October 4, 1957, the USSR launched Sputnik and astounded Americans with their missile technology. On January 7, 1958, concern over secure military communications prompted President Eisenhower to appropriate $520 million for Space and Strategic Missile Research in the Department of Defense (DOD). Although the majority of the money was given to NASA (a division of DOD), the remaining $150 million was allocated for the Advanced Research Projects Agency (ARPA). The purpose of ARPA/DARPA was to define research topics and manage programs that fund these areas. The United States military strategists acknowledged a need for a decentralized communications network that could withstand a nuclear attack. Other governments (Germany, Britain, and Japan) also funded network development, but none of these produced anything like the Internet.

The Rand Corporation did a series of studies for the U.S. Air Force, which outlined the theoretical basis for a packet-switching network. In 1962, the report titled, "On Distributed Communications Networks," by Paul Baran was published by the Rand Corporation. Packet switching (which was originally coined by Donald Davies of the U.K.) would permit messages to be routed around damaged parts of the network. Also, in the early 1960's, Len

Kleinrock did his thesis on packet switching at Massachusetts Institute of Technology (MIT). At the same time, Ted Nelson (who later invented and coined the term "hypertext") wrote a plan for "Xanadu," a conceptual prototype of today's World Wide Web (WWW). In October 1962, ARPA hired J.C.R. Licklinder (from Bolt, Beranek and Newman—BBN or BB&N) to investigate how to utilize military investment in computers in the Control and Command Research (CCR) subdivision of ARPA. He introduced the idea of the computer as a communications medium between people, rather than as an arithmetic engine. This very "unbureaucratic" subdivision of ARPA was officially renamed the Information Processing Techniques Office (IPTO) in 1965, when "Lick" became the director. The IPTO researched the idea of networking computers together in order for ARPA's principle investigators to share resources. In early October 1967, the IPTO (where Lawrence Roberts was the chief scientist, later to become the director) proposed these ideas which included the creation of an Interface Message Processor (IMP) subnetwork. Later in 1967, ARPA gave the contract (to write IMP specifications) out to Stanford Research Institute (SRI). From this research (largely done by Elmer Shapiro), Lawrence Roberts and Barry Wesler (ARPA) wrote the final version of the IMP specs.

The four original host-sites of this mainframe connecting experiment (now known as ARPAnet) included UCLA, SRI, UCSB, and the University of Utah. They were picked because they provided either network support services or unique resources. Researchers at each site were responsible for writing the software that was necessary to connect their computers to the IMP at their site. Each site had a different kind of computer to connect. By trusting that the programmers at each site would be capable of modifying their operating systems, ARPA made the academic computer science community an active part of the ARPAnet development team. In 1968, a group met to establish agreed-upon standards (protocols) and discuss programming technical problems. This group, called the Network Working Group (NWG), documented their commentary, called Request for Comments (RFC), to be used by the mainframe computers. The first node was installed at UCLA (where Prof. Kleinrock was now teaching) on September 1, 1969. The second node was installed at SRI (where Prof. Engelbart taught) on October 1, 1969. The first message sent was: "Log-in. Crash." The UCSB and the University of Utah were connected respectively, at one-month intervals.

In March 1970, ARPANET spans the US, connecting BBN (node #5) into the net. By 1971, 15 nodes were connected. Basic email programs (called SNDMSG and READMAIL) by Ray Tomlinson (BBN), were created in 1972, along with the first email management program (developed by Larry Roberts). Also in 1972, a time slotted radio network, ALOHAnet (developed by Norman

Abrahams at the University of Hawaii) was connected to the ARPAnet. Robert Kahn, of BBN, organized the first public ARPAnet demo in October of that same year at the first International Conference on Computer Communications. There was much skepticism about packet switching and little enthusiasm for this "government" project (outside of the technical groups) until this successful computer conference. Vinton Cerf was chosen to be the Chairman of the International NWG. In July 1972, File Transfer Protocol (FTP) specifications were released by Jon Postel (a request for comments editor), and Abhay Bhushan (NWG chairman). Later that same year Robert Khan was hired by Larry Roberts into ARPA, and the following year, he and Vinton Cerf outlined the TCP/IP (transmission control protocol/Internet protocol). In May 1974, they published the report titled, "A Protocol for Packet Network Interconnection" (originally known as Kahn-Cerf protocol). In 1975, the ARPAnet was transferred to the control of the Defense Communications Agency (DCA). In July 1976, Vint Cerf joined ARPA as the program manager of the packet radio and packet satellite network. By March 1977, there were 62 hosts connected to the ARPAnet. In March 1978, TCP split into TCP and IP. This was also the year that the Apple II was launched; it was the first retail mass market of personal computers (PCs).

Although the "no strings attached" federal defense dollars undoubtedly made it easier for the early networking pioneers to concentrate on the technical details of their work, private-sector activity was slowed because of the absorbed tax dollars and resources. In October 1973, Larry Roberts left ARPA to join Telenet (a private packet switching network) as the CEO. In 1975, Telenet experienced many regulatory barriers to entry, and thus did not show a profit until the early 1980's. Also USENET (a private network among universities) sprang up and supposedly resembled today's Internet much more then the ARPAnet did (USENET still exists today—it is a collection of newsgroups). Barriers to entry, not a lack of entrepreneurial activity slowed the efforts to build private networks.

During the 1980's, 3 major technologies gained public visibility: the IBM PC, the Ethernet local area network (LAN) designed by Robert Metcalf, and the UNIX operating system (originally designed by Ken Thompson). In 1980, Tim Bernes-Lee wrote a program called "Enquire Within," which was the predecessor of his World Wide Web (WWW) program. In September 1981, there were 213 nodes on ARPAnet. The National Science Foundation (NSF) organized the Computer Science NETwork (CSNET), taking over 70 sites by 1983, and integrating most computer science sites by 1986. CSNET provided network services (email) to non-ARPAnet sites. Also in 1981, the IBM PC was introduced and the following year Drew Major (and his SuperSet colleagues) networked an IBM PC. In 1983, ARPAnet began using TCP/IP (which is today's Internet standard protocol) enabling every ARPAnet computer to connect with every other one (regardless of operating system) and technically became the "Internet." The DCA split MILnet (military network—45 nodes remaining) from ARPAnet (68 nodes remaining). As well, the domain naming system (DNS) was designed by Jon Pastel, Paul Mockapetris, and Craig Partridge. Carlton Amdahl tossed his "smart bridge" idea around with SRI professors, who refined the idea. In 1983, Cisco Systems was founded by Sandy Lerner (who named Cisco from "San Francisco") and Len Bosack. In 1987, Sequoia Capital invested $2 million in Cisco for 1/3 of their stock. In 1990, soon after Cisco went public (valued at $288 million), Lerner was fired and Bosack resigned. Approximately 73 percent of the routers in the Internet today are produced by Cisco. In 1985, the National Science Foundation (NSF) organized the NSFNET backbone to connect 5 supercomputing centers and interconnect all other Internet sites. This was the same year that bulletin boards were invented using the GUI (Graphic User Interface). In 1986, there were 5,000 hosts on the Internet; in 1989, there were 100,000.

1990 marked the year that the ARPAnet was "reinstalled," and traffic shifted to a mix of public and private transmission lines by 1995. It was also a big year for Tim Bernes-Lee, who created the WWW (web) at the Counsel European for Particle Research (CERN). The following year, CERN published a copy of the WWW code on the Internet. In 1992, commerce was permitted to flourish. The Web really began to grow in 1993 with the development of the mosaic browser by Marc Andreessen (and others) at UCLA. He was 24 years old then; the following year he joined Jim Clark to become the co-founder of Netscape Communications and produced the popular Netscape Navigator. In July 1994, Odyssey Homefront studies found that 27 percent of U.S. households had home computers and 6 percent of these were on-line. It is estimated that the Web was growing at a rate of 341,000 percent/year. Even the White House and the United Nations were on-line at this time. Commercialization of the Internet had begun. In 1994, Microsoft licensed technology from Spyglass for their net browser (Internet Explorer) to be used with Windows 95. A great deal of political controversy was created when Microsoft bundled their browser together with their Win95 software; legislation forbids tying of the purchase of one product to another. In August 1998, Microsoft asked a federal court judge to dismiss the antitrust lawsuit, but the request was denied. Zona Research has published survey indicating that in July 1998, 54 percent of business offices used Netscape's browsers (including both the Navigator and Communicator versions), while the remainder used the Internet Explorer.

2.5.2 Internet Commercial Landscape

Today the Internet backbone, now known as the very high-speed backbone network services (vBNS) is maintained by IBM corp., MCI Comm., and Merit (a non-profit organization owned by 11 public universities in Michigan). Outside the U.S., Internet access is provided by international carriers (AT&T, British Telecom, MCI WorldCom, etc.). The U.S. accounts for approximately half of the world's total Internet users (Computer: Consumer Services & the Internet). In July 1998, Homefront studies indicated that 45 percent of U.S. households had home computers and 27 percent of these were on-line.

This new communication technology is quickly replacing traditional forms. The emergence of the Internet has produced a 32 percent decline in television use, and a 25 percent decline in telephone use ("Evolution of the Internet," by Vinton Cerf, December, 96). More messages are sent through the Internet then the U.S. post office. According to Science magazine (1998), 320 million web pages exist (with millions more added each month) as net use doubles every 100 days. More than 70 million adults in the U.S. used the Internet regularly as of the second quarter of 1998 (Nielsen Media Research). Its popularity has influenced the appearance of support groups, such as "Internet-eers Anonymous" and "Webaholics" to address the concern over excessive use (IAD—Internet Addiction Disorder; discussed at the 1996 Convention of American Psychology Association). This spontaneously growing virtual community ignores national borders; it is present in over 186 countries. Annual growth rates (percentage of connected hosts) range from 63 percent/year in Europe, 90 percent in North America, 111 percent in Africa, 136 percent in Asia, and 152 percent in Latin America (ISP Survival Guide, 1999).

On-Line Activities - Odyssey 1996

Chat lines / chat room (7.5%)
Entertainment (17.9%)
Email (30.2%)
News/Information (20.8%)
Research (23.6%)

The commercialized race to be "on-line" has had many interesting effects. The *Financial Times* reported in March 1998 that of the 25 most valuable Internet companies (with a combined value of $37 billion), 20 of these companies are still operating at a deficit. But their soaring high stock prices don't seem to have been affected by these financial reports. The optimistic mood pervading in the Internet industry has provoked some financial investors to view these phenomena as "investing in potential"; others see these stocks as "high risk." The market will determine which company's become multibillion dollar corporations and which go bankrupt. Financial advisors predict that the telecommunications industry will continue to grow 10 percent/year for the next 10 years. The Internet puts the customers in direct contact with the suppliers, thereby eliminating the need for the distributor in many cases. Small businesses value the Internet; they can make their mom and pop proprietorships look just as large as big corporations.

Portals are navigation companies that provide customers with search engines to locate websites. Today these portals offer more competitive services such as email, chat, stock quotes, news headlines, on-line games, and apartment listings. Although their main source of revenue is advertising, some portals are also looking into on-line billing and buying services to increase their revenue. They will also probably realize revenues through electronic commerce, such as when a customer buys a product through an advertisement seen on a portal. In 1998, $4.8 billion was invested in on-line shopping; in 2001, that number is projected to become $17.3 billion. The "www" acronym has sarcastically been described as the "world wide wait" as users anxiously await for the search engines to retrieve data, often only to find thousands of possible websites to choose from. There is some controversy about the fact that some of these search engines retrieve their list of websites in the order of how much money they were paid by that site (High-tech Stocks and Mutual Funds, p. 127). As of July 1998, Yahoo has amassed 18 million registered users—up 50 percent from January 1998. However, their competitors are closing in on their lead. Odyssey's Homefront study indicated that 63 percent of on-line U.S. households occasionally use Yahoo, followed by Excite with 38 percent and Infoseek with 32 percent.

2.5.2.1 Top 15 Most Visited Websites

1. yahoo.com	29,519
2. aol.com	23,088
3. netscape.com	18,470
4. microsoft.com	17,784
5. excite.com	15,904
6. geocities.com	14,820
7. infoseek.com	11,528
8. lycos.com	10,081
9. msn.com	9,615

10. altavista.digital.com	9,116
11. tripod.com	8,048
12. hotmail.com	6,533
13. angelfire.com	6,202
14. zdnet.com	5,859
15. amazon.com	5,664

Customers interested in Internet access have a choice between using an Internet Service Provider (ISP), or an Online Service Provider (OSP). ISPs normally offer the www, newsgroups, and email. They have a direct connection to the Internet, and tend to be cheaper and faster. OSPs usually provide other services—some of which cost extra to activate. Odyssey's Homefront study reveals that America Online's (AOL) lead in the market of commercial online services is continuing to widen. According to IDC/Link, AOL (acquired CompuServe) has penetrated 54.8 percent of the market while Microsoft has 9.2 percent, Prodigy has 6.0 percent, AT&T WorldNet has 6.5 percent, and Internet MCI has 1.7 percent within the US (as of December 1997). After these, the remaining 4,500 service providers compete over the remainder of the market. Dataquest (San Hose, CA) estimates that the service provider market will expand from $11 billion in 1997, to $38 billion in 2002. There are over 30,000 service providers worldwide (ISP Survival Guide). For an international view of the access providers, download a daily updated, color-coded Internet map (courtesy of Bell Labs) from: http://ww.cs.bell-labs.com/~ches/map/db.gz.

2.5.3 Current Internet Structure

The Internet is an astonishingly successful outcome from the networking research that the Defense Department's Advanced Research Projects Agency (ARPA) sponsored 25 years ago. It is not what one would naturally expect from a "government" project. Its success was not only the technology developed, but the manor in which the privatization has handled.

Initially, ARPAnet tied together the computers of military researchers and universities. The goal was to share computer resources and supported applications such as remote log in and file transfers. Later email, usenet discussion groups and the World Wide Web application were added. As traffic grew, the military portion was split from the civilian portion run by the National Science Foundation (NSF) known as NSFNET. At that point, the NSF allowed

of users to access NSFNET according to an "acceptable use
~~~~~rally forbade commercial traffic. Various commercial re-
~~~~ grew up around the NSFNET that provided them a back-
~~~~~ercial traffic. Eventually the NSF discontinued NSFNET,
~~~~~~mber of <u>Network Access Points (NAPs)</u> interconnected
~~~~~~one called the vBNS.

~~~~re of the Internet is shown in Fig. 2.17. Users access
~~~~ an Internet Service Provider (ISP) of which there are more
~~~~ today. Typically, individuals access ISPs via modems over dial up
~~ phone lines or basic rate ISDN from the telephone company. Businesses often connect to ISPs via T1 or even DS-3 dedicated private lines. The
ISPs provide access to regional IP networks. These regional IP networks interconnect according to bilateral agreements via a number of Network Access
Point (NAPs). The NAPs are interconnected via various commercial backbones and via the vBNS, a high-speed backbone for non-commercial use
funded by the NSF.

The Internet differs from the PSTN in many ways. Instead of carefully
planned routing tables downloaded into switching systems, the routers of the
Internet use a self-discovery process to determine the connectivity of the
network and decide themselves where to send traffic (often not over the most
ideal route). Rather than use redundancy throughout the network, the Internet achieves reliability via a mesh topology rich with alternate routes. Rather
than strict quality of service guarantees, the Internet provides a best-effort
service of widely varying quality. Rather than charge by distance and traffic
sent, the Internet is characterized by flat rate, distance insensitive pricing.

From its data application beginnings, the Internet has evolved to host a
number of multimedia applications. RealAudio streaming audio letting users

Figure 2.17
Internet Architecture

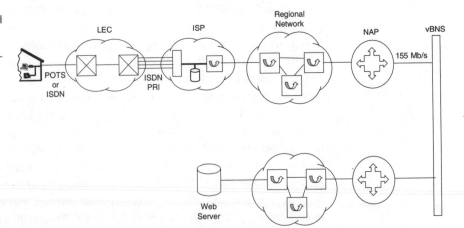

listen to remote radio stations, play music, or listen in on interviews. Rich Music Format (RMF) lets users interact with the music such as changing the tone or pitch. The MBone (Multicast Backbone) allows video broadcasts to a wide audience. Streaming video allows video files to be viewed over the Internet while it is being downloaded. Various videoconferencing applications have been developed for the Internet including NetMeeting and Eye2Eye. Animations are available via Shockwave, a multimedia authoring language. Even virtual reality applications are now available using the Virtual Reality Modeling Language (VRML).

The current collision between the Internet and PSTN was the result of the addition of telephony applications on the Internet such as the Internet Phone from VocalTec. The quality is only fair but the price is great—free if you already have Internet access.

2.5.4 The Next Generation Internet

For all its success, the Internet is not without its problems. Performance in particular can suffer during busy periods. Universities and research labs complain that because of this, the Internet is no longer a good tool for networking research. Recognizing this, the U.S. Government would like to see if lightning can strike twice and has begun a research project known as the Next Generation Internet (NGI) initiative. This is complemented with a university-led effort known as Internet2.

2.5.4.1 NGI Initiative. The NGI initiative was created with a broad agenda and the ability to involve government, research institutions, and the business sector. It is coordinated by the NGI Implementation Team under the Large Scale Networking Working Group of the Subcommittee on Computing, Information, and Communications (CIC) of the White House National Science and Technology Council's Committee on Technology. The Federal government believes that it has a unique role to play in stimulating technological progress. Through its NGI initiative, the U.S. Federal Government believes it will help create an environment in which advanced networking R&D breakthroughs are possible.

The NGI initiative began October 1, 1997, with the following participating agencies:

- DARPA: Defense Advanced Research Projects Agency
- NASA: National Aeronautics and Space Administration
- NIH: National Institutes of Health

- NIST: National Institute of Standards and Technology
- NSF: National Science Foundation

The goal of the NGI initiative is to conduct research and development (R&D) in advanced networking technologies and to field trial those technologies in test-beds that are 100 to 1,000 times faster than today's Internet. The NGI goal is to also test revolutionary applications that meet important national needs and that cannot be achieved with today's Internet.

The NGI initiative was initially conceived by Vice President Al Gore with full support from President Clinton. In his 1998 State of the Union address, President Clinton sought Congressional support for the Next Generation Internet. A portion of his speech is quoted below:

> We should enable all the world's people to explore the far reaches of cyberspace. Think of this—the first time I made a State of the Union speech to you, only a handful of physicists used the World Wide Web. Literally, just a handful of people. Now, in schools, in libraries, homes and businesses, millions and millions of Americans surf the Net every day. We must give parents the tools they need to help protect their children from inappropriate material on the Internet. But we also must make sure that we protect the exploding global commercial potential of the Internet. We can do the kinds of things that we need to do and still protect our kids. For one thing, I ask Congress to step up support for building the next generation Internet. It's getting kind of clogged, you know. And the next generation Internet will operate at speeds up to a thousand times faster than today.

As a result, on January 27, 1998, the 105th Congress passed resolution H.R. 3332: the Next Generation Internet Research Act of 1995.

> To amend the High-Performance Computing Act of 1991 to authorize appropriations for fiscal years 1999 and 2000 for the Next Generation Internet program, to require the President's Information Technology Advisory Committee to monitor and give advice concerning the development and implementation of the Next Generation Internet program and report to the President and the Congress on its activities, and for other purposes.
>
> Be it enacted by the Senate and House of Representatives of the United States of America in Congress assembled

The makeup of the congressional appropriation authorization is as shown in Fig. 2.18.

The Next Generation Internet (NGI) initiative is a multi-agency Federal research and development (R&D) program to develop, test, and demonstrate advanced networking technologies and applications. The rational for investing in NGI is highlighted as below:

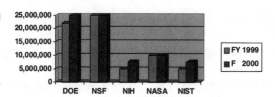

Figure 2.18
Congress Appropriation Authorization for FY 1999 & 2000

- Internet traffic has been growing 400 percent per year.

- By the year 2000, more than half of the U.S. population is expected to have access to the Internet.

- The Federal government, universities, and businesses are developing medical, environmental, manufacturing, educational and defense applications that require new high-capacity networks to make them fully functional and widely available.

- NGI-developed applications and technologies will be available to the business sector for incorporation into services for schools, work places and homes.

- The NGI initiative, together with investment by academia and industry, is laying the foundation for networks that are more powerful and versatile than the current Internet.

- The NGI has virtually unlimited potential to help Americans live better and work smarter.

2.5.4.2 NGI Goals.   There are three goals for NGI:

1. Goal 1: Experimental research for advanced network technologies
2. Goal 2: NGI testbed
3. Goal 3: Revolutionary applications

2.5.4.2.1 Goal 1: Network Research and Implementation Strategy.   The NGI initiative will focus network research in three areas:

1. Network growth engineering
2. End-to-end quality of service
3. Security

The strategy for implementing Goal 1 is to fund federal, industrial, and university R&D organizations to develop and deploy the various network infrastructure and applications. This will be done in an open technology collaboration as exemplified by the IETF, ATM Forum, and others.

Beginning 1999, DARPA will lead the Goal 1 effort with active participation from NSF, NASA, NIST and DOE. This strategy is similar to the one that contributed to the success of the original Internet.

2.5.4.2.2 Goal 2: NGI Test Bed. The networks developed in Goal 1 will connect at least 100 sites (universities, federal research institutions, and research partners) at speeds at speeds from 100 Mb/s to 1 Gb/s end-to-end—100 times faster than those of today's Internet. This network will be large enough to provide a proof-of-concept test bed for the hardware, software, protocols, security and network management that will be required in the future commercial Internet.

Goal 2 efforts begin in 1999 and will be led by NSF, NASA and DOE with active participation by DOD. The strategy for achieving this goal is for Federal agencies to build high performance, collaborative networks in partnership with the telecommunications and Internet provider industries and federal research institutions. A sub-goal is to interconnect the top research universities and federal research institutions (100 NGI sites) in the country through a fabric delivering 100 Mbps end-to-end in an interoperable mesh interconnecting federal networks, such as vBNS, ESnet, NREN, DREN, and other appropriate networks, and to demonstrate a few high-end applications at a small subset of these above sites.

2.5.4.2.3 Goal 3: Revolutionary Applications. Goal 3 focuses on demonstrating new applications that meet important national goals and missions. These applications will include federal agency mission applications, university and other public sector applications, and private sector applications. These applications are those that could improve U.S. competitiveness in existing business areas and will demonstrate the potential for entirely new business areas based on commercializing the technologies developed within this initiative.

NGI applications under consideration are:

- Education: Distance education, digital libraries
- Emergencies: Disaster response, crisis management
- Environment: Monitoring, prediction, warning, response
- Government: Delivery of government services and information to citizens and businesses
- Health care: Telemedicine, emergency medical response team support
- National security: High performance global communications, advanced information dissemination
- Scientific research: Energy, earth systems, climate, biomedical research

Many of these areas are mission critical and fundamental to the success of the next generation commercial Internet. Two classes of applications have already been accepted as foundation-applications.

1. Distributed computing applications recognize that the network seeks to provide very high bandwidth with low latency. Examples of specific applications are: global ocean-atmosphere climate models, quantum mechanical materials models, etc.

2. Collaborative applications that require moderate to high bandwidth and also the ability to reserve a piece of the network pipe for high quality video and audio streams. Examples of specific applications are distance learning and collaborative engineering design.

The strategy for achieving Goal 3 is to demonstrate a number of applications that are perceived as important by the private sector and the general public.

Sponsoring organizations, from academia and industry, will choose applications to demonstrate within the NGI such advanced functionality and performance. The proof-of-concept opportunities provided by this initiative will give substantial visibility to new approaches for meeting important federal missions as well as those of other institutions.

2.5.4.3 Technical Resources. Most major U.S. universities are directly or indirectly involved with the NGI initiative. And most major communication companies are also represented to some degree with the NGI. Participating governmental agencies directly involved and their technical contributions include the following:

- DARPA: Pioneer of long term networking research; developed cutting-edge network technologies
- DOE: With long term experience in managing production and research networks; specialized in networking technology
- NASA: Experience in network management and in specialized network test beds; strength in mission driven applications involving high data rates
- NSF: Excellent relationship with the academic community
- NIST: Long experience in standards development, networking research, computer systems security, and in test beds involving many industrial partners.
- National Institute of Health (NIH): Bring their experience in medical research, and health care applications.

2.5.5 Internet2

In the 1960s and 1970s, the federal R&D agencies, major universities, and private companies worked together to develop today's Internet. That partnership created the Internet's phenomenal success story. Internet2 was created to resemble the original Internet model with the goal of developing and implementing new technology needed by all network users, thereby ensuring continued U.S. leadership in the application of computers and communications. Internet2 is a collaborative effort by over 135 U.S. universities chartered to develop advanced Internet technology and applications. It is a project of the University Corporation for Advanced Internet Development (UCAID) working with industry and the federal government to replicate Internet success for the new millennium. UCAID's university members have committed over $50 million per year in new funding for Internet2. Each university has also committed significant resources towards upgrading campus networks, connecting to regional gigaPoPs, and providing the facilities for advanced application development. UCAID's corporate members have thus far committed $20 million to the project.

2.5.5.1 Internet2 and NGI Relationship. Internet2 and NGI initiative are working together in partnership with NSF on various research programs. Today, over 50 Internet2 institutions have already received competitively awarded grants to connect to the vBNS (very high performance Backbone Network Service) developed by the NSF and MCI. Internet2 members are expected to receive more funding in the next two years in the form of competitively awarded grants from the NSF and other federal agencies participating in NGI. As Internet2 develops and NGI programs go forward at federal agencies, it expected that network services will be available on interoperable backbone networks. Over the next three to five years, Internet2 expects to provide an opportunity to reduce the number of low speed network connections now supported with federal agencies.

2.5.5.2 Internet2 Goals. A key goal of Internet2 is to transfer technologies into the commercial sector, thus creating the basis of a next generation network that will continue U.S. leadership in this area.

Below are areas the Internet2 community is addressing:

- Recreate a leading edge research and education network capability
- Transfer new capabilities to the global production Internet
- Demonstrate new applications that enhance researchers' ability to collaborate and conduct experiments

- Delivery of education and other services such as:
  - Health care
  - Environmental monitoring
- Development of a communications platform to support QOS based applications

2.5.5.3 Internet2 Working Groups. Internet2 created nine Working Groups to address the goals and develop/support the advanced applications for higher education's research and education missions. Working groups include representatives from Internet2 universities and corporate members. Nine topics that are being explored include:

1. IPv6
2. Measurement
3. Multicast
4. Network Management
5. Network Storage
6. Quality of Service
7. Routing
8. Security
9. Topology

There are several initiatives which were spin-off of the working groups. One of the more important one is QBone (Quality of service Backbone test bed) that was a spin-off of the QOS working group.

2.5.5.4 Abilene and QBone. The Internet2 Abilene network, like the vBNS, is a high-performance backbone network that meets the demanding bandwidth requirements of the research and higher-education (R&E) communities. Abilene is a project of the University Corporation for Advanced Internet Development (UCAID) and is based on the Qwest corporation's nationwide fiber network. Unlike the ATM-based vBNS, Abilene will provide nationwide OC-48 (2.4 Gbps) packet-over-SONET (POS) connectivity through the Cisco 12000 Series (GSR) routers, with GigaPoPs and universities connecting at OC-3 (155 Mbps) and OC-12 (622 Mbps) speeds.

Abilene will peer directly with the vBNS and with advanced federal agency networks at the emerging NGIXs. One of its crucial differentiating goals is to construct a backbone network supporting QoS and IP multicast.

The QBone proposal is to provide Diff QOS service over components at Abilene PoPs. The expedited forwarding (EF) per-hop behavior (PHB) will be supported between two or more Abilene PoPs. Aggregate policing at the edge will be exercised.

The schedule for Qbone is as follows:

January 1999:

- Classifying and policing with CAR at several OC-3 and OC-12 access points attached to other QBone participants.
- Use of WRED on the OC-48 GSR line cards.
- Use of MIB data and Surveyor one-way delay measurements between Abilene PoPs and between Abilene and other Surveyor boxes deployed throughout the QBone to audit Abilene's performance and the performance of end-to-end QBone services.

June 1999:

- Test new WFQ (or WRR) features over OC-48 POS.
- Semi-automated admissions control and edge device.
- Evaluate automated resource management tools.
- Test QBone inter-BB signalling protocol.

End 1999:

- Test Production-quality interdomain QoS

2.6 Cable Networks

2.6.1 A Case for Cable Network Modernization

The cable network architecture was developed for a specific service and hence was optimized to efficiently carry broadcast video services to the home. Today's cable network model cannot cope with, nor was designed to provide these new interactive services. This picture gets even more complex because of the unpredictability of future applications. CATV networks of today needs a major overhaul in order to cop with applications requiring integrated bidirectional communication.

One could imagine cable network operators forgoing this phase of turmoil and taking a wait-and-see approach as they have often done in the past. Yet the situation is different today. Cable companies are now facing significant competition, already from DBS operators and now from telephone companies as well. If the cable operators take a defensive strategy, they could rapidly loose market share to others. Cable companies need differentiators to meet this competition. Taking the offensive by providing these demanding network services is the safer route. The challenge, of course, is how the MSOs can do so without going broke in the process.

Today's users are becoming very sophisticated regarding what the network services must provide in order to satisfy bandwidth hungry applications. Cable modems can play the major role in providing these bidirectional interactive services.

Modernizing the cable network to provide high speed interactive service appears to give cable operators an advantages over the other networks. HFC bandwidth capacity is enviable. Deploying a cable modem over a modernized HFC might be all it takes to be able to provide today's demanding high speed interactive applications. A well engineered HFC with cable modems can provide the full range of analog TV cable channels, high speed Internet access, voice, and high quality interactive video.

Cable companies are finding the Internet access market a compelling reason for upgrading their networks to support cable modems. Unlike ILECs who are required to allow any ISP to have access to their broadband infrastructure, the cable companies can restrict their cable modem service to their own imbedded ISP service.

Voice traffic can be carried over cable modems using Voice-over IP technology. The larger cable operators may opt to invest and deploy digital switches to directly provide telephony services. This will elevate these cable operators to telephone service provider. Some large cable operators have already tested this approach and are deploying it today. Smaller cable operators could provide only the access portion of the network and sell this to CLECs who would offer the telephony service.

Some cable operators believe that their current lack of a class 5 switching infrastructure is a blessing because of the opportunities it presents. While the RBOCs struggle to re-engineering their legacy network, a process that could take many years, the cable industry has a blank drawing board. Whatever the cable companies deploy today will at least for now be the latest switching technology.

2.6.2 HFC: The Next Generation Cable Network

The overall HFC topology as recommended by CableLabs is shown in Fig. 2.19. It was specifically developed for the client MSOs accommodating various economic models and network needs.

Figure 2.19
Regional Hub HFC
Network Architecture

Hybrid Fiber
Cox s cable
system

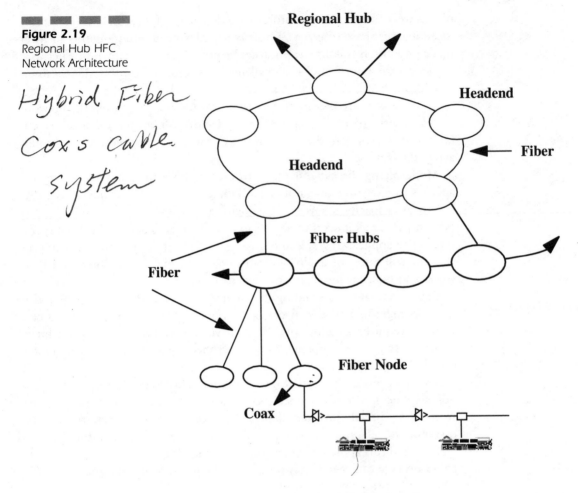

The architecture includes the following attributes:

- Centralized regional hub with common equipment among multiple operators within the region
- Fiber ring connecting hubs and headends providing routing capabilities
- No amplifiers in the fiber rings
- Maximum distance between furthest end user node and fiber hub of 80 km

Future cable systems will be limited in size. A region may have ten or 20 different cable systems. The other functions as depicted in the reference model include telephony switches, video-on-demand regional video servers, and access to information databases.

2.6.3 HFC Access Shortfalls

HFC is evolutionary from the current tree and branch networks and this upgrade can be done in a stepwise manner. Nonetheless, compared to competitive systems being deployed by the telephone companies, HFC has certain shortfalls that will need to be addressed. Primary among these are reliability, security/privacy, and operations, administration and maintenance (OA&M).

2.6.3.1 Reliability. The one-to-many/many-to-one architecture of HFC, although cost effective, is not ideal from a service reliability standpoint. Single failures in the system can disable service for large numbers of customers. This contrasts with the equipment practice of the telephone companies who typically require redundancy for any equipment affecting more than 16 users. Indeed, the outage history of cable networks has caused many to question whether customers will trust their lifeline telephone service to the cable company. Potentially, cable companies could target the second and third line market, letting the phone companies keep the basic line.

2.6.3.2 Security/Privacy. The shared medium topology of HFC networks retains the security problems that have always plagued cable networks. Since the same signal is sent to all users, theft of service remains a concern. By contrast, the VDSL system envisioned by the ILECs sends individualized video channels to each home. Similarly there is a concern for privacy if my conversation is being transmitted down the coax to all my neighbors. Whether or not this is a realistic threat, it could be perceived as such and this perception could impede user acceptance. The best solution appears to be encryption. Cryptography technology is being developed in the industry for cable systems that should address this issue. Indeed, once encryption systems are in place, cable companies could advertise that their telephony service is much less vulnerable to eavesdropping than is the phone company's.

2.6.3.3 Operation, Administration and Maintenance. All network operators from time to time experience network failures. The way service providers respond to these problems will have a major impact on the loyalty subscribers have to their service provider. Surveys consistently reveal that cable companies have a poor reputation for network quality, reliability and customer service. These are the top attributes that telecommunications managers consider when selecting a telecommunications service provider.

This reputation must change if cable operators hope to survive in the competition resulting from the convergence of telephony services, data services, and broadcast entertainment services. The ILECs have invested heavily in operations support systems and are required by the public utility commissions to measure and satisfy strict measure on reliability and customers service including how long a customer must wait when calling in to report a problem. Customer service calls should take no longer than two to three minutes in total. Technologies such as computer telephone integration (CTI), automatic number identification (ANI), and voice response units (VRU) enable a customer's account information to appear on the screen before the customer service representative answers the call. These technologies do more than satisfy customers—they also reduce operations costs.

The MSOs are drafting new support infrastructures tailored to accommodate the multi-service HFC platform. If that support infrastructure is built correctly, the cable industry stands a good chance to survive and indeed prosper in the competition with the DBS and ILEC competition.

2.6.4 HFC Business Case

Planning and deploying new platforms for the interactive services must include such factors as the geographic area, ease of access, existing network capacity, etc. An important question facing planners will be whether to integrate additional capabilities into their current networks or opt to build an overlay network with a high speed fabric. The ultimate decision will be determined by the business plan of the particular cable company.

The potential revenue to a cable company upgrading to a two-way HFC network appears to be very attractive. At the minimum, the following market opportunities exist:

- The access market for POTS (Plain Old Telephone Services) including long distance access revenue is about $70 billion. That is about $475 dollars per household per year.
- CATV broadcast services included charges for premium channels is about $390 per year.
- Internet access charges are about $240 per year for narrowband access. Best estimates of willingness-to-pay for broadband Internet access is nearly twice this.

- Video-On-Demand (VOD) market has been puzzling and predictions of its demise have been exaggerated. Market trials suggest that subscribers will use such a service if included as part of a package, but may not pay for it as a separate service. Best estimates are around $200 dollars per year per household.

2.6.4.1 Cost of Modernization. The upgrade of a conventional cable network to HFC including support high speed Internet access, voice telephony and other interactive services will require considerable investment. An economic analysis by David P. Reed showed that the cost of building an HFC network from scratch is not much more than the cost of a traditional cable network. Installing fibers instead of coax in the trunk without the need for special amplifiers accounts for the low cost differential. Reed estimated that the system total cost of an HFC network was about $450 per home passed based on 60 percent penetration rate and 500 home nodes. Major cable operators have recognized this and are basing their strategy on building HFC networks even without accounting for new interactive services. In this way, their networks will be two-way ready should they decide to add these interactive services at a later time.

Upgrading the traditional cable network is a more complex economic model. The cost of upgrading depends on how much rebuilding needs to be done. In order to fully upgrade a traditional cable network to an HFC two-way transmission, the following upgrades will typically be needed:

- At the minimum, the trunk cable (from the headend to the distribution network) must be replaced with fiber.
- Upgrading or replacing the amplifiers in the distribution network enabling them to transmit in both directions.
- Rebuilding the headend.

The range of system upgrade costs was estimated in various studies to be between $100 to $240 per home passed. This is based on 60 percent penetration with 500 subscribers per node

2.6.5 HFC and Cable Modem Penetration

The HFC business case will be a complicated task for some small cable operators and a challenging one to larger MSOs. There are over 11,000 cable systems in the United States. Large cable operators claim to have revamped

TABLE 2.3

HFC
Modernization/
Penetration

Year Passed	% of Home by Upgrades
1995	18%
1997	50%
1999	77%
2001	85%

their systems and a good percentage is HFC ready today. Table 2.3 is Merrill Lynch's estimate of the HFC modernization effort.

MSO will be modernizing the cable network on a case-by-case bases. Estimates of the number of cable modems served by these networks vary considerably. Table 2.4 shows deployment estimates from Foster Research.

Paul Kagan Associates per Table 2.5 estimates were more pessimistic.

Even the lower number are still impressive; however, the cable companies recognize that they are in a race with ADSL being deployed by the phone companies. Once a customer has invested in either a cable modem or an ADSL modem, it may be difficult to get them to switch to the other technology.

2.6.6 HFC Deployment Issues

The HFC topology does not lend itself to dynamic reconfigurations or salability. The HFC systems, of today, seem to be hardwired to provide that specific service. For example, if more upstream or downstream channels are

TABLE 2.4

Cable Modem
Forecast (Source:
Forester Research
Inc.)

Year Modems	Cable
1996	90,000
1997	550,00
1998	2 million
2000	7 million

	Year Shipped	Cable Modems in Million
TABLE 2.5	1996	0.1
Cable Modem Forecast (Source: Paul Kagan Associates)	1997	0.4
	1998	1.4
	1999	2.2
	2000	4.0

needed to accommodate an increase in traffic or due to introduction of new applications, the analog amplifiers in the HFC distribution part of the network will have to be manually modified or may be replaced. If video telephony becomes a popular service, then increasing the upstream bandwidth will be required (allocate more channels in the RF spectrum). Reducing the home passed per node would be a costly alternative. The cable modems and possibly the headend will also need to be modified to handle tuning to these new channels.

Going digital all the way is another parameter in the equation that must be evaluated. A 6 MHz analog channel with QAM modulation (see chapter 5) can deliver over 30 Mb/s of digital bandwidth and possible more in the future. A good quality video picture requires 4 to 6 Mb/s.

Competition from other large cable companies is an unlikely scenario. According to the FCC (CS Docket # 95-61), cable television is available to 95 percent American household, yet only 1.5 percent of American homes are passed by multiple systems. Therefore the competition will exclusively be from non-traditional network operators such as the telephone companies.

The uncertainty of the technology use and present and future service offering is a fundamental difficulty facing the cable operators. Most of the upgrades that are needed will be sunk cost (unrecoverable). They cannot afford to make a strategic mistake.

Service differentiation is an important tool the modem service is something the DBS operators cannot offer. Hence an effective competitive strategy would be for the cable operators to offer service bundles of CATV to win and hold onto customers. Service bundling is packaging multiple services offered to subscribers at an attractive price. A cable operator may be able to attract those customers who want to buy several services together, even if a competitor offers some of these individual services at a lower price.

2.6.7 Voice Over HFC Solutions

Architecturally, handling voice in an HFC environment is not all that complex. There are three principal approaches to providing telephony services over HFC:

1. Traditional STM technology (POTS over HFC)

2. Voice over IP

3. PCS (Personal Communication Services)

4. Voice over ATM

All three solutions will most likely find deployment in various HFC networks. Which approach if any dominates over time is difficult to predict. A brief descriptions of the three solutions is given below.

2.6.7.1 STM Approach to Voice Over HFC.   The initial approach used to carry voice over an HFC network was via a telephony modem. Part of the HFC bandwidth was devoted to 64 kb/s channels. In this approach, a switch or Digital Loop Carrier system is located at the headend. The headend controller receives the voice channels from cable modems located in the home and passes these channels to the DLC or the switch for further processing. If the headend is equipped with a DLC, then T1 trunks will carry the voice circuits to nearby class 5 switch for call handling. Alternatively, the headend could be equipped with a fully featured class 5 switch.

2.6.7.2 Voice Over—IP Over HFC.   In this approach, the IP service provided by cable modems is used to carry voice. The headend would be equipped with IP routers for providing IP services and providing connectivity to the Internet. The headend controller will separate IP traffic and hand it to the routers for further processing. A gateway would be used to provide connectivity with the PSTN.

An advantage of this approach is that it is directly compatible with the emerging voice over Internet. Also, nothing need be added to the HFC network itself beyond that required for data services over cable modems. A disadvantage is that the packetization will result in added delay, thereby necessitating echo cancellers. Also, if compressed speech is used for voice-over-IP, then the quality may not be as good as the ILEC voice service. The additional electronics required at the modem for cable modem service and for voice-over-IP increases power consumption, thereby making network powering difficult.

2.6.7.3 PCS Over HFC Solution.

An attractive means of providing voice in HFC is to introduce PCS. PCS is a digital micro-cellular service.

PCS seems well-matched to the HFC infrastructure and to cable plant in general. CDMA (call division multiple access) technology at 1900 MHz can be installed cost effectively by deploying the mini low power antennas in the HFC plant. The remote antenna can be installed using a tap in various locations in the distribution network. The subscriber signals are picked up by these antennas and converted into the upstream band (between 5-40 MHz) and sent towards headend. The PCS signal is separated from the TV signals at the headend and then relayed to the switch and the rest of the landline network. This is illustrated in Fig. 2.20. When terminating a call to an HFC subscriber, the reverse will apply. The Downstream signal is transported in one of the 6 Mhz bands.

PCS over HFC has several advantages over cellular or PCS over telco-based network infrastructure. The antennas of a cable-based PCS system can be hung on the aerial cable, powered much like a cable amplifier, and can be located in the distribution plant to accommodate that geographical area. Additional antennas can be added just as easily to increase capacity and serve growth in that market. Since the HFC infrastructure is already in place, cable operators have an important advantage over other network infrastructures where tower siting, installation costs, power connections and network connections to the cell sites are difficult challenges.

Another significant advantage is the shared medium in PCS over cable. There is a single connection and common electronics at the head end. New

Figure 2.20
PCS in the HFC Infrastructure

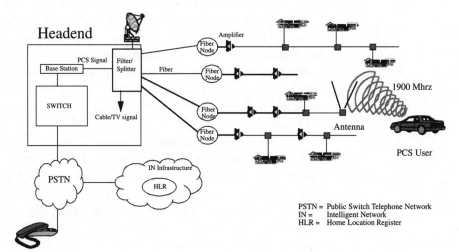

PSTN = Public Switch Telephone Network
IN = Intelligent Network
HLR = Home Location Register

radio channels can be assigned or reassigned dynamically to meet capacity and market growth.

Nonetheless, building a PCS network is a risky proposition and capital-intensive. Recently, the FCC auctioned the licenses for the PCS spectrum band for a total of 120 MHz. A 30 MHz license costs approximately $15.00 per population. To serve a city of one million licenses can cost as much as $15 million. The license winners have already begun deploying systems using conventional cellular architectures.

Cox cable recently joined Sprint, Comcast Corporation, and Tele-Communications Inc. in a venture that offers service bundling of a variety of telecommunications and video services. Services include local and long distance; wired and wireless telephone, cable television, and high-speed internet services. Cox is promoting a plan called Alternate Access to reduce billing to their customers. This allows businesses to directly connect to their preferred long distance carrier, bypassing the local phone company's access charges.

2.7 Layer 2/Layer 3 War

There are many in the industry who are still skeptical and are questioning the complexities and maturity of ATM as the technology that can meet the market demand, especially in the near term. The skeptics want to see ATM prove itself in the marketplace first before investing heavily in it.

The phenomenal growth of the Internet has brought about a very heated debate in the industry. It is common knowledge that the Internet is in trouble. Stories about Internet traffic jams have become a staple of major media technology reports. The backbone Internet Service Providers (ISPs) are scrambling to upgrade the capacity of backbone links, which is straightforward enough. However, the major problem of limited capacity of switching equipment remains.

The conventional approach to native IP routing, as embodied in routers by Cisco Systems which presently dominate the Internet, is no longer adequate. Routing IP packets is a rather complicated process, requiring traversal of a tree-like routing table, processing IP options, and other relatively complicated actions. The reason for the complexity is that every IP datagram is routed separately. Implementation of complete IP routing in hardware at best can be called experimental.

Routing ATM cells, on the other hand, is very simple because all cells have a small fixed size and because they are not independent. Rather, virtual circuits are formed, and all cells sent by parties in a virtual circuit follow the same path. This means that intermediate ATM switches keep tables of virtual circuits, so routing of a cell is reduced to a simple retrieval from an array indexed by the virtual connection identifier. The simplicity allows ATM switching to be implemented completely in hardware, thus making ATM switches outperform existing IP routers by an order of magnitude.

However, ATM has problems that are not usually understood, or discussed, by anybody but a few computer scientists and backbone engineers. Those obscure problems are not evident in lab tests, benchmarks or "pilot projects"; they affect the quality known as scalability. In fact, most leading Internet backbone engineers agree that all-out ATM replacement for the Internet simply won't fly.

The following sections explain the concepts and argue that native IP routing is the only currently existing technology that can hold a global data network together. The debate discusses only the applicability of the different technologies for global communications; the suitability of technological solutions for private networks is certainly different. For example, ATM can be a perfectly good technology for LANs and medium-size corporate backbones, providing that it can compete on a price-performance basis with other approaches, such as gigabit Ethernet.

2.8 Layer 2 Switching vs. Layer 3 Switching War

For the past two decades, industry leaders and researchers have sought a telecommunications technology that could accommodate all forms of information. In contrast to the multiple networks and technologies of today, this technology would unify all forms of information in a common format—voice, data, video and whatever is over the horizon. This unification could reduce networking costs versus the alternative of multiple overlay networks. It would simplify network deployment, operations, provisioning, engineering, billing and marketing.

Some researchers had proposed multi-rate circuit switching as the grand unification approach. Since all forms of information can be converted into bits, then an $n \times 64$ kb/s circuit switch, for wide ranges of n, would accommodate virtually any type of traffic. The problem with circuit switching, we now understand, is that it is poorly matched for bursty data. Also many forms of information are simply not a natural multiple of 64 kb/s.

Others have suggested WDM optical switching could achieve the sought unification. By providing a transparent optical path from end-to-end for a given wavelength, any protocol could be transported: STM, ATM, IP, etc. The problem with this approach is that the burden of dealing with multiple protocol types is merely pushed to the end points. It relegates the network to the role of a "dumb pipe."

The consensus of the industry today is that the grand unifying protocol will be one of two protocols: ATM or IP. Basically this comes down to a question of whether switching at the core should be done at layer 2 or layer 3.

ATM had the early lead in public telecommunication networks. The goal of ATM from its beginning was to accommodate disparate forms of information and to operate at very high speeds. Adaptation protocols have been defined that allows circuit-oriented traffic of arbitrary bit rates to be carried over ATM. ATM allows information to be burst at high rates, but does not waste network resources between bursts thereby allowing it to carry data traffic efficiently. At this point, most major U.S. interexchange carriers have announced their intention to build their backbone network using ATM.

The underlying protocol upon which the Internet is built, TCP/IP, is a packet switching protocol. Unlike ATM that switches cells at layer 2, IP routes packets at layer 3. IP supports variable size packets versus ATM's reliance on fixed-sized cells. ATM is circuit-oriented vs. IP which is a datagram protocol.

The Internet gold rush and its phenomenal growth have inspired a continuing social transformation. Currently over 30 percent of the U.S. population reported that they now watch less TV and about 25 percent reported less telephone use. Fifty percent of the Internet traffic is going toward the World Wide Web.

Presently, the Internet is plagued with access delays and can be slow when downloading files or video clips. This slow speed access is turning off a lot of potential users to the point that it may threaten its future growth. ADSL, cable modems, and others will play a major positive role in remedying this frustrating experience. These high-speed access technologies multiply bit delivery to the user by 200 times. This will put a lot of pressure on the backbone and so the debate of layer 2/layer 3 switching intensified recently to address this dilemma.

The argument among the telecommunications experts, academia, and intellectuals seems neverending over layer 2 versus layer 3 switching. The outcome of the debate, in most cases, never resulted in any change of heart and this issue becomes more philosophical than technical. This long debate created two cultures, the "netheads" and the "bellheads" with nethead rooting for layer 3.

Layer 3 switching is analogous to driving coast-to-coast using well known interstate highways and changing routes in major hubs based on traffic patterns, speed, radio announcement, electronic signs/detours. Layer 2 switching, on the other hand, is when one loads his or her car on an express train and travels from coast-to-coast. In the later case, the network must be intelligent and one must negotiate before hand with an operator who is able to guarantee and reserve a slot on the express train. This example depicts the fundamental concept of the two scenarios and hints toward the engrained bias that shaped these views. These most talked about issues are:

1. Cost, in general

2. End-to-end reliability

3. Dumb versus smart networks

4. Switching techniques and efficiency/scalability

Below is a brief and objective description of the issues being debated.

2.8.1 Cost

Networks using layer 3 routers, it has been argued, are more cost effective than ATM alternatives. The cost differential primarily has to do with the state-of-the-art of the technology choices and the stringent requirements that were placed on building ATM switches. Public ATM switches provide QOS performance guarantees and are built to be fully redundant. The popularity of the Internet has resulted in high production volume, off-the-shelf components for building IP networking equipment. This resembles, in our example, building passenger vehicles that are competitively priced versus building a train.

The cost of overhead is also playing its role in the ATM vs. IP debate. If the traffic being carried originates as IP, then carrying this on ATM in the access or backbone imposes an "ATM cell tax" due to the addition of the ATM headers. Why use IP over ATM if IP already has adequate addressing for packet delivery? In other words, why load the passenger car onto the train if the car is already capable of making its own way to the destination?

2.8.2 End-to-End Reliability

The debate over end-to-end reliability is more philosophical in nature than technical. The original goal of ARPA was to create a network survivable in times of war. It was believed that an IP-based network based on a mesh architecture without fixed routes could achieve high reliability on an end-to-end basis. In an IP world, the reliability is viewed from a network perspective rather than a component (switch) view. ATM switches are expected to be highly available and are designed so that no single board failure will result in the failure of switch.

TCP/IP was designed from the outset to handle packet delivery in a failure-prone network structure. TCP manages the packet delivery such as packet numbering (reordering and duplicate detection), positive acknowledgment, retransmission on timeout, etc. on an end-to-end basis, while IP layer is responsible for low level functions such as addressing, duplication, disorderly arrival, and alternate routing, etc.

2.8.3 Dumb Versus Smart Networks

This dumb vs. smart network is fundamental in the layer 2 layer/3 switching argument. It is generally accepted that intelligence is migrating to the edge of the network and hence to the user. Indeed, the Internet's success has been achieved from major advancements in PC hardware and software/applications.

The strength of the telco network infrastructure is predicated on building an intelligent network (IN) delivering rich features and services to the end user. Its network-embedded databases provide significant advantages. The ubiquity and success of 800-900 number services can be directly traced to this approach. This IN concept has been instrumental in delivering rich services to dumb home terminals such as cheap phones. Today, users can subscribe to rich features such as call waiting, caller ID, last call return, call tracing and other fancy CLASS™ features. These complex features add costs and complexity to the network and erode efficiency but without adding costs or complexity to the terminal. Telco advocates, on the other hand, proclaim that a PC is a $2,000 phone.

The IN concept has several soft spots. Network complexity and therefore cost plays result from centralized intelligence. Common databases result in bottlenecks. From a reliability standpoint, centralization is "putting all your eggs in one basket." The SS7 network failure several years ago that disrupted telephone service in much of the U.S. East Coast is a case in point.

The Internet was intended to be a dumb network and that is reflected in its design choices:

- No global control of operations
- Network element independence

- Best effort communication
- Stateless machines
- Interconnections of gateway and routers in a mesh topology

This concept is very appealing to ISPs who are not eager to deploy a complex network or deploy even more complex operations systems to maintain that network. "Make it simple" is the philosophy advocated by most the netheads. The current effort to provide a smarter network with a notion of QOS plays against the idea of "keep it simple." Some experts promote the concept of solving the QOS problem by simply providing more bandwidth. This is a novel solution, especially since the cost of bandwidth is dropping dramatically. However, sooner or later, in the multimedia environment the concept of discipline will prevail and "pay for what you get" becomes attractive even at the risk of building a more complex network: "What Intel giveth, developers taketh away" creating sophisticated services requiring fat bandwidth and QOS aware networks.

To the application users, dumb network meant that control of the application is at the end points that they control. They can use their creativity when developing applications and platforms to run over the Internet backbone. Only basic knowledge of TCP/IP was required to develop products for the Internet.

2.8.4 Switching Techniques and Efficiency/Scalability

Below are some of the critical technical issues that are used by many to make the case for or against ATM playing the key role in the Internet network. The authors are not attempting to draw any conclusion other than state the facts without going into excessive technical details. What is puzzling is that experts are reading the same fact sheet and build the case for or against IP/ATM.

The critical parameters identified for building the pro and cons for ATM are summaries as follow:

- In native IP, datagrams are individually routed using a tree routing algorithm table. This is difficult to implement in hardware.
- ATM uses a virtual circuit identifier in the header to route cells. Each cell in a virtual circuit follows the same path. This path is established via signaling, and the connection stays active for the duration of the session. This allows ATM to switch packets in hardware, thereby outperforming existing IP routers by an order of magnitude.

■ The traffic pattern of the PSTN where 80 percent of calls are local (stong local community of interest) does not fit Internet traffic. Internet servers are distributed all over the United State and even globally. One finds himself/herself connected to Germany by a click of an icon. This traffic behavior has a profound impact on network scalability. The backbone router pipes have to be fat enough to accommodate this traffic.

■ In surfing the Internet, sessions to a given server are typically short. Use of a connection-oriented protocol like ATM would result in a heavy volume of call set ups and tear downs. For downloading a file (long session), the overhead of establishing virtual circuits would be justified.

■ Ipsilon's IP switching protocol or the Cisco's tag switching technologies add an optional circuit switching capability to IP networks. This increases network capacity at the cost of greater complexity. The need for this may have been overtaken by promised terabit IP routers.

■ ATM switches deployed as a hub providing virtual paths between routers can provide rich connectivity and high-speed paths between connected routers. IP's mesh topology is cost effectively maintained while providing link bandwidth flexibility.

■ ATM virtual paths provide traffic metering, traffic measurement, distant sensitive billing, etc.

■ IP proponents advocate flat rate billing because of the huge billing infrastructure that would otherwise be needed. Many question whether users accustomed to "all you can eat" flat rate service would accept metered billing. Nonetheless, the current revenues of many ISPs do not cover their investment costs. Thus far, this has not bothered investors who continue to see share price appreciation despite operating losses.

■ As more and more traffic demands faster and faster packet switching, ATM may become less scalable. The time required to switch each small 53-byte ATM becomes limiting. The ATM forum is working on a Frame ATM work item which may alleviate this "beat the clock" limitation. IP packets are variable in length with most packets much longer than 53 bytes. This gives more time to route a packet.

■ Responding to network failures is more complex in an ATM network. New routing tables need to be downloaded or pre-stored in the switch. Existing calls would be torn down or automatically re-established (for PVCs) on a different route.

- IP networks do not guarantee that packets are rec
  order transmitted.

2.8.5 And the Winner is . . .

There is no doubt that as an end-user service, IP has won. IP-based LANs are deployed in virtually every business. The ATM-25 interface defined by the ATM Forum never caught on in North America. Nonetheless, ATM has found application in Wide Area Networks. AT&T, MCI Worldcom and Sprint are each deploying ATM as a core network due to its high capacity. The Internet's vBNS backbone is based on ATM. In the access arena, most ADSL deployments have been based on ATM. ATM's selection was due to its support of QoS that network operators believe will be important to support voice and video applications over ADSL.

The major advantages of ATM, capacity and QoS, may not last. Techniques such as TAG switching and IP switching will allow IP networks to take advantage of packet flows to reduce routing overhead at intermediate routers. Protocols such as DIFFSERV and RSVP will add QoS capabilities to IP. In some ways, as these capabilities are added to IP, the simplicity of IP begins to erode and IP begins to look more and more like ATM.

3

The Fiber Solution: ATM Passive Optical Networks

3.1 Fiber's Manifest Destiny

Most network operators would assert that the ideal broadband access network would be one based on optical fibers. Versus copper pairs, coax or wireless technologies, the bandwidth capacity of an optical fiber is nearly limitless. State-of-the-art optical transmission systems transmit 10 Gb/s on a single wavelength and new Wavelength Division Multiplex Systems have been announced allowing 64 wavelengths to be carried on the same fiber–and even these systems exploit less than 1 percent of the theoretical capacity of an optical fiber. Optical fibers carry signals long distances without regeneration. For example while T1 lines require repeaters every 1.7 km and while CATV networks typically require amplifiers in every 500–700 m of coax, fiber optic transmission systems can carry signals 100 km or more between repeaters or amplifiers.

Optical fibers have much longer operational lives versus copper cables that are subject to the inevitable corrosive effect of water infiltration. When cable cuts occur, it is much easier to repair a single glass fiber than the individual pairs of a 2000 pair copper cable. Adding the fact that optical fiber systems are on a much faster technological learning curve than other transmission technologies, and one easily understands the continued interest in fiber access networks.

The greater the traffic to be carried and the greater the distance over which the traffic must be carried, the greater the advantage of optical fibers. For this reason fiber optic transmission systems quickly displaced other alternatives in long distance backbone networks. Then with continued cost improvement, optical fibers began (and have now largely displaced) microwave and T1 for interoffice trunks in metropolitan areas. More recently optical fibers are being used in the feeder portion of access networks and between CATV head ends and electronic nodes in the field. Figure 3.1 illustrates this migration of fiber towards the end user suggesting a "manifest destiny" of fiber eventually entering the distribution network and perhaps premises wiring as well. Indeed, both ILECs and CLECs are already running fiber to large office buildings in downtown areas. However, aside from a few well-publicized trials, extending fiber networks to small business and residential customers remains a distant goal.

3.2 Impediments of Fiber Access Networks

If optical fiber access networks are so desirable, what are the impediments that have prevented their use up to now? First, fiber cables are more expensive than copper. The cost of a single mode optical cable is today about 20¢ per fiber-meter. This is significantly higher than the 10¢ per pair-meter of twisted-pair copper cable. The cost of optical splices adds significantly to installation costs. The cost of splices ranges from about $14 for a mechanical splice, to about $20 for the lower loss fusion splice. These costs can be expected to decrease significantly with deployment volume.

The above comparison is for new construction. In most cases where installation of fiber would be considered for advanced services, a twisted pair or coax network would already exists. The cost of overbuilding this existing

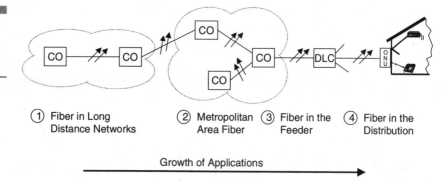

Figure 3.1
Migration of Fiber
Applications Toward
the User

① Fiber in Long   ② Metropolitan   ③ Fiber in the   ④ Fiber in the
   Distance Networks   Area Fiber   Feeder   Distribution

Growth of Applications →

network with fiber is considerable. The situation is easier in downtown metropolitan areas where fiber cables can be pulled through existing underground ducts. In suburban and rural areas, the fiber cables must be hung on telephone poles or buried in the ground—after the roads, trees and bushes are in place.

Another significant impediment is the cost of the optoelectronic components–the lasers, detectors and diplexer at each end of the fiber. Although the costs of these components are rapidly decreasing, they are still much more expensive than the corresponding electrical drivers and receivers. The costs are highly volume dependent with a optical costing initially around $300 by decreasing to under $150 with volume.

A third impediment is the cost of optical connectors and patch panels required between the cables and the central office equipment. The cost of connectorizing a fiber ranges from $8 to $25.

A final impediment is providing power for the network termination equipment at the customer's premises and power for the user's terminal. Since the beginning of the 20th century, powering for the user's telephone has been provided from the network over the same twisted pair used to carry telephone traffic. Hence no local connection to the electric utility has been required, and lifeline telephone service is maintained in the event of power outages. With fiber to the premises, no means is available to deliver power from the network. One possibility is to provide an additional copper pair in the cable to be used for powering. Those advocating this approach envision "Siamese" drop wires to the premises containing both an optical fiber and copper twisted pair or coax in the same cable. Another alternative is to utilize local powering, although this raises the issue of providing a connection to a power outlet on the premises and requires providing some form of battery backup to handle power outages. Both of these alternatives significantly raise the cost of a fiber-to-the-home or fiber-to-the-business system.

Despite these obstacles, the advantages of fiber-to-the-business and fiber-to-the-home systems appear irresistible with the result that engineers continue to attempt to devise means to overcome these obstacles.

3.3 Types of Fiber Access Networks

The most straight-forward technique for deploying fiber access networks is to deploy the same SONET transmission systems designed for interoffice lightwave facilities. This is done today for large business customers. SONET facilities are extended to office buildings in either a point-to-point configuration to each business or in a ring add/drop configuration passing through several business locations. This is illustrated in Fig. 3.2. While the volume of traffic from large business customers may be sufficient to justify a SONET transmission system, the same is not true for small business or residential customers. Other architectures providing a lower cost per user is required.

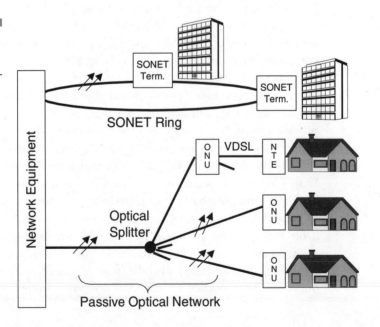

Figure 3.2
Optical Access
Networks

In the mid-1980's, the major network operators began to develop a Fiber-to-Home specification known as Broadband ISDN. B-ISDN was viewed as the next logical step beyond ISDN over copper wires and was to be driven by video entertainment to the home. At that time some advocated a TDM approach for B-ISDN. Their view was that B-ISDN should consist of narrowband ISDN plus some number of high-bit-rate pipes. What finally emerged was a specification of a 155.52 Mb/s (or 622.08 Mb/s) fiber operating with SDH (or SONET in ANSI nations) and carrying ATM cells. Completed in 1988, such a residential broadband system has never been deployed.

In 1986, BellSouth began two field trials of video delivery over Fiber-to-the-Home in the Orlando, Florida area, one in a subdivision called Heathrow with AT&T (now Lucent) and the other in a subdivision called Hunter's Creek with Nortel. While both provided high-quality video service, the cost of both systems were far too high for commercial deployment.

Despite these early disappointments, the industry continues to seek a Fiber-to-the-Home approach that would be cost-effective for residential customers. A promising architectural approach known as passive optical networks is shown in Fig. 3.2. Here, a single network interface is shared by multiple network users. A passive optical splitter connects the shared network fiber to multiple users fibers. Although a single optical splitter is shown, multiple splitters could be used up to the total allowed splitting ratio.

Two types of optical splitters have been proposed. One type is the WDM splitter. Here, multiple signals of different wavelengths are carried on the shared fiber. The WDM splitter directs each wavelengths to a corresponding user fiber. In the upstream direction, each Optical Network Unit (ONU) transmits a signal at a different wavelength; these upstream signals are simply combined onto the shared fiber towards the network. From the network equipment, individual lasers and detectors are required for each wavelength. The economic savings come about through minimizing the number of fibers that must be brought into the network. A difficulty with this approach is ensuring that each ONU transmits at its assigned wavelength. The output wavelength of semiconductor lasers is highly temperature sensitive. Precise temperature control that would be required is expensive to achieve at each customer's premises.

A second type of splitter proposed for passive optical networks is a power splitter. Here, the network laser transmits a single, high-intensity signal. The power splitter sends $1/N^{th}$ of the power of this signal to each of the N users. In the upstream direction, each ONU transmits signals with the same wavelength. A Time Division Multiple Access (TDMA) method is used to ensure that each user's traffic arrives at different times at the splitter to ensure that there is no overlap of signals. Advantages of power splitting versus WDM are that only a single laser and detector are required in the network, and also that identical network interfaces can be used by each user. This technique, however, introduces a number of other difficulties. Each ONU must operate at the total multiplexed data rate. Since the multiplexed traffic is sent towards each user interface, techniques such as encryption must be used to prevent users from intercepting traffic destined for other users. A faulty ONU could disrupt communications for all users sharing the same passive optical network, and remote fault isolation may be difficult to achieve. Finally, power splitting itself will tend to limit the total operational length or splitting ratio of the system.

Note that it is possible to combine the WDM and power splitting approaches to avoid some of the difficulties of each technique. For example, one could employ WDM downstream for privacy and to achieve maximum operational length while using TDMA upstream to obviate the need for distributed temperature control of user interface lasers.

Currently, the pure power splitting approach appears to more viable, and it is around this approach that industry standards are being written.

As shown in Fig. 3.2, an APON can be used for Fiber-to-the-Home or Fiber-to-the-Business where the fiber extends all the way to the customer's premises or in a Fiber-to-the-Cabinet or Fiber-to-the-Curb architecture where the fiber terminates at a street-side terminal. From this terminal, VDSL can be used over copper wires to provide broadband services.

3.4 The Full Services Access Network Industry Group

In 1995, a group of seven telecommunication operators formed a collaborative effort to identify technologies and network architectures that could cost-effectively support a wide range of narrowband and broadband services.[1] This effort has now evolved into a forum know as the Full Services Access Network (FSAN) and includes network operators from around the world representing over 350 million access lines. The members of the FSAN concluded that a cost-effective broadband access network could only be achieved with significant manufacturing volume. This in turn would only be achieved with world-

---

[1] The original members of the FSAN initiative were British Telecom, Deutsche Telecom, France Telecom, NTT, CSELT and Telefónica.

wide industry consensus around a common set of specifications. FSAN is representative of a recent trend in the telecommunications industry to initially bypass formal standards bodies to develop specifications when time to market is critical, but instead organize an *ad hoc* forum of interested parties. The specifications developed may later be brought to an appropriate standards body for formal standardization.

The work of the FSAN initiative have taken place in three phases. The initial phase (July 1995 to June 1996) identified technical and economic barriers to low-cost broadband access networks and identified ATM Passive Optical Network (APON) as the most promising access architecture for overcoming these barriers. It concluded that components for broadband access networks could be manufactured at low cost only with production volumes of a million units or more. It also concluded that while some factions of the industry favored Fiber-to-the-Cabinet and others favored Fiber-to-the-Home, these two approaches could share a significant number of components leading to high manufacturing volume. The second phase (July 1996 to February 1997) developed a set of specifications for the APON. These specifications have now been presented to ITU-T and standardized in recommendation G.983 "High speed optical access systems based on Passive Optical Network (PON) techniques." The third phase, ongoing at this writing, is to conduct field trials of APON networks. These would include encouraging early equipment production and would focus on demonstrating services and examining practical field issues such as installation, powering, testing and OAM&P (Operations, Administration, Maintenance and Provisioning).

During the latter phases, the work of FSAN has been organized in two sub-teams called "chapters," one focused on Fiber-to-the-Cabinet and Fiber-to-the-Curb, while the other focused on Fiber-to-the-Home and Fiber-to-the-Business. The FTTH team consisted of the members listed in Table 3.1. Their goals are to ensure that cost-effective FSAN-compliant FTTH systems are available from multiple manufacturers, to prove this technology in realistic

TABLE 3.1

FSAN FTTH Chapter

BellSouth	BT
France Telecom/CNET	Deutsche Telekom
NTT	KPN
Swisscom	SBC
Telecom Italia/CSELT	

TABLE 3.2

FSAN FTTCab
Chapter

Bell Canada	British Telecom
Deutsche Telekom	France Telecom
GTE	Swiss PTT
Telecom Italia	

field environments, to minimize the number of technical solutions supported by network operators and manufacturers, and to drive toward equipment interoperability. A particular interest of the FTTH Chapter has been addressing the ONU powering question. Privacy and security are also particular concerns in the FTTH environment.

The FTTCab Chapter consisted of the network operators listed in Table 3.2. An area of unique interest to the FTTCab Chapter has been VDSL. In carrying out their work, the members of the FTTCab developed a common baseline RFP (for field trials) which each member modified and issued to its suppliers. Several APON field trials were started as a result of this as described in section 3.7.

3.5 ATM Passive Optical Networks

A reference architecture for the APON specified by the FSAN consortium is shown in Fig. 3.3.

Figure 3.3
APON Reference
Architecture

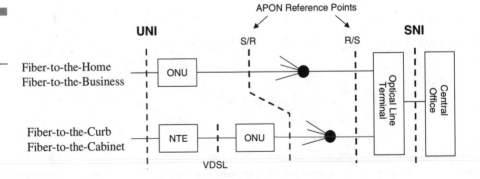

At the Service Node Interface (SNI), various existing narrowband protocols would be used such as V5.1 or V5.2 (TR-08 and GR-303 in North America) or a new broadband specification known as VB5.1 or VB5.2.

At the User-Network Interface (UNI), an ATM-based interface would be specified such as ATM Forum 25.6 Mb/s or ATM-155, although in typical implementations, various terminal adaptation functions could be integrated with the ONU. Hence the interface to the user from such an integrated box could be POTS, ISDN (I.430, I.431), frame relay, 10BaseT, 100BaseT, analog video, etc. In the case of Fiber-to-the-Cabinet or Fiber-to-the-Curb, data is sent over twisted copper pair using the VDSL specification from the ONU to the Network Termination Equipment (NTE) on the customer premises. The data rate of the VDSL is largely a function of the length of the copper facility. Note that in early implementations, some network operators may deploy ADSL and HDSL as well.

The Passive Optical Network specified by FSAN, illustrated in Fig. 3.4, is based on an ATM Passive Optical Network (APON). The APON utilizes a power splitter with information sent in each direction in ATM cells. Each APON interface on Optical Line Termination (OLT) in the network connects to up to 32 Optical Network Units (ONUs) located close to the customer's premises (for Fiber-to-the-Cabinet or Fiber-to-the-Curb) or at the customer's premises (for Fiber-to-the-Home or Fiber-to-the-Business). The proposed APON system transmits downstream information at either 155.52 Mb/s or 622.08 Mb/s and upstream information at 155.52 Mb/s. The symmetrical configuration (155.52 Mb/s downstream and upstream) would typically be used for business applications, while the asymmetrical configuration (622.08 Mb/s downstream and 155.52 Mb/s upstream) would typically be used for residential applications supporting video entertainment services.

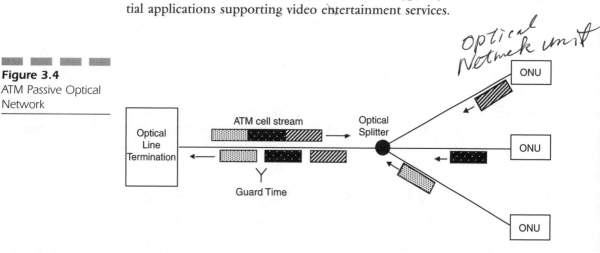

Figure 3.4
ATM Passive Optical Network

Figure 3.5
Upstream Cell
Overhead

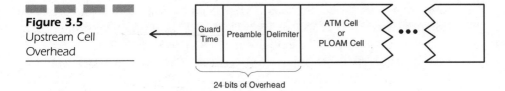

Figure 3.5
Upstream Cell
Overhead

Upstream and downstream information can be sent on two separate fibers, or wave-division multiplexing (downstream at 1550 nm and upstream at 1300 nm) can be used to allow a single strand of fiber to transmit information in both directions. Cost studies indicate that the single-fiber WDM approach will be less expensive for large-scale deployment.

Internal message formats allow up to 64 ONUs to share the same APON. However, transmit power limitations will typically limit an APON to 16 to 32 ONUs with the operational range up to 20 km. The average transmission quality should have a very low bit error rate, less than 10^{-9} across the entire APON system with an error rate of 10^{-10} for the optical components.

The sharing of the bandwidth on the APON is controlled by the OLT. Cells transmitted in upstream slots require additional overhead for the Time Division Multiple Access Function. The format of these overheads is shown in Fig. 3.5. The OLT specifies which ONU is allowed to transmit in a given cell slot. This procedure is made more difficult by the fact that the various ONUs will typically be located at varying distances from the splitter. In order to ensure that one user's cell arrives at the splitter almost immediately after the previous cell is completed, a ranging protocol is employed so that each ONU can determine its distance from the splitter and hence know when to launch its cell. A small guard period is needed between each cell, the length of which depends on the accuracy of the ranging protocol. For practical systems operating at 155.52 Mb/s, guard periods of at least four bits are required. Upstream cells also require a preamble to enable the OLT to recover the phase of the bits of the cell and adjust for signal intensity. Finally, a delimiter is required to denote the start of the ATM cell. The total overhead required for each upstream ATM cell, guard period + preamble + delimiter, is 24 bits. The guard time length, preamble pattern and delimiter pattern are programmable under OLT control and are set via downstream messages.

3.5.1 Physical Layer Operation

The operation of the APON is largely controlled though use of Physical Layer OA&M (PLOAM) cells. The downstream cell stream is divided into frames,

■■ ■■ ■■ ■■
Figure 3.6
PLOAM Cell Format

1	IDENT	25	GRANT19
2	SYNC1	26	GRANT20
3	SYNC0	27	CRC
4	GRANT0	28	GRANT21
5	GRANT1	29	GRANT22
6	GRANT2	30	GRANT23
7	GRANT3	31	GRANT24
8	GRANT4	32	GRANT25
9	GRANT5	33	GRANT26
10	GRANT6	34	CRC
11	CRC	35	MESSAGE_PON_ID
12	GRANT7	36	MESSAGE_ID
13	GRANT8	37	MESSAGE_FIELD0
14	GRANT9	38	MESSAGE_FIELD1
15	GRANT10	39	MESSAGE_FIELD2
16	GRANT11	40	MESSAGE_FIELD3
17	GRANT12	41	MESSAGE_FIELD4
18	GRANT13	42	MESSAGE_FIELD5
19	CRC	43	MESSAGE_FIELD6
20	GRANT14	44	MESSAGE_FIELD7
21	GRANT15	45	MESSAGE_FIELD8
22	GRANT16	46	MESSAGE_FIELD9
23	GRANT17	47	CRC
24	GRANT18	48	BIP

56 cells per frame on the 155.52 Mb/s system and 224 cells per frame on the 622.08 Mb/s system. Every 28th cell transmitted in the downstream direction is a PLOAM cell. There are 53 slots, able to fit an ATM cell plus overhead, on the upstream. An ONU will transmit a PLOAM cell upstream as requested. These PLOAM cells are used to set operational parameters, provide status information, and allocate upstream bandwidth among the ONUs. The format of the downstream PLOAM cell is given in Fig. 3.6.

The use of the various fields is as follows:

PLOAM IDENTification	Used to identify the first PLOAM cell of each frame and hence the start of the frame. Also used to identify future versions of the PLOAM format.
SYNCronization	Used to transmit the 8 kb/s clock from the OLT to the ONUs. The ONU keeps a counter of the number of bytes transmitted (every 4 bytes for the 622.08 Mb/s interface) and resets this counter every 125 μsec. The value of this counter is transmitted in the SYNC field.

GRANT   Used by the OLT to grant the ONUs access to upstream slots. The 53 GRANT bytes across the initial two PLOAM cells each correspond to an upstream slot and indicate which ONU can transmit in that slot. It is also used to initiate the ranging protocol with that ONU.

CRC   Used to protect each group of 7 GRANTs. An ONU will not use a GRANT if the CRC is bad.

MESSAGE_PON_ID
MESSAGE_ID
MESSAGE_FIELD
CRC   Used by the OLT to send a message to the ONU identified by MESSAGE_PON_ID. The type of message is indicated by MESSAGE_ID with message content contained in MESSAGE_FIELD. The message is acted upon only if the CRC is correct.

Bit Interleaved Parity (BIP)   Used to monitor the bit error rate of the downstream link.

The format of the upstream PLOAM cell is shown in Fig. 3.7.

The use of the various fields is as follows:

Figure 3.7
PLOAM Cell Format

1	IDENT	25	LCF10
2	MESSAGE_PON_ID	26	LCF11
3	MESSAGE_ID	27	LCF12
4	MESSAGE_FIELD0	28	LCF13
5	MESSAGE_FIELD1	29	LCF14
6	MESSAGE_FIELD2	30	LCF15
7	MESSAGE_FIELD3	31	LCF16
8	MESSAGE_FIELD4	32	RXCF0
9	MESSAGE_FIELD5	33	RXCF1
10	MESSAGE_FIELD6	34	RXCF2
11	MESSAGE_FIELD7	35	RXCF3
12	MESSAGE_FIELD8	36	RXCF4
13	MESSAGE_FIELD9	37	RXCF5
14	CRC	38	RXCF6
15	LCF0	39	RXCF7
16	LCF1	40	RXCF8
17	LCF2	41	RXCF9
18	LCF3	42	RXCF10
19	LCF4	43	RXCF11
20	LCF5	44	RXCF12
21	LCF6	45	RXCF13
22	LCF7	46	RXCF14
23	LCF8	47	RXCF15
24	LCF9	48	BIP

PLOAM IDENTification	Used to identify future versions of the PLOAM format.
MESSAGE_PON_ID MESSAGE_ID MESSAGE_FIELD CRC	Used by an ONU to send a message to the OLT. The ONU sending the message identified by MESSAGE_PON_ID. The type of message is indicated by MESSAGE_ID with message content contained in MESSAGE_FIELD. The message is acted upon only if the CRC is correct.
Laser Control Field (LCF)	Used for maintaining the nominal specified mean optical output power and to control the extinction ratio.
Receiver Control Field (RXCF)	Used to recover the threshold level for regenerating the data from the incoming analog signal.
Bit Interleaved Parity (BIP)	Used to monitor the bit error rate of the downstream link.

3.5.2 APON Bandwidth Management

A wide variety of traffic types is envisioned to be served by the APON including mixtures of both isochronous traffic, such as telephony and entertainment video, which must be reproduced as a smooth clocked bit stream, and isochronous traffic, such as web surfing, which is bursty and more delay tolerant. The management of the available bandwidth must be carefully managed to ensure proper performance of each class of traffic. Further complicating the situation, the upstream and downstream bandwidth must be shared in a fair way among the users of the APON. The responsibility of this bandwidth management is given to the OLT. The OLT manages internal downstream queues for the various users and traffic types and priorities, and services these queues in ways to provide appropriate priority among traffic classes and equal quality of service among users. The required performance for various classes of service as specified in FSAN requirements are given in Table 3.3.

Access to the upstream bandwidth is controlled by a media access control function in the OLT. The ONUs notify the OLT of their need for bandwidth via Request Access Unit (RAU) messages indicating the length of its internal queues. The medial access controller in the OLT indicates each frame of 53 slots in the upstream direction which ONU can transmit a cell in each slot. The ONU can transmit either a data cell or a PLOAM cell in the slot granted to it.

TABLE 3.3

Performance
Requirements of
Various Services

Service	Traffic Type	Peak Bandwidth	Access Transmission Delay/Access Delay/ Response	Cell Loss
IP Routing	UBR	10 Mb/s (typical)	1.5 ms/<1 s/−	Cell discard on IP packet basis (one cell loss results in one IP packet loss)
ATM SVC	CBR/VBR/ ABR	<150 Mb/s	1.5 m/−/−	$<10^{-5}$
Video on Demand	CBR	6 Mb/s (typical)	1.5 ms/<3 ms/− (cell deviation value)	$<10^{-8}$
Switched Video Broadcast- ing	CBR	6 Mb/s (typical)	1.5 ms/<3 ms/<500 ms (cell deviation value)	$<10^{-8}$
POTS/ISDN	CBR	<2 Mb/s	1.5 ms/−/−	$<10^{-5}$

3.5.3 APON Ranging Protocol

Because ONUs will be located at varying distances (from 0 km to 20 km) from the optical splitter, the times will vary when they must launch a cell in order for it to fall within an upstream cell slot. To accomplish this, each ONU has a programmable equalization delay and a ranging protocol has been defined to measure end-to-end delay and to set this equalization delay. Part of these procedures are also used to specify to the ONU the format of the three overhead bytes added to upstream cells and to specify to the ONU its address on the APON.

The procedure is initiated by the OLT sending an UPSTREAM_OVER-HEAD message to indicate to the ONUs which overhead format they must to use. Assume for the moment that the equalization delay in one ONU is being adjusted. The OLT will send the ONU a message giving it a Ranging Grant and asking it to acknowledge in a specified upstream slot. The OLT will also free up slots near the one assigned since the ONU's equalization delay has not yet been adjusted. The ONU will respond with ranging PLOAM cell. The OLT measures the arrival of this cell versus the ideal time in the upstream slot, and based on this difference, sends a RANGING_TIME message to the ONU setting the equalization delay. During normal data transfer operation, the OLT monitors the arrival time of cells versus the target time.

Each ONU on the APON has an assigned PON_ID that is used by the OLT to address messages to a specific ONU. This is achieved by the OLT assigning the ONU a PON_ID with an ASSIGN_PON_ID message which it addresses to the ONU via its unique serial number. These serial numbers are assigned by the manufacturer. The OLT could discover these by the network operator manually entering them. It is highly desirable, however, for the OLT to be able to "self discover" ONUs. This is done by the OLT periodically initiating the ranging procedure, addressing the Ranging Grant to "all ONUs without an assigned PON_ID." If there is only one new ONU, that one will respond with a SERIAL_NUMBER_ONU message giving its serial number which the OLT will use to then assign it a PON_ID.

What if more that one ONU has not been assigned a serial number as would typically be the case when the APON is first provisioned? In that case, multiple ONUs will respond to the above Ranging Grant and their responses will collide. This is handled by the OLT recognizing the collision and sending another Ranging Grant with a SERIAL_NUMBER_MASK message. Only those ONUs whose serial number matches the grant will respond. The OLT will repeat the procedure while varying the mask in a binary tree algorithm until only one ONU responds.

3.5.4 Use of Encryption on APONs

A complication of power-splitting passive optical networks is that all the downstream traffic is sent to every ONU. It is the responsibility of the ONU to filter the traffic so as to send to the user only that traffic destined to that user. However, since the ONU (for Fiber-to-the-Home or Fiber-to-the-Business) is physically located at the user's premises, it is conceivable that a sophisticated user could tap in at the APON and intercept traffic destined for other users. A business customer would be very uncomfortable knowing that his or her competitor might be eavesdropping.

To address this problem, an encryption mechanism has been defined for APON. The OLT initiates the process by sending a New_Key_Request message in a PLOAM cell to an ONU. That ONU responds by sending an encryption key to the OLT in a New_Encryption_Key message in a PLOAM cell. The OLT acknowledges receipt of the key with an Encryption_Key_Update message to the ONU and, to protect against message loss, repeats this message a known number of times with a known delay after it begins using this new encryption key. The ONU completes the procedure by replying with an Acknowledge message.

Note that with the addition of encryption, APONs, far from being susceptible to eavesdropping, become one of the most secure broadband access platforms being deployed.

3.6 Prevention of Babbling ONUs

The operation of the upstream data transmission depends upon each ONU transmitting only during its slots as assigned to it by the OLT. Under normal operation, this would be the case. However, one can imagine various ONU failure modes where an ONU could interfere with the operation of other ONUs. In the extreme, the laser of an ONU could become stuck in the "On" state causing the failure of the entire APON. Such faults might be difficult to isolate. Obviously one would want to avoid the cause of having to send a craftsperson to each ONU site to isolate a failure.

A basic requirement is that the ONU must be designed to detect failures of its laser and to take actions to shut off power to this device. This could be done by detecting that there is no nominal backfacet photocurrent or that the drive current is beyond the maximum specification. Other conditions that would result in the ONU shutting off its laser are these:

- Loss of signal on the downstream
- Startup procedure failure
- Loss of three consecutive PLOAM cells
- Loss of cell delineation (7 consecutive cells with invalid header error checks)
- Loss of downstream framing
- Receipt of a Deactivate PON_ID message from the OLT

For applications requiring greater reliability, the best solution is to provide redundant APONs. Requirements for duplex operation are included in the APON specifications.

3.7 Trials and Deployments of APON Systems

With the completion of APON specifications, network operators have announced or begun early trials and deployments of APON systems. Of the

Figure 3.8
Bonaparte Turin
Network

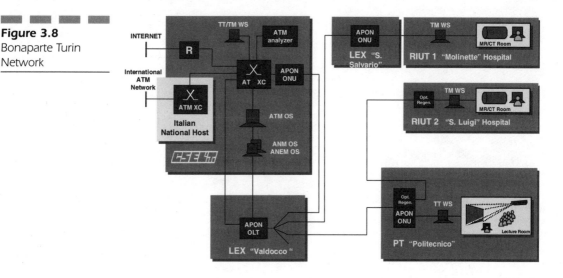

companies that have indicated their intention of going forward with APON trials, GTE began a field trial in August 1998 of APON in a Fiber-to-the-Cabinet configuration. VDSL would provide the final connection to the customer's premises. This would be followed by commercial deployments in 1999. According to GTE, the prices in the APON bids received compare favorably with the current prices for the baseline HFC approach. According to GTE, the price of serving a customer with HFC is approximately $766 (includes inside wiring and an average of 2.3 Set Top Boxes). Serving these same homes with APON/VDSL would cost $675–$700 (including residential gateway system). The limitation is that the APON/VDSL approach can only provide three to four simultaneous video channels, a problem for some TV-centric homes.

CSELT has announced an APON trial as part of their Torino 2000 project, an interactive multimedia services trial and application development testbed targeting hospitals, research institutes, etc. This is shown in Fig. 3.8.

Other trials are about to begin of APON systems. Both BellSouth and NTT are about to begin Fiber-to-the-Home trials for data and entertainment video. SBC has expressed interest in APON for fiber to small businesses.

The success of these trials will determine whether APON plays a major roll in future broadband access networks or simply joins the long list of ill-fated Fiber-to-the Home trials.

4

ADSL and VDSL— The Copper Highway

4.1 Introduction

The telephone network is the only network providing near universal access to residential and business customers. The number of telephone subscribers worldwide is estimated to be of the order of 700 million. From its genesis, the telephone access network has consisted of twisted pairs of copper wires strung on poles or buried in the ground to the customer's premises. These copper wires were designed for the transmission of analog voice signals. Nonetheless, technologists later discovered that it was possible to transmit data through these same voice channels. From early modems that could only squeeze 75 b/s through a voice channel, this technology has advanced to the point where modern inexpensive modems can transmit nearly 56 kb/s through a voice grade connection.

Modem technology has at last reached its limit. Since the telephone network internally encodes voice connections at 64 kb/s, further increases of modem speeds on dial connections are not even theoretically possible. This limit, however, is imposed by the telephone switching systems and interoffice facilities; the lowly twisted pair copper wire to the home has yet more capacity to yield. In the 1980's, ISDN became the first system to exploit that capacity. ISDN transmits 144 kb/s in each direction over a single twisted pair of wires over 6000 m. At the central office switching system, this 144 kb/s is broken into two 64 kb/s switched channels and a 16 kb/s advanced signaling channel. Clever data communications engineers determined ways to dial the same end point with the two channels and combine their capacity to form a 128 kb/s connection. This may have been the savior of ISDN in North America. While in France and Germany ISDN was widely deployed for conventional telephone service, in North America ISDN languished awaiting a driving application, especially for residential customers. Many claimed that ISDN stood for "I Still Don't Know." Recently, with the growing enthusiasm for the Internet, residential customers in North America have started ordering ISDN for fast Internet access.

The announcement of ISDN's success at last may have been premature. The demand has grown for even higher access data rates for residential customers to support new services such as web pages with video and multimedia components. Telcos still aspire for a means to offer switched Video-On-Demand services. All this is occurring in the context of increasing competition from CATV companies and new players in the telecom market brought about by the current telecommunications liberalization and deregulation. The telephone companies are again looking at the twisted pair copper wire to see if there isn't more capacity to exploit.

The amount of analog capacity on a twisted pair of copper wires is fundamentally related to its length. The majority of telephone loops are less than 4 km in length and yield a usable analog capacity of about 1 MHz. Shorter loops have even more capacity. Exploiting this capacity has been enabled by recent advancements in Digital Signal Processing technology.

ADSL (Asymmetric Digital Subscriber Line) provides one answer to this need. Recognizing that residential customers have a greater need for fast download speeds than for transmitting upstream data, ADSL devotes the majority of the capacity on the loop to the downstream channel. Depending upon the length of the loop, ADSL can achieve downstream bit rates up to about 7 Mb/s and upstream rates of several hundred kb/s. In this way, ADSL transforms the twisted pair from one limited to voice and low-speed data to a powerful wideband bit pipe. ADSL does all this while leaving the lower 3 kHz intact for support existing (analog) telephone service.

Although ADSL is a vast improvement over modems and ISDN and appears well-matched for high-speed Internet access, there remains a need for even higher capacities to support services such as Switched Digital Video Broadcast and even switched HDTV (High Definition TV). A conventional broadcast quality video signal, MPEG-2 encoded, requires 6 Mb/s. Given that many homes have multiple TV sets, a capacity of at least 12 Mb/s or 18 Mb/s is required. A single HDTV signal requires about 20 Mb/s. It turns out that even this is not too great for the copper loop. Using a technique known as Very High Speed Digital Subscriber Line (VDSL), 26 Mb/s can be transmitted on 3 km of copper loop and 52 Mb/s transmitted up to 1 km. This is much shorter than most subscriber loops. However, it can provide the "last mile" connection to Fiber-to-the-Cabinet (FTTCa) or Fiber-to-the-Curb (FTTC) architectures where optical fiber extends from the local exchange to the cabinet or the curb unit. Note that even for the high speeds of VDSL, the underlying analog POTS or ISDN service is maintained on the same twisted pair.

This chapter starts with an overview of the evolution of the capacity on copper twisted pairs. After a description of the main impairments of the medium, the chapter concentrates on ADSL and VDSL, covering the topology, the system requirements, the transmission techniques used and the current status in standardization.

4.2 Evolution of Capacity on Twisted Pairs

Figure 4.1 depicts the evolution of transmission speeds on twisted pairs.

Since the early 1960's, voice band modems have been developed to transport digital data over the telephone network. With these modems, data is modulated at the transmitter, transported (bidirectionally over one pair or over two pairs) transparently through the telephone network and demodulated at the receiver. Different modulation techniques have been used, starting with FSK (Frequency Shift Keying) in the ITU-T standard V.21, DPSK (Differential Phase Shift Keying) in V.26, V.27 and V.29, to QAM (Quadrature Amplitude Modulation) in V.22bis, V.33 and V.34. Different duplexing techniques have been applied: full duplex transmission over a single pair with separated (FDD, Frequency Division Duplexing) or overlapping (EC, Echo Canceling) frequency bands for both directions of transmission, dual simplex transmission over two pairs and half duplex transmission over one pair. The

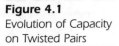

Figure 4.1
Evolution of Capacity
on Twisted Pairs

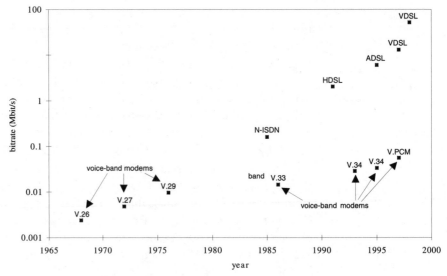

bit rates achieved range from 300 b/s in V.21 to 33.6 kb/s in V.34. More recently (in 1997), bit rates of 56 kb/s have been achieved (V.PCM [1]). However, these very high rates are only possible in the downstream direction and only in particular configurations, e.g., where an Internet service provider has a digital connection to the central office and where only one A/D converter exists in the connection.

Key for voice band modems, as the name implies, is that the data is modulated in the voice band (from 300 to 3.4 kHz). This is a main difference with DSL modems whose spectrum is not limited to this band. The direct consequence is that DSL signals cannot be carried transparently over the POTS (Plain Old Telephone Service) network end-to-end. They have to be terminated before entering the local telephone exchange. Figure 4.2 depicts as an example of Internet access by means of voice-band modems (top) and ADSL (bottom). In case of the ADSL, the transmission on the twisted pair is terminated in the LT (Line Termination) on the local exchange and the NT (Network Termination) the subscriber side. The "POTS-splitter" in the LT separates the analog telephone signal from the digital data signal. The first is sent to the narrowband network (telephone exchange), the second to the broadband network (e.g. ATM switch or router).

A first step towards DSL was the introduction, in the 1980's, of basic rate ISDN that offers full duplex (EC) data transport at 144 kb/s (data plus sig-

---

[1] V.PCM is still in the process of standardization.

Figure 4.2
Voice Band Modem
Versus DSL

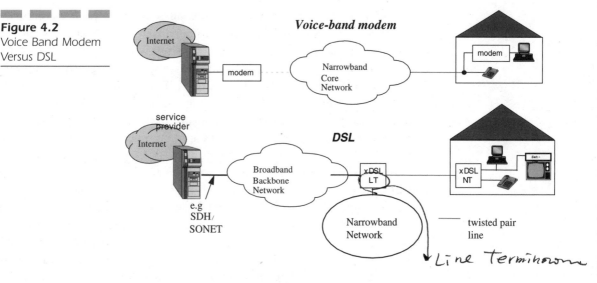

naling) over a single pair. The modulation format is 2B1Q (in North America) or 4B3T (in Europe). During the past years, the evolution in Digital Signal Processing techniques has enabled the design of several new modem technologies, such as HDSL, ADSL and very recently VDSL, boosting the transmission capacity of the existing twisted pair infrastructure. HDSL transports 1.5 Mb/s (North America) or 2 Mb/s (Europe) over 1, 2 or 3 pairs. The transmission on each pair is full duplex (EC). The modulation is 2B1Q. ADSL and VDSL are described in more detail later in this chapter. While HDSL is more suited for business applications, ADSL and VDSL primarily address the residential and SOHO (Small Office Home Office) markets.

It should be noted that besides those listed above, other DSL systems have been defined such as HDSL-2 (a single pair version of HDSL that is currently being standardized by ANSI/T1E1.4, sometimes referred to as SDSL) and RADSL (Rate Adaptive DSL, being the single carrier version of ADSL). Also, some proprietary systems have been announced: MDSL (Multi-rate DSL or Moderate speed DSL), UDSL (Unidirectional DSL) and IDSL (Integrated DSL).

4.3 Twisted Pair Impairments

The telephone access network consists primarily of twisted pairs of copper wires. (For a brief period aluminum wires were used in some countries, but this conductor was not well-suited for this purpose as is a poorer conductor, is less pliable, and is more susceptible to corrosion). The twist in the twisted pair ensures equal coupling into the two wireless, thereby introducing few differential signals (longitudinal signals will not pass through coupling transformers at the end of the twisted pair). This also reduces electromagnetic radiation from signals carried in the copper loop.

In the past, paper was used as an insulator between the wires. Today polyethylene is common. These wires pairs are combined in cables of different sizes ranging from a couple to a few hundreds of pairs. The cable structure can vary considerably. In Europe, the pairs are often combined in quads that consist of two pairs twisted around each other. These quads constitute binders of some tens of pairs. Finally, several binders are grouped in a single cable. In the U.S., this hierarchy is often non-existent and pairs are grouped in cables without forming quads. Between different wires in the same cable there exists capacitive and inductive coupling. The coupling increases as the wires are closer together. It causes unwanted crosstalk between the pairs. The crosstalk is typically worse between two pairs in the same binder than for wires in adjacent binders. It can be reduced by the optimization of the twist of the individual pairs and of the topology of the cable.

The transmission channel capacity depends highly on the twisted pair characteristics and suffers from a number of impairments.

- Since the early days of telephony, load coils have been added to long loops (typically longer than 6 km) to boost and flatten the frequency response of the line at the upper edge of the voice band. In so doing, these coils greatly increase attenuation at frequencies above the voice band making them unusable for DSL. They must be removed to allow for DSL services.

- The frequency dependent attenuation and dispersion leads to pulse distortion and inter-symbol interference. In multi-carrier systems, inter-carrier interference can occur. Figure 4.3 illustrates the attenuation of the signal as a function of frequency after propagation over 1 km of 26 gauge twisted pair.

- Crosstalk between wire pairs in the same binder or adjacent binder groups is a principal noise contributor. Two types of crosstalk can be distinguished: Near-End crosstalk (NEXT) occurs at a receiver that is collocated with the disturbing source while Far-End crosstalk (FEXT) occurs at a remote receiver. FEXT is attenuated by propagation through the loop while NEXT is not. Therefore, NEXT dominates FEXT by far for echo canceled systems. Further, NEXT increases at a typical rate of 4.5 dB/octave ($f^{3/2}$) while FEXT increases by 6 dB/octave (f^2).

Figure 4.3
VDSL Signal and
Noise Spectra

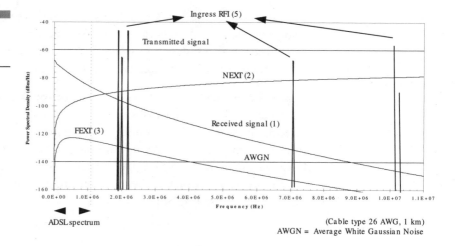

(Cable type 26 AWG, 1 km)
AWGN = Average White Gaussian Noise

- In the process of connecting and disconnecting portions of the loop plant, sometimes a portion of open-circuited wire pair is left connected to a working wire pair. This is called a bridge tap. The existence of bridged taps in the loop plant differs from country to country and depends upon the cabling rules used in the past. Their presence causes reflections and affects the frequency response of the cable leading to pulse distortion and inter-symbol interference.

- A loop can also be built up of wires with differing diameters (referred to as gauge transitions), leading to reflections and distortion.

- Copper transmission suffers from impulsive noise that is characterized by high amplitude bursts of noise with a duration of a few microseconds to hundreds of microseconds. It can be caused by a variety of sources such as central office switching transients, ringing, ring trip, dial pulses and lightning.

- The impedance mismatch between the line impedance and the hybrid transformer (used to interconnect the bi-directional transmission of the loop to the separate transmit/receive path in the network) causes signal reflections. This problem can be resolved either by echocancellation or by separation of upstream and downstream transmission on the loop by means of Frequency Division Multiplexing (FDM).

- As a result of the cable unbalance, RF (Radio Frequency) signals can be picked up during propagation over the wire and interfere with the transmitted data at the receiving side. The balance of the cable decreases as the frequency increases. The aerial drop wires, the vertical cables in high-rise buildings and the in-house wires are the most vulnerable for RF ingress. This ingress can come from a variety of sources such as AM broadcast (LW, MW and SW), radio amateur communication and public safety and distress bands. The radio amateur is the most troublesome interferer for many reasons.

Because of the wide differences in network topology, installation practices and cable types, the impact of these impairments can vary significantly from operator to operator. Also, some DSL systems will suffer more than others from given impediments. As an example, RF interference is a much greater threat for VDSL than for ADSL because of its spectrum allocation (see section 4.5). In any event, a highly adaptive transmission system is needed to cope with the above imperfections.

4.4 ADSL

4.4.1 System Requirements Reference Model

Figure 4.2 (bottom) depicts a network reference model for ADSL. A somewhat more detailed view is given in Fig. 4.4.

ADSL has to coexist with POTS on the same pair. A POTS-splitter consisting of a low-pass filter (LPF) and a high-pass filter (HPF) separates the analog telephone signal from the digital data signal. The high-pass filter may be integrated with the ATU (ADSL Transceiver Unit) at central office side (CO) or the remote terminal (RT) side, i.e. the customer side. At the customer's premises, the low-pass filter is typically installed at the entrance of the home, i.e. in the basement or in the NID (Network Interface Device).

Figure 4.4
ADSL System Reference Model

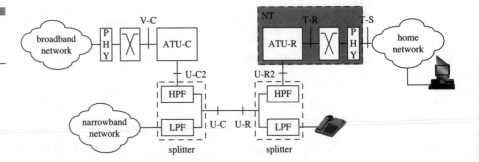

4.4.2 Performance

The ADSL transmission system offers an asymmetric capacity to the subscriber. In the downstream direction (towards the subscriber), it provides a capacity up to 7 Mb/s, while in the upstream direction it provides up to 640 kb/s. In general, the maximum ADSL data rate depends upon the distance covered, wire gauge, and interference as shown in Table 4.1 (for uniform cable sections). The values in this Table are for DMT based modems. Single carrier modems have a different performance.

4.4.3 Transfer Mode

Three types of data transport are provided for in the ADSL standard:

- bit synchronous data such as DS1 (1.544 Mb/s) or E1 (2.048 Mb/s)
- packet data (e.g. making use of the HDLC protocol)
- ATM (Asynchronous Transfer Mode) transport

Provisions are made in the standard for the transport of each of these data types but the support of all three modes is optional.

4.4.4 DMT Transmission

DMT transmission for ADSL has been standardized in ANSI/T1E1.4 and is supported by ETSI/TM6.

TABLE 4.1

ADSL Performance Figures

Reach	Cable (AWG)	Downstream Data Rate	Upstream Data Rate
18,000 ft	24	1.7 Mb/s	176 kb/s
13,500 ft	26	1.7 Mb/s	176 kb/s
12,000 ft	24	6.8 Mb/s	640 kb/s
9,000 ft	26	6.8 Mb/s	640 kb/s

4.4.5 Spectrum and Bit Allocation

The basic principle of DMT is to transmit the information bits in parallel over a large number (256) of carriers (tones), each of which is QAM modulated. The carrier frequencies are multiples of the same basic frequency (4.3125 kHz). For the separation of the up- and downstream transmission, two bandwidth allocation policies are included in the standard.

The first one uses overlapping spectra for up- and downstream transmission and applies echo canceling (EC). The second option uses frequency division duplexing (FDD) in which case no tones are shared by the up- and downstream bands. The latter is depicted in Fig. 4.5.

The upstream band ranges from about 25 kHz to 138 kHz (carriers 6 to 32) while the downstream band extends up to 1.104 MHz (carrier 256). The start frequency of the downstream spectrum can be anywhere above 25 kHz and is manufacturer discretionary. The lowest carriers are not modulated to avoid interference with POTS. The transmit power spectrum is almost flat over all used tones. For up- and downstream transmission, the average nominal psd (power spectral density) is respectively −38 dBm/Hz and −40 dBm/Hz across the whole band. In downstream direction, a power cut-back is applied on short lines to avoid saturation of the remote receiver. An optional power boost (of 6 dB) for long lines was provided in T1.413 Issue 2 to reduce the crosstalk to other services in the same cable. The passband ripple shall not be greater than ±3.5 dB. The number of bits that is assigned to a tone and its precise transmit power is determined during system initialization as

Figure 4.5

Transmit Spectrum of DMT-based ADSL Systems

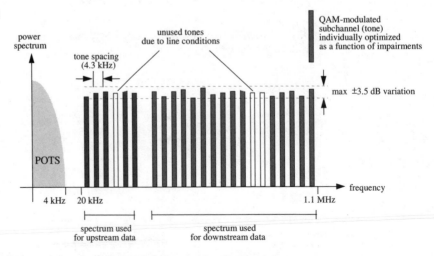

a function of the SNR (Signal to Noise Ratio) on that tone and the requested overall bitrate. During operation, adaptation of the bit assignment or corrections to the transmit power are possible to compensate for alterations in line conditions, due to a variation of the noise or a (slow) drift of the cable transfer function (e.g. because of temperature variations). These on-line adaptations do not interrupt the data flow.

4.4.6 Error Correction

In order to improve the Bit Error Rate (BER) or (equivalently) to increase the system performance, expressed as an increase in capacity for a given BER, Forward Error Correction (FEC) is applied. ANSI specifies the use of Reed-Solomon (RS) coding combined with interleaving. The additional use of Trellis coding is optional but may further reduce the BER or increase the SNR margin.

A distinction is made between delay sensitive or "fast" data, for applications as video conferencing or TCP/IP sessions, and delay insensitive data, such as for Video On Demand (VOD). Delay sensitive data is not interleaved and is transmitted within less than 2 ms (one-way). Delay insensitive data is interleaved to make it more robust against impulsive noise at the cost of increased latency. The ANSI standard allows for simultaneous transport of "fast" and interleaved data.

4.4.7 Bit Rate Adaptation

In T1.413 Issue 2 compliant modems, the up- and downstream bitrates can be programmed in any multiple of 32 kb/s. This small granularity is characteristic for DMT transmission. An increment (or decrement) of the bit rate with 32 kb/s is achieved by the allocation of one extra (respectively one less) bit on a single carrier.

At startup of the modem, two strategies are possible:

1. Manual rate selection (mandatory capability): The system starts up at the rate fixed by the operator. This can be done in accordance with the type of service the user is requesting (and paying) for.

2. Automatic rate selection (optional capability): At startup, the modem *itself* determines the transport capacity of a specific line and initializes at this rate. This concept has also been referred to as "Rate Ad-

aptation at Start-up'' or ''Available Bit Rate'' ADSL as it maximizes the throughput of each individual line. The operator keeps control over essential parameters as delay, BER and SNR margin.

In an informative annex, T1.413 Issue 2 describes Dynamic Rate Adaptation (DRA). Its purpose is to allow reconfiguration of the modem during data transport and to avoid a lengthy restart procedure if the channel conditions or the service requirements would change over time. The mechanism allows rate modifications (up- and downgrades) for both up- and downstream as well as a redistribution of capacity between the fast and interleaved paths. However, these adaptations might involve a service interruption of the order of tens of milliseconds.

4.4.8 Characteristics of ADSL

The most important feature of ADSL is that it can provide high speed digital services on the existing twisted pair copper network, in overlay and without interfering with the traditional analog telephone service (plain old telephone service: POTS). ADSL thus allows subscribers to retain the (analog) services to which they have already subscribed. Moreover, due to its highly efficient line coding technique, ADSL supports new broadband services on a single twisted pair.

As a result, new services such as high speed Internet and On-line Access, Telecommuting (working at home), VOD, etc., can be offered to every residential telephone subscriber. The technology is also largely independent of the characteristics of the twisted pair on which it is used, thereby avoiding cumbersome pair selection and enabling it to be applied universally, almost independent of the actual parameters of the local loop.

The asymmetric bandwidth characteristics offered by the ADSL technology (64–640 kb/s upstream, 500 kb/s − 7 Mb/s downstream) fit in with the requirements of client–server applications such as WWW access, remote LAN access, VOD, etc., where the client typically receives much more information from the server than it is able to generate. A minimum bandwidth of 64–200 kb/s upstream guarantees excellent end-to-end performance, also for TCP/IP applications. These basic characteristics are reflected in two important advantages of the ADSL technology:

- No installation is required for laying new cables, making it useful solution in advance of fiber deployment in the local loop.

■ ADSL can be introduced on a per-user basis. This is important to the network operators for it means that their investment in ADSL is proportional to the user acceptance of high speed multimedia services.

The mature ADSL product combines the benefits of the DMT and ATM technologies, resulting in:

■ Full bandwidth flexibility: upstream and downstream bit rates can be chosen freely and continuously up to the maximum physical limits. At initialization, the system automatically calculates the maximum possible bit rate, with a predetermined margin. The service management system can then set the bit rate to the level determined by the customer service profile, thus maximizing noise margin and/or minimizing transmit power.

■ Full service flexibility: a random mix of services with various bit rates and various traffic requirements (guaranteed bandwidth, bursty services) can be supported, within the available bit rate limits.

4.4.8.1 ATM Over ADSL. In addition to the total capacity of an ADSL line varying due to loop length and other factors, the bit rate required by the various services supported over ADSL will vary as well. Rather than attempt to define fixed capacity channels within ADSL matched to the envisioned services, it is better to use a flexible bandwidth management mechanism that is future safe; that can adapt to the changing needs of the services carried; that can provide the flexibility needed to support new services with bit rates other than those presently envisaged; and that can take advantage of changes in required bit rate resulting from improved compression technologies. A good example of the latter is MPEG 2 video services which can be sent with variable quality at bit rates from 1.5 to 6 Mb/s.

ATM offers the flexibility sought. Indeed, ATM is envisioned as the protocol base for future fiber-based access networks (such as APON described in chapter 3). By utilizing ATM on each of the advanced access network types, a uniform structure is created towards the user with the specific type of access network utilized affecting only the overall ATM capacity. This allows a clean evolution from the point of view of the subscriber and the service providers, from ADSL now, to VDSL in the near future, and eventually to fiber-to-the-home.

The ADSL units at the Central Office side and subscriber side are commonly referred to as ATU-C (ADSL Transceiver Unit, central office) or LT (Line

Termination) and ATU-R (ADSL Transceiver Unit, remote terminal) or NT (Network Termination) respectively. ADSL segment is typically operated as an independent part in the access network. In this way, the ADSL system can interface to an ATM switch in the Central Office or to a fiber-based access network with ATM transport as depicted in Fig. 4.6.

4.4.8.1.1 Description of the ADSL Termination Units. The functional blocks of both the ATU-C and ATU-R are shown in Fig. 4.7. A splitter is used to separate the ADSL data from the underlying analog POTS channel. The DMT modem consists of an analog front-end circuit and a digital signal processor. The ATM processing part is independent of the DMT modem. The potential of receiving a channelized bit stream is also shown. The following paragraphs focus on the ATM functional blocks.

4.4.8.1.2 ATM Mapping. Although the current version of the ANSI standard allows for a channelized interface with one of these channels used for ATM, for the rest of non-ATM uses, this approach is far from optimal. Further, the ANSI standard puts unnecessary restrictions on the up- and downstream

Figure 4.6
Access Network Configurations

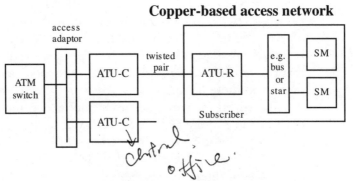

Copper-based access network

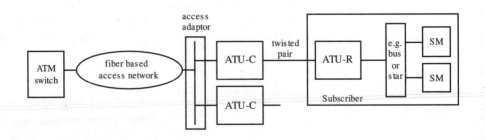

Hybrid fiber & copper-based access network

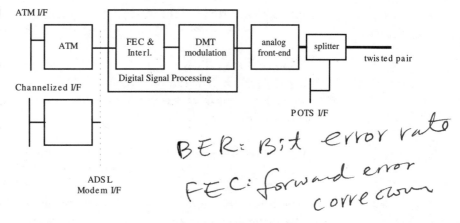

Figure 4.7
Interfaces Provided in the ADSL System

BER: Bit error rate
FEC: forward error
correction

ATM rates in this approach. It is much more desirable to avoid the channelization altogether and offer the ATM functional block a direct interface to the digital signal processing block.

The ADSL modem provides a transparent transmission of data over one unconditioned twisted pair. It provides two alternative transmission methods referred to as the Slow channel and the Fast channel as shown in Fig. 4.8. These might have been better named Low delay and Medium delay. The data stream on the slow channel is interleaved after Forward Error Correction, before it is passed to the DMT modulator. This improves the error correction capability at the expense of extra delay. The Fast data is not interleaved and thus a lower delay, while the error correction capability is smaller.

The ATM functional block (de)multiplexes the received ATM cell stream to and from the Slow and Fast transmission channels. Both cell streams are treated in an independent way. The (de)multiplexing function is based on the ATM header content, for which a specific context table is provided, that can be updated dynamically. From an overall ADSL service standpoint, two Quality of Service (QOS) classes can be defined on the ADSL transmission system: the Fast channel with a low delay but larger BER, and the Slow channel with

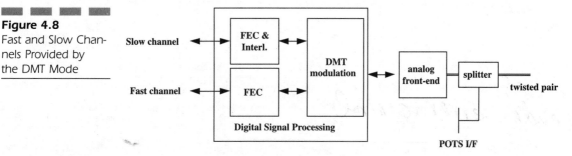

Figure 4.8
Fast and Slow Channels Provided by the DMT Mode

better BER performance but larger delay. At connection set-up, a choice will be made between Fast and Slow channel depending on the QOS requirements of the invoked service. The above mentioned multiplex/demultiplex function of ATM cells is based on the full header (including the VP). Therefore, the ADSL system can be incorporated in an access network where the QOS management is VP based. An overview of the ATM functional block is provided in Fig. 4.9 and Fig. 4.10.

4.4.8.1.3 ATM Transport. The ATM transport on the ADSL transmission system is cell based. In the direction towards the DMT modem, the bit rate adaptation to the DMT modem characteristics is provided by means of idle cell insertion. The same chip set is able to operate on both terminations: at the Residential Subscriber (ATU-R) and at the Central Office (ATU-C). The functions to be performed by the ATM interface are essentially the same for both ADSL terminations, but the transmit/receive bit rates are different. Therefore, the Fast and Slow buffers are able to serve the downstream as well as upstream bit rates. Cell delineation on the received ATM stream is based on the HEC field in the ATM header and conforms to ITU recommendation I.432. Therefore, the correct HEC field is inserted in the ATM header at the transmit side (direction to DMT modem).

4.4.8.1.4 Delineation. In order to ease the delineation process, a payload scrambling is performed at the transmit side. Specific performance monitoring is done on both the Fast and Slow transmission channel by the HEC

Figure 4.9
ATM Functional Block (from ATM Interface to DMT Modem)

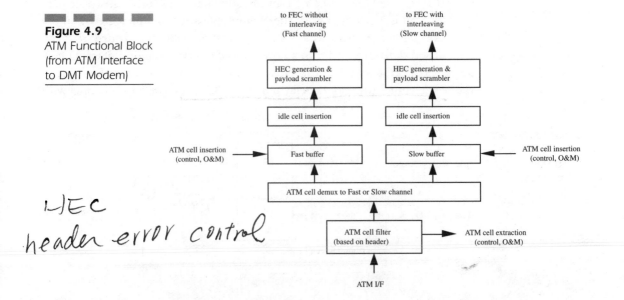

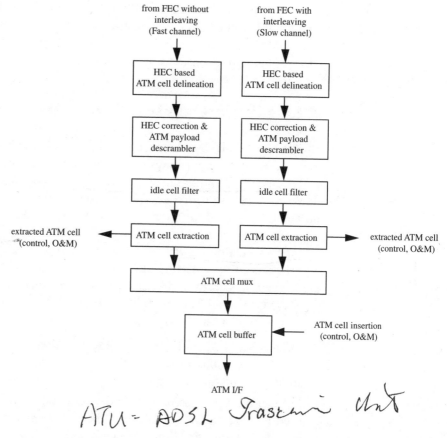

error detection and correction block. After idle cell filtering and cell extraction, both ATM cell streams from the Fast and Slow channels are multiplexed on a single ATM stream. An ATM cell buffer serves the output ATM interface.

The ATM interface can be used in a bus configuration: either as a slave component in the ATU-C, or as a master component in the ATU-R. The former allows connection to several ADSL systems on a single network termination (e.g., an SDH STM-1 or SONET OC-3c interface), while the latter permits several service modules to be connected to the ADSL system.

4.4.8.1.5 OAM&P. For operation and maintenance, the ATM interface provides insertion and extraction of ATM cells in both directions (from and to the DMT modem), on the Fast as well as the Slow channel. Cell extraction is based on the ATM cell header. The extracted cells are processed in an On-Board Controller (OBC). In this way, operation and maintenance of the ATM transport in the ADSL system can be performed autonomously. The cell insertion to and extraction from the ATM interface is also used for signaling purposes (e.g., for allocation of VP/VC resources in the ADSL system). Specific VP/VCs can be defined dynamically.

The ADSL system provides the means to perform tests in various loopback configurations on the system in operation.

4.4.9 Single Carrier Transmission (RADSL)

4.4.9.1 CAP and QAM. In the early days of ADSL standardization, two types of modulation techniques were offered for standardization: DMT and CAP (Carrierless Amplitude and Phase). In early testing, prototypes of DMT outperformed CAP with the result that the current T1E1.413 standard is based on DMT. Backers of the alternative CAP method have not given up and have been joined by backers of another technique called QAM (Quadrature Amplitude Modulation) and have requested standardization by ANSI/T1E1.4. The backers of these techniques claim that chips based on their techniques are simpler, less costly and dissipate less power than DMT.

While CAP and QAM modulation techniques are closely related, the two proposals differ in many aspects:

- Modulation: CAP and QAM respectively. Although different, conversion from one modulation technique to the other requires only an additional rotation of the transmitted or received data symbols.

- FEC: In the CAP system, the downstream data is RS-encoded, interleaved and scrambled before being passed through a Trellis encoder. In the upstream direction, RS-coding and interleaving is optional. The interleaver uses an implied convolutional interleaving method. During data transfer, the scramblers at transmitter and receiver are locked (synchronized scrambler). In the QAM system, a distinction is made between "fast" and interleaved data. The interleaved data for up- and downstream is scrambled *before* being RS-encoded and interleaved. No Trellis coding is applied. The "fast" data is not coded against channel errors. The interleaver is a Ramsey Type-2 interleaver. The scrambler is self-synchronizing.

- Precoding: In the CAP system, the data is passed through a Tomlinson (channel) precoder before modulation. In the QAM system, no precoding is applied.

- Framing: The framing in both systems is entirely different.

4.4.9.2 Spectrum and Bit Allocation. As in DMT, both the CAP and QAM proposals place the upstream channel is at frequencies below the downstream channel. Both channels use nominal square-root raised cosine shaping with an excess bandwidth of 15 percent for CAP and 20 percent for QAM. In the CAP system, the start frequencies for the up- and downstream channels are respectively 35 kHz and 240 kHz. The stop frequencies depend upon the selected symbol rate. For the upstream, two mandatory symbol rates are predefined: 85 and 136 kBaud. For the downstream, five mandatory symbol rates are predefined ranging from 136 to 1088 kBaud. Besides the predefined symbol rates, variable symbol rates may be provided. The upper limit for the psd in the passband region is -40 dBm/Hz nominal for the downstream and -38 dBm/Hz nominal for the upstream. For the QAM system, two different spectral allocations classes exist. Both classes define the range of frequencies over which up- and downstream transmitters are allowed to transmit significant amounts of energy. The class A spectral allocation is consistent with that defined in T1.413. The limits for the upstream band are 34 and 134 kHz; those for the downstream band are 148 and 1104 kHz. In class B, the upstream extends up to 189 kHz and the downstream up to 1366 kHz. In both classes, the psd is limited to -39 dBm/Hz for the downstream and -37 dBm/Hz for the upstream.

Both systems, CAP and QAM, support all constellation sizes up to 256 with an integer number of bits per symbol (with the exception of the 8 point constellation for QAM).

4.4.9.3 Future of CAP and QAM Proposals.   Whether or not the arguments put forth by the CAP and QAM proponents have merit, it appears that these alternative ADSL techniques have lost in the marketplace. The great majority of the local exchange carriers worldwide have selected DMT for their network deployments. The few that had selected one of the other techniques have now announced that they will be switching to DMT. Where CAP has enjoyed some success is among the data CLECs. Whether they too switch to DMT is to be seen.

4.4.10 ADSL-Lite

With conventional ADSL, a passive splitter is installed at the customer's premises, typically on an outside wall, to separate the telephone service from the data service. The telephone service is connected to the normal inside wiring of the premises, and the data service is sent on a separate wire pair to an ADSL modem located inside the premises, typically near a PC. This requires the network operator to dispatch a craftsperson (that is, to initiate a "truck roll") to install the splitter and, if a spare inside wire is not available, to install the wire to the ADSL modem. This trunk roll adds to the cost of installing

ADSL and (due to the limited number of craftspeople) limits the rate at which ADSL can be deployed.

A new version of ADSL, entitled "ADSL-lite" has been proposed to address this problem. With ADSL-lite, no splitter is installed and the telephone and data service share the same wire pair in the premises. An ADSL modem would be installed exactly like a voice-band modem, by plugging it into any available voice jack on the premises. Although obviously desirable, this approach raises the problem of the telephone (as well as fax machines and answer machines, etc.) interfering with the data service, or of the data service introducing noise onto the telephone service. Unfortunately telephone instruments are not characterized by the impedance or noise they present at frequencies above the voice band. Potentially, these instruments could present non-linearities at high frequencies and high signal levels causing intermodulation distortion with the result that some ADSL energy falls into the voice band resulting in audible noise. Nonetheless, ADSL-lite is generating considerable excitement in the industry. Early measurements by a couple of manufacturers have indicated that splitterless operation might be possible at the cost of reduced performance.

Without the splitter, it was assumed that ADSL would not operate at as high a bit rate, but this was viewed as an indirect advantage. By reducing the bit rate and complexity of regular ADSL, ADSL-lite could be implemented on (advanced) general purpose DSPs (Digital Signal Processors). The advantage of such a DSP approach is that it allows quick changes (through downloadable software) to modifications in the standard or to respond to demands for new features. It would further allow the modem at the customer premises to be programmed in different modes, e.g. as a voice band modem initially then converted to a ADSL modem when that service was available in the customer's area. Contributions were brought to T1E1.4 on low complexity ADSL with reduced features or a limited set of parameters (e.g. a reduced number of tones, limited FEC capabilities, etc.).

The ADSL approach was initially advanced by a Special Interest Group (SIG) founded by Intel, Compaq Computer, and Microsoft and now consisting of most of the major ADSL manufacturers and network operators. This group has proposed requirements of upstream rates in the range of 32 to 512 k/s and downstream rates between 32 kb/s to 1.5 Mb/s with a minimum of 384 kb/s upstream and 1.0 Mb/s downstream on an 18 kft, 24 AWG loop.

Recent field measurements of equipment based on the ADSL-lite specifications indicate that in most cases telephones will interfere with ADSL-lite operations especially during ringing and on-hook/off-hook transitions. However these impairments can be largely eliminated by placing small in-line low-pass filters in front of each telephone. These filters could be installed by the homeowner, thereby avoiding the truck roll and allowing the existing telephone pair in the home to be shared by telephone and ADSL service.

The initial ADSL deployments have been based on standard ADSL. Most network operators have announced plans to switch to ADSL-lite, at least for residential applications, as soon as this technology is available.

4.5 VDSL

While ADSL may be quite sufficient for most data applications, for entertainment video even higher bit rates are needed. VDSL could play a significant role in the delivery of these services.

4.5.1 System Requirements Reference Model

Because of its short reach, VDSL will not extend to the central office, but must be terminated at a node 1 to 3 km from the customer's premises. Figure 4.11 shows the network reference model for VDSL. It is essentially a fiber to the node architecture with an Optical Network Unit (ONU) located in the copper access network. Like ADSL, VDSL also has to coexist with existing narrowband services. A service splitter accommodates shared use of the physical transmission line for both VDSL and either POTS or ISDN BRA.

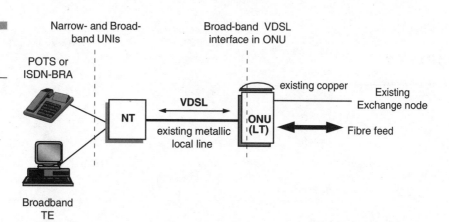

Figure 4.11
VDSL System Reference Model

4.5.2 Transfer Mode

While delivery of video entertainment over ATM has been considered by both the telecommunications and CATV industries, the more likely delivery mode for video entertainment signals may be STM. Hence, VDSL transceivers are required to transport ATM and STM (SDH) with associated network timing reference(s).

VDSL provides two or four channels with a bit rate under control of the network operator, consisting of one or two downstream and one or two upstream channels. These channels have a programmable latency. The VDSL transceiver is required to transport delay sensitive services (e.g. POTS/video conferencing) as well as services which are sensitive to impulsive noise (e.g. digitally encoded video signals). Low latency services are transported over a fast channel (with no or shallow interleaving), while impulse noise sensitive services are better served by the slow channel (with deep interleaving). The maximum fast-path delay is 1 ms and the maximum interleave-path delay is 10 ms.

4.5.3 Performance

VDSL should be designed to accommodate both asymmetric and symmetric transmission between the ONU and the customer. Table 4.2 lists the target ranges for several combinations of upstream and downstream bit rates. Ideally, a VDSL system should be able to handle a mix of symmetric and asymmetric rates in the same binder group, depending upon the demand for spe-

TABLE 4.2

VDSL Performance Figures

Reach	Cable (AWG)	Downstream Data Rate	Upstream Data Rate
1,000 ft	26	52 Mbit/s	6.4 kbit/s
1,000 ft	26	26 Mbit/s	26 kbit/s
3,000 ft	26	26 Mbit/s	3.2 kbit/s
3,000 ft	26	13 Mbit/s	13 kbit/s
4,500 ft	26	13 Mbit/s	1.6 Mbit/s

cific services. The ability to provide such flexibility is closely related to the selected duplexing mode.

The VDSL system is required to meet its distance and quality of service requirements with a 6 dB margin at a 10^{-7} BER, taking into account impairments such as crosstalk from other xDSL systems (both NEXT and FEXT), impulsive noise, RF ingress noise, system noise and broadband environmental noise. Network operators insist that ADSL and VDSL systems be spectrally compatible in the same multi-pair cable and that both services can be deployed without complicated and excessive planning restrictions. Furthermore, VDSL must to be spectrally compatible with other xDSL systems like ISDN-BRA, ISDN-PRA (Primary Rate Access) and HDSL.

4.5.4 Transmit Spectrum

The VDSL band is located between 300 kHz and 30 MHz with the maximum wideband transmitted power set to 11.5 dBm. Dual requirements exist for the psd of the VDSL transmitter. In cases where EMI (i.e. RF egress noise) is not problematic (e.g. buried cables), the psd is upper limited to -60 dBm/Hz in the entire VDSL band. In case where EMI must be tightly controlled (e.g. aerial cables), the psd is upper limited to -60 dBm/Hz except in the internationally standardized amateur radio bands where the psd should be less than -80 dBm/Hz.

A power back-off is required in the upstream (and downstream) direction to reduce self-FEXT.

4.5.5 Power Consumption

Since the ONU will be located outside, stringent limits are imposed on its power consumption and heat dissipation. Indeed, although the ONU will be powered over a separate cable (note that the ONU is connected to the central office via an optical fiber), the power that can be provided to the ONU is limited. Also, fan assisted cooling is often unacceptable, thus limiting the allowable heat dissipation. The power consumption of the VDSL line card located at the ONU should be on average restricted to 3 W per line when delivering broadband service. Considerably less is required when the VDSL transceiver is in a quiescent mode. Therefore some power-down facilities must be provided.

4.5.6 Transmission Techniques

Basically four different transmission techniques have been proposed in ANSI/T1E1.4 for VDSL:

- SDMT (TDD-DMT): multi-carrier transmission with Time Division Duplexing
- Zipper (FDD-DMT): multi-carrier transmission with Frequency Division Duplexing
- CAP/QAM: single-carrier transmission with Frequency Division Duplexing
- MQAM (Multi-QAM): multiple single-carriers with Frequency Division Duplexing

In the following sections, these techniques are briefly described.

4.5.6.1 SDMT. Synchronized discrete multitone transmission is a time-division duplexed (TDD) implementation of DMT. Up- and downstream data are alternatively transmitted on the copper wire (''ping-pong'') and use the full frequency band (Fig. 4.12). The ratio of the up- and downstream capacity is determined by the duty cycle of the ping-pong frame, which allows a flexible division of the channel capacity among up- and downstream transport, from highly asymmetrical to entirely symmetrical. Between the symbols, silent periods are inserted in the frame to allow the channel echo response to fade out before reception starts in the other direction.

To prevent co-located transceivers from injecting NEXT into each other, all VDSL transmission frames must be synchronized. Although this requirement limits bit rate flexibility, SDMT still provides the capability to initialize all VDSL systems in the same binder at any (but the same) up/down ratio.

An advantage of the SDMT technique, as compared to FDD DMT, is that in this time multiplexed mode, main parts of the VDSL transceiver (e.g. the FFT block) are shared by the transmitter and receiver, resulting in a considerable reduction of the modem complexity. Another advantage of SDMT is the possibility to change the system bandwidth (5.52, 11.04 or 22.08 MHz)

Figure 4.12
SDMT Frequency and Time Domain Schemes

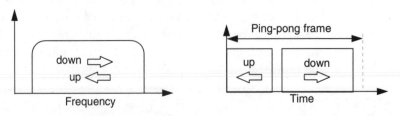

while preserving the number of tones (256) to achieve better performances for low bitrate services.

To avoid RF egress in the amateur radio bands, the tones that overlap with these bands could simply be suppressed. Additional techniques such as windowing the transmitted symbols or the use of dummy tones have been described in order to realize steep edges and deep nulls. Windowing reduces the spectral spillover and thus improves the frequency selectivity of the FFT. Dummy tones do not carry data but are modulated with a linear combination of neighboring carriers to shape the spectrum.

Windowing can also be applied at the receiver to lower the sensitivity for RF ingress. Additional techniques to reduce the residual RF ingress are suppression by means of an adaptive filter in the time domain or RF canceling in the frequency domain. The first technique creates notches in the spectrum at the location of the RFI (Radio Frequency Interference). The second technique makes use of an antenna tone that is not modulated with data to measure the actual amplitude and phase of the RF interferer. With this information, the RF contribution on the other carriers can be canceled. To avoid saturation of the AFE (Analog Front End), analog RF canceling can be done.

4.5.6.2 Zipper. The Zipper technique makes use of DMT modulation where different carriers are used for up-and downstream transmission. Figure 4.13 represents a possible carrier allocation for a symmetric service.

This carrier per carrier assignment makes the Zipper very flexible. The ratio of up- and downstream transmission can be chosen in almost any proportion. Moreover, the Zipper VDSL system can easily coexist in the same cable with asymmetric systems like ADSL as NEXT from ADSL into VDSL (and vice versa) can be avoided by a careful selection of the up- and downstream carriers.

The separation of up- and downstream data is not based on filtering, but on the orthogonality of the tones in both transmission directions. To ensure this orthogonality, the Zipper system requires the following:

■ The extension of the DMT symbol with a cyclic suffix (in addition to the cyclic prefix) to compensate for the propagation delay.

Figure 4.13
The Zipper Duplexing Scheme

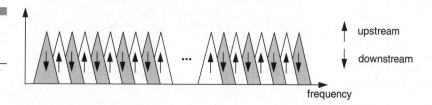

■ Time and frequency synchronization between all transmitters at both ends.

With these two conditions, NEXT from adjacent pairs and echo from transmission in the opposite direction on the same pair are also avoided as illustrated in Fig. 4.14.

Synchronization is required between all transmitters to keep the symbols sufficiently aligned in time. Some type of ranging ("timing advance") ensures that the VTU-O (VDSL Terminal Unit at ONU side) and the VTU-R (VDSL Terminal Unit at Remote terminal side) start transmission of a DMT symbol at the same time. Also synchronization between the transmitters and the receivers is needed to maintain the orthogonality. Notice that similar requirements for synchronization exist in SDMT although the reasons differ.

The suffix introduces redundancy and thus decreases the transmission efficiency. In principle, the suffix needs to be dimensioned after the longest pair in the cable. A high duplex efficiency is obtained if a large number of carriers (1024 or more) is used. The drawback is the need for large (I)FFT blocks. Furthermore, in contrast to SDMT, the Zipper technique requires the calculation of 2 (I)FFTs per DMT symbol, resulting in a more complex transceiver and increased power consumption.

Figure 4.14
Ensuring Orthogonality with Zipper

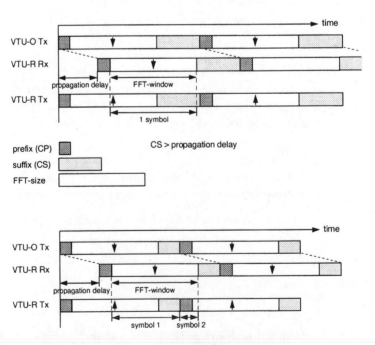

In principal, the same techniques can be applied here to cope with RF interference as described above for SDMT: use of dummy tones, windowing, frequency domain canceling and suppression in the time domain by adaptive filtering. However, due to the large number of tones, the Zipper system exhibits a better frequency selectivity. Therefore, the required psd suppression in the RF bands might already be achieved by only zeroing the tones located in these bands.

Because of the simultaneous up- and downstream transmission and the absence of TDD frames, the latency is less than for SDMT.

4.5.6.3 CAP/QAM. The single carrier VDSL specification allows for both point-to-point and point-to-multipoint transmission between the network and the user equipment. The down- and upstream transmitter may be implemented as either a CAP or a QAM transmitter. The down- and upstream receiver should be capable to detect and decode either CAP or QAM incoming signals. For downstream transmission, the data is differentially encoded in a 16-point constellation. For upstream transmission, the data is differentially encoded in a 256-point constellation in case of a point-to-point configuration. Reed-Solomon coding, combined with interleaving, is applied. The RS coding parameters (N and K) are programmable. The interleaver is a triangular convolutional interleaver as specified by DAVIC for FTTC. The downstream spectrum is located above the upstream spectrum. Figure 4.15 depicts the location of the up- and downstream spectra for several symmetric and asymmetric profiles.

Figure 4.15
VDSL Single Carrier
Spectrum Allocation

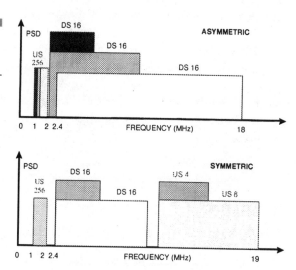

Figure 4.16
VDSL MQAM Spectrum Allocation

4.5.6.4 MQAM. Up to six carriers are used in parallel for up- and downstream transmission (Fig. 4.16). The frequency bands are placed in between the amateur radio bands. As the SNR at these frequencies is very low, the loss in capacity due to the frequency gaps is limited. Each band has a fixed center frequency and symbol rate and is allocated to up- or downstream transmission. The lowest band that ranges from 300 kHz up to 1.8 MHz is used for downstream transport. This allows the MQAM VDSL system to co-exist with ADSL. Band splitting between up- and downstream bands is based on analog and digital filtering. Downstream (and upstream) bands are synchronized. Sufficient flexibility is provided to support both symmetrical and asymmetrical services.

4.6 Conclusions

DSL technology has been the subject of a great deal of media attention recently, both in the technical magazines and in the non-technical press. It is the weapon the telecommunication operators will be using to win the hearts and pocketbooks of Internet users against the competition of cable modems. It is expected both ADSL and VDSL will play an important role in the coming years, for the customer at home, for the telephone operators worldwide and for the telecommunications and microelectronics industry. xDSL has been recognized worldwide as being among the top 5 product concepts driving the microelectronics industry in the next decade. In this chapter, an overview was given of both technologies covering, amongst other things, reference models, system requirements, performance, modulations techniques and standardization.

Hybrid Fiber Coax and Cable Modem

5.1 Overview

Developing a high performance Global Information Infrastructure (GII) for the information age is a national and international goal. A key element identified of this global "information highway" is the on-ramps: high-speed access links available to all at affordable rates. In North America, the cable industry is in a strong position to build these on-ramps. CATV networks provide the only broadband infrastructure reaching nearly every home in the United States and Canada. By adding cable modems, a powerful data access network is created enabling not only high-speed data services, but telephony services as well. The attractiveness of this architecture was demonstrated by AT&T's decision to acquire TCI and to form a joint venture with Time Warner with the stated goal of letting CATV networks be the local portion of AT&T's future telecommunications strategy. The situation in countries outside of North America is currently less favorable, but this situation is rapidly changing with telecommunications liberalization.

In this chapter, the technologies and deployment issues associated with cable modems are described. Deploying cable modem service requires a costly modernization effort by the cable operator to enable bi-directional communication. Hence, to fully understand the technology of the cable modem, the CATV environment will be briefly described to give a better appreciation of the various constraints placed on the developers of this technology.

This chapter is organized in two parts:

1. The first part describes the network environment in which the cable modem must operate

2. The second part describes the two cable modems the industry is specifying:

   - ATM-centric cable modems

   - IP-centric cable modems

   - Effort to converge IP and ATM based cable modem

5.2 Market Pull/Technology Push

Today's networks are service specific. Cable TV networks were optimized for one-way video broadcasting. Telephone networks, wireline and wireless were designed to efficiently handle telephony; its transmission and switching systems were optimized for the needs of voice traffic. Neither network was designed to accommodate the transport needs of the Internet nor emerging interactive multimedia services. To meet these needs, both telecommunications and cable operators will need to make radical changes to their network infrastructure. At the same time, the operators of these networks will be in fierce competition with each other and with emerging operators who do not have the constraint of trying to evolve their current infrastructure. This "battle of the bundles" will not just be for the right to carry emerging services, but over basic telephony and video distribution as well.

In this looming battle, the telcos will be fighting with ADSL and VDSL modems over copper lines. The primary weapon the cable multiple system operator (MSO) will be fighting back with is the high-speed cable modem. The free market will determine the winners and losers.

5.3 Cable Network and Evolution to HFC

5.3.1 History of the Cable Network

The cable network was originally deployed to perform a very simple task. Reception of TV signals could often be poor, especially in suburban areas where high willingness-to-pay consumers began moving in the 1960's. The cable network provided a solution: good reception, greater channel selection, and no ugly antenna on the roof.

The CATV network used coax shielded cable to deliver clear, equal strength TV signals to the home. A good quality antenna tower received TV channels from the airwaves or from satellites and mapped them in the cable spectrum. In North America, the band 50–550 MHz is reserved for NTSC analog cable TV broadcasts as shown in Fig. 5.1. The range of frequencies from 50 to 550 MHz is divided into 6 MHz channels (8 MHz for Europe). TV analog signals are modulated in each of the 6 MHz channels.

The TV signals in the cable coax are replicas of the one broadcaster send through the airwaves; hence no modifications were needed to the television set. CATV brought about yet another advantage: the ability to provide more channels with signals of equal strength to the end user. TV signals traveling through the air do not have the aid of periodic amplification as does the CATV coax, and TV receivers and tuners cannot cope with the interference of more powerful adjacent TV signals.

Channel allocation in the radio spectrum is regulated by the Federal Communications Commission (FCC). The FCC also regulates the frequency location, and signal power used by TV broadcasters. These FCC rules guarantee that stations that use the same TV channel are far enough apart as not to interfere with one another. That, coupled with the rapid attenuation of TV

Figure 5.1
CATV Cable Spectrum

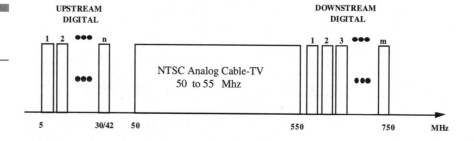

signal in the air, provided distinct disadvantages to broadcasters versus cable operators.

Cable operators, by virtue of the market demand for program variety, flexibility, and premium channel availability, were elevated from CATV operators to broadcasters and later they became content providers. During that period, the cable network remained a one-way broadcast medium. More recently, HFC modernization plans have accelerated not only to set the stage for providing bi-directional communication, but also to increase channel capacity to meet the new competition from Direct Broadcast Satellites. These modernizations include fiber optics installation in the trunk plant yielding higher quality signals through a reduction in amplifiers, increased system reliability, and reduced operating and maintenance costs. With that as prerequisite, the *cable modem* became practical since the number of amplifiers that would need to be replaced for two-way operation was reduced, just in time to meet the demand for high-speed interactive services.

5.3.1.1 Regulation. Over the past three decades, Cable operators have become a powerful force in the market. More and more homes subscribed to the service to the point that the majority of homes now receive their television service from a cable company. Channels brought in from "Super Stations" began to compete heavily with local channels. By that time, cable operators became content providers as well. The result was that the FCC and local franchise authorities began to regulate the industry and dictate what channels must be carried to serve the local community. "Must Carry" rules were issued to protect local broadcasters.

The Federal government has been schizophrenic with regard to regulating CATV prices. The Cable Communications Policy Act of 1984 eased price control to encourage growth. In 1992, the U.S. Congress passed the Cable Television Consumer Protection and Competition Act and reenacted price control regulation with some exceptions. Barely four years later in the Telecommunication (Reform) Act of 1996, Congress deregulated this industry with the proviso that the telecommunications industry can now compete for video services and, *quid pro quo*, the cable operators can also enter the local telephone service market.

5.3.2 Legacy Cable Network

The technologies of the 1960's and 1970's were adequate to provide one way CATV broadcasting services. Networks were built in a simple and somewhat organized topology of a branch and tree architecture. The broadcast nature of the services allowed a high degree of sharing of the coax. Figure 5.2 illustrates a traditional cable network. These were the major functional elements:

1. Cable TV headend
2. Long haul trunks
3. Amplifiers
4. Feeders
5. Drops

5.3.2.1 Headend. The Cable TV headend is mainly responsible for the reception of TV channels gathered from various sources such as broadcast television, satellite, local community programming and local signal insertion. These 6 MHz TV analog channels are modulated using a frequency division multiplexing technique, and are placed into the cable spectrum as shown in Fig. 5.1. This central control headend can serve thousands of customers using a simple distribution scheme. To achieve geographical coverage of the community, the cables emanating from the headend are split into multiple cables. When the cable is physically split, part of the signal power is sent down each branch. The content of the signal, however, remains constant.

5.3.2.2 Trunks. High quality coax cables are used as trunks to deliver the signals into the distribution network and finally to its intended destination. The trunk can be as long as 15 miles. Lower quality coax is commonly used in the distribution and drop portions of the plant.

Figure 5.2
Cable System Topology

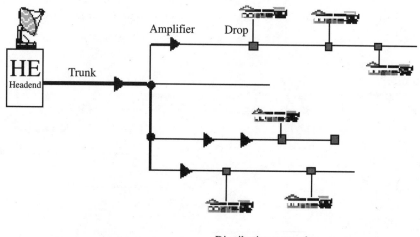

5.3.2.3 Amplifiers. TV signals attenuate as they travel several miles through the cable network to the subscriber's homes. Therefore, amplifiers have to be deployed throughout the plant to restore the signal power. The more times the cable is split and the longer the cable, the more amplifiers are needed in the plant. Excessive cascade of amplifiers in the network creates signal distortion. Amplifies are also located in the distribution network (sometimes referred to as the last mile). These amplifiers, used in the traditional cable network, are one way (amplifying signal from the headend to the subscriber). This scheme introduces several potential problems when a network needs to be upgraded to provide bi-directional communication. In such cases, these amplifiers need to be replaced with new two-way amplifiers.

5.3.2.4 Feeders. Feeders are sometimes referred to as the distribution network. The term "homes passed" usually refers to homes that are near the distribution network. The coax cables in the distribution network (branch/tree) are usually short and are in a range of 1 to 2 miles.

5.3.2.5 Drops. Drops are usually located on telephone poles or more recently buried in the customer's lawn. A lower quality coax is used to connect from the drop to the home.

5.3.3 HFC Network

The HFC (Hybrid Fiber Coax) is the next generation cable network (shown in Fig. 5.3). Evolution to HFC is the first step needed to provide bi-directional communications paving the way for cable modem deployment.

An HFC network is a bi-directional shared-media system consisting of fiber trunks between the headend and the fiber nodes, and coaxial distribution

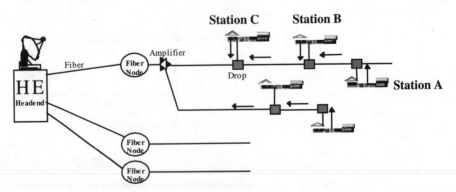

Figure 5.3
Hybrid Fiber Coax Topology

from the fiber nodes to the customer locations. The video signals are carried in analog format on the optical fiber. The fiber node interconnects the fiber trunk with the coaxial distribution. It typically serves 500 to 2000 subscribers via coaxial cable drops. These connected subscribers share the same cable and thus the available capacity and bandwidth. Compared with traditional tree-and-branch cable networks, HFC greatly reduces the number of amplifiers between the headend and the residence, thereby improving signal quality and improving reliability (amplifier failure is the leading source of service outage on CATV networks). In some HFC designs, no more than two or three amplifiers are in the coax between the fiber node and the home. The shorter coax allows more usable bandwidth up to 1 GHz. This greater bandwidth plus the low amount of sharing of the coax connecting any given subscriber, provides the potential of individualized communications to each subscriber for Video-on-Demand or two-way services like telephony and Internet surfing. However, since several subscribers share the same downstream and upstream bandwidth, special privacy and security measures have to be taken.

Because of the sharing, some form of medium access control (MAC) scheme is required in the upstream direction. Essentially, the MAC acts as a traffic cop. The MAC controls and mediates information flow, i.e., to prevent collision of information transmitted from users to the headend.

There are several other advantages of this HFC topology. The fiber truck no longer needs amplifiers. Fiber is less immune to noise, and has very low signal attenuation. Amplifier failures, when they do occur, affect fewer customers and are easier to locate.

5.3.4 Upstream/Downstream Cable Spectrum

Referring back to the Frequency Division Multiplexing (FDM) shown in Fig. 5.1, the range of frequencies from 5 to 42 MHz is dedicated to upstream digital transmission. The cable modem as transmitter uses this range to transmit digital information from the users to the headend. In the downstream direction, the frequency range from 450 to 750 MHz is reserved for downstream digital transmission.

Cable modems must be able to tune its receiver across the range 5 to 42 MHz to receive data digital signals. This digital data is modulated and placed into 6 MHz channels (as in a traditional TV signal). A cable modem, therefore, functions as a tuner. The QAM modulation scheme was selected by the industry for the downstream direction. In the upstream direction, the data is modulated using the QPSK modulation technique and placed in a 6 MHz channel in the 5 to 42 MHz band. At this frequency range, the spectrum is very polluted and noisy because of interference from such sources as CBs and amateur radios, and impulse and ingress noise from home appliances. QPSK is more robust in terms of its immunity to noise, but at the cost of delivering data at fewer bits-per-Hertz than other modulation techniques.

5.3.5 Digital Cable Network

All forms of communication today have either migrated or are migrating into digital format e.g., CDs, cellular, voice, video MPEG. Most, if not all, future services will likely be in digital format. The cable companies are under competitive pressure from the DBS industry (such as from the digital video service—DirecTV) to go digital. Digital transmission results in a noticeably better quality picture, an important differentiating factor for the consumer.

While current HFC networks carry the video signal in analog format on both the fiber and coax portion of the network, there is nothing inherent in the design of the network that prevents signals from being carried in digital format. Indeed, today's cable system can carry digital signals without modifications as long as the modulated signal fits within the bandwidth and power constraints that the cable system will carry. Digital communication can co-exist with analog TV signals as long as the digital signals are contained in their own 6 MHz band.

In upgrading a cable network for digital transmission, analog amplifiers in the system will be augmented with digital repeaters, much like what the telephone network uses today to transmit T1 digital signals on copper lines. The advantage of going digital, beyond improving signal quality due to noise, is the increased capacity of the cable system. Digital signals are more robust that analog and are better able to the utilize the high attenuation portion of the CATV spectrum above 550 MHz.

5.3.5.1 Potential Capacity. The capacity of a cable system will increase enormously if digital signals are transmitted instead of analog TV signals. Typical cable networks are built to support 50 or more television channels. A Signal-to-Noise (SNR) of 48 to 50 dB SNR must be maintained in each 6 MHz channel. With this SNR, modulation techniques such as QAM encoding can achieve 43 Mb/s capacity in a 6 MHz channel. The compression scheme used in the MPEG-2 standards for audio and video dramatically reduced the data rate required for transmission. A digitally compressed video signal of 3 to 6

Mb/s can deliver an excellent quality broadcast video. Hence, a single analog video channel can carry 6 to 10 digital channels and the digital capacity of the cable system can easily reach over 500 channels with existing cable bandwidth.

5.3.6 Cable Network Modernization Effort

Modernizing the cable network to provide high-speed interactive service is underway by most MSOs. HFC appears to be the network architecture of choice. HFC bandwidth capacity is enviable. An HFC network with well-engineered cable modems can provide not only bi-directional data transmission, but all the TV cable analog channels, digital channels, high speed Internet access, voice, and high quality interactive video that customers are demanding.

5.3.7 HFC Access Shortfalls

HFC is evolutionary and can be accommodated in a step-wise approach. However, two-way HFC networks introduce a host of technically challenging problems and suffer a number of shortfalls relative to the point-to-point networks of the telecommunications carriers. The most crucial issues are reliability, security/privacy, operations and maintainability.

The HFC architecture, although potentially more cost effective than other approaches, have a number of challenges to overcome:

- Component failure in an amplifier in the distribution network can render an entire neighborhood out-of-service.

- Failure of AC power (powering the amplifiers) is an even more serious problem, potentially rendering the entire serving area out-of-service. Historically cable operators did not concern themselves with power outages at remote equipment (who cares if CATV service is out if there is no power for the TV anyway). However, as cable operators go after telephony market and business data communications, backup powering for amplifiers and other equipment must be addressed.

- Because of the shared medium topology, the action of a malicious user can effect the operation and communication of all other users in the same branch or tree. A failed cable modem may have the same effect (of disrupting the shared bus), but it is expected that cable modem will be designed to make such failure modes rare.

■ The upstream transmission path is highly prone to noise ingress. The entire cable network must be well maintained to ensure that noise does not leak into the system. This is a particular problem with user inside wiring.

5.3.8 Factors Influencing Cable Modem Operation

5.3.8.1 Amplifiers Bi-directional Issues. Modern cable systems (HFC) with bi-directional communication must use amplifiers that work in both directions. To accomplish this, back-to-back amplifiers with filters are arranged so that downstream signals are first filtered and then amplified. Similarly, the upstream signals will also be filtered and then amplified. The upstream path has an inherent disadvantage because of the branch and tree topology. During amplification of the upstream, the splitter outputs become its input. The splitter simply combines the incoming signals and noise, hence both are amplified. In the downstream direction, the signals passing through a splitter are attenuated on the splitter outputs, but the noise carried downstream is also attenuated.

5.3.8.2 Frequency Agility. Frequency agile capable modem simply means that it can tune into any one of the downstream or upstream frequencies. The cable modem, in the upstream, is able to transmit on whatever frequency the cable system is equipped to handle. This gives cable operators the tools to change the upstream and downstream bandwidth allocation spectrum in their system due to changing traffic demand, without user intervention or, even worst, having to change the terminal equipment. Excessive noise due to ingress (temporarily or long term) of an upstream channel can be dynamically isolated by re-tuning the cable modem to other downstream and associated upstream channels. Modems with frequency agility beyond the 5 to 42 MHz range might be more flexible, but may not be cost justifiable.

5.3.9 Noise

The upstream channel in HFC networks has been the source of great concern. The frequency band assigned to the cable modem positions is in a very hostile noise environment. Ingress noise in the upstream direction is the main cause of impairments in an HFC system. This noise comes in different flavors and severity.

The industry developed a channel model that mathematically defined the nature and physics of the cable network noise. This model was used to refine the specifications of the physical and MAC layers for the cable modem.

Noise Roots. A cable system may be viewed as giant antenna for various noises and impairments. This is especially true in the 5 to 42 MHz band of the RF spectrum. Each type of noise must be fought at its source before it is allowed to propagate into the network and mutate. Just as challenging is the fact that the noise phenomena in the cable network are time dependent. What is measured in the morning is quite different from measurements made during the peak TV viewing hours. Moreover, these measurements are different from one region of the country to the next. The age of the cable plant and the drop wires in particular, the humidity of that region, number of subscribers on the drop, inside home wiring, and past maintenance practices—all play a part in the how the network behaves under different loads. To say that the system must be developed for a worst case scenario is not the optimal solution. In many situations, a field technician can enhance video signal quality and reduce noise markedly simply by mechanically and electrically securing the cable plant.

This presents a special problem for the industry. Noise solutions, are, to great extent, based on field measurements. A cost-effective solution in one region may not work at all in another region.

In general, network noise problems come from three areas:

1. Within the subscriber's home (70 percent)

2. Drop plant (25 percent)

3. Rigid coaxial plant (5 percent

Troubleshooting intermittent problems is costly and time consuming. Finding the problem does not always mean it can be fixed.

5.3.9.1 Noise Characteristics in the Upstream Direction. In the upstream direction, there are several noise sources that can impair communications. A channel model was developed by the industry identifying these sources. These noise sources are briefly described below.

Hum Modulation. Hum modulation is amplitude modulation due to coupling of 60 Hz AC power through power supply equipment onto the envelope of the signal.

Microreflections. Microreflections occur at discontinuities in the transmission medium which cause part of the signal energy to be reflected.

Ingress Noise. Ingress noise is the unwanted narrowband noise component that is the result of external, narrowband RF signals entering or leaking into the cable distribution system. The weak points of entry are usually drops and faulty connectors, loose connections, broken shielding, poor equipment grounding, and poorly shielded RF oscillators in the subscribers household. Since the upstream transmission is at the lowest frequency of the network's passband, the noise summates at the trunk.

Ingress noise contribution includes most if not all FCC conforming RF power levels such as hair dryers, power line interference, electric neon signs interference, electric motors, vehicle ignitions, garbage disposals, washers, nearby passing airplanes, high voltage lines, power systems, atmospheric noise, bad electrical contacts and any open air RF transmissions such CB and Amateur radio transmission, leaky TV sets and computers, civil defense, aircraft guidance broadcasts, international short-wave and AM broadcasters.

Common-Path Distortion. Common mode rejection is due to nonlinearities in passive devices and corroded connectors in the cable plant.

Thermal Noise. White noise is generated by random thermal noise (electron motion in the cable and other network devices) of the 75-ohm terminating impedance.

Impulsive Noise/Burst Noise. Burst noise is similarly impulse noise but with a longer duration. It is a major problem in two-way cable systems and is the most dominant peak source of noise. Impulse noise is mainly caused by 60 Hz high voltage lines and any electrical and large static discharges such as lightning strikes, AC motors starting up, car ignition systems, televisions, radios and home appliances such as washers. Loose connectors also contribute to impulse noise. There are two kinds of impulse noise: Corona noise and Gap noise.

Carona noise is generated by the ionization of the air surrounding a high voltage line. Temperature and humidity play a major role in contribution of this event.

Gap noise is generated when insulation breaks down or via corroded connector contacts. Such failures pave the way to discharge in 100 KV lines. Other sources are automobile ignitions and household appliances using electric motors. This discharge or arc has a very short duration (of μsec duration) with a sharp rise and fall time period.

Phase Noise and Frequency Offset. Phase noise arises in frequency-stacking multiplexers which occur in some return path systems.

Plant Response. The cable plant contains linear filtering elements that are dominated by the diplex filters that separate upstream frequencies from downstream frequencies.

Nonlinearities. Nonlinearities include limiting effects in amplifiers, laser transmitters in the fiber node, and laser receivers in the headend.

5.3.9.2 Noise Characteristics in the Downstream Direction

There are several noise sources that can impair downstream communications. The noise sources are additive and are described below.

Fiber Cable. The fiber effects the digital signal in two ways:

1. Group delay is due to the high modulation frequency of the signal in the fiber.

2. White Gaussian noise is added to the power.

Plant Response. Impulse response is defined as "tilt" and "ripple". The tilt is a linear change in amplitude with frequency and is an approximation to the frequency response of the components in the network. The ripple is a sum of a number of sinusoidal varying amplitude changes riding on top of the tilt and is a measure of the effect of Microreflections in the network.

AM/FM Hum Modulation. AM/FM hum modulation is amplitude/frequency modulation caused by coupling of 110 Hz AC power through power supply equipment onto the envelope of the signal or shift both up and down in frequency.

Thermal Noise and Intermod. Thermal noise is modeled as white Gaussian noise with power defined relative to the power at the output of the plant response. Intermod is caused by nonlinearities in the system generating harmonics of other channels.

Burst Noise. Burst noise is due to laser clipping which occurs when the sum total of all the downstream channels exceeds the signal capacity of the laser.

Channel Surfing. Channel surfing causes microreflections to appear and disappear. Because the significant sources of channel surfing are close to the receiver, a large but slowly changing ripple in the frequency domain will appear and disappear.

5.3.10 Approaches to Noise Suppression

There are many approaches to suppress or avoid noise ingress in HFC networks. Since these approaches are not mutually exclusive, they could be combined to improve the performance of the network. The guidelines shown below will have to be made on a plant by plant basis. Network performance is effected for both the upstream and downstream channels, although the upstream channel is more pronounced in the overall performance.

- Aligning the amplifiers properly in the reverse direction.
- An important aspect of cable plant installation or modernization is to ensure the system is both mechanically and electrically sealed. For it to be otherwise will invariably cause a significant contribution to Ingress and Impulse Noise within the system.

■ Electrically, all powered devices and the cable plant must be grounded appropriately. This may prove difficult in arid and/or rocky climates due to the inability to establish a good electrical ground.

■ Almost 70 percent of the source of this ingress noise is generated mostly at the subscriber drops and within the home. Low quality coax is used for the subscriber drops. Radial cracks and cracks in the shield's foil are the main source of leaks and hence ingress noise. Do-it-yourselfers are also doing their share to contribute to system leakage when installing their in-home wiring using older, bad, or loose connections. One effective approach to improving the network performance is through the upgrade of adequate coaxial residential wiring and adding good connector and good grounding practices. This, however, may prove costly.

■ Reducing the channel bandwidth adds robustness to the system, since it reduces the group delay distortion and enables the use of higher order modulation schemes. This approach may not be economically feasible in some regions.

■ A frequency agile cable modem (in a multi-tone carrier) is one method used to reduce (skip) noise impairments. The approach is to select only those carrier frequencies in the return path where noise is minimal. This means that portions of the return path spectrum will be marked as not usable. Fine frequency agile systems function well by avoiding noise. This approach, if fully adopted and deployed, may become a liability in the future when interactive services demand more and more upstream bandwidth. Frequency Agility at sub-carriers is most appropriate if the noise sources are narrowband and not broadband. Ingress noise is the only type of noise that meets these criteria. Frequency agility is not an effective strategy for dealing with Impulse or Amplifier Noise as this noise is broadband in nature.

5.4 Cable Modem

The previous section provided an overall picture of the cable environment, and outlined the ground-rules for the modernization plan of developing and preparing requirements for the *Cable Modem*. This section focuses specifically on the cable modem and the competing specifications as given in IEEE 802.14 and MCNS/SCTE.

As discussed in chapter 2, there is debate in the industry over whether ATM or IP will provide the kernel for network convergence. ATM, with its support for multiple qualities of service, promises high-speed integrated services across heterogeneous applications. Given that we cannot fully predict how systems will be built, the quality of service needed, costs, and the market demand for applications and services (including many that have yet to be devised), the approach taken in IEEE 802.14 was to develop a cable modem standard based on ATM.

Others view that since the Internet is an engine that is driving high-speed data networking to the home, the cable modem should be based on the common language of the Internet: IP. As a result, MCNS/SCTE developed a cable modem specification that is IP-centric. These specifications are referred to in the industry as DOCSIS℗ (Data Over Cable Service Interface Specifications).

Hence two types of cable modems are under development by the industry:

1. Cable Modem that is ATM-centric.
2. Cable Modem that is IP-centric.

5.4.1 ATM-Centric vs. IP-Centric Cable Modem

Architecturally and philosophically, IEEE 802.14 and DOCSIS 1.0 modems are similar. The physical layer of both cable modems is in fact identical and is based on ITU J.83 (with some exceptions for North America). The fundamental difference between DOCSIS and IEEE 802 cable modems is in the development of the MAC and the layers above. Unlike DOCSIS, the ATM-based cable modem MAC layer contains the segmentation and reassembles (SAR) necessary for ATM end-to-end operation. For IP services, the IEEE specifies IP over ATM using the AAL-5 segmentation and reassembly. DOCSIS on the other hand, uses the variable size IP packet as the transport mechanism per ISO 8802-3.

Other differences between the ATM-based cable modem and IP-based cable modem lie in the funtionalities of the upper layer services and security, maintenance and management messages (e.g., registration and initialization).

Such differences may be considered major, but there are some vendors who are toying with the idea of providing glues to accommodate both ATM-based or IP-based cable hardware in the same silicon. Although this is possible, especially since the two physical layers are similar, it is unlikely we will see cable modems with dual personality in the market, as we expect the market will eventually force one of the types to dominate.

The differences between the IP-centric cable modem and ATM-centric cable modem will be noted as we describe the various functional components.

5.4.1.1 ATM-centric/IP-centric Modem Convergence.

This confusion in the industry is not unique to cable modems, i.e., should the underlying technology be ATM or IP based. Some refer to it as the fight of the century. The fundamental questions, regarding cable modem, which the industry is struggling with are:

- Is there room to market two types of cable modems: IP-based and ATM-based?
- Will the MSOs ever embrace ATM as the technology solution for broadband applications?
- Will DOCSIS version 1.1 include QOS IP friendly applications in time to compete with the available high-speed access alternative?
- Will the service platform be ATM or IP based and hence will native ATM service becomes ubiquitous?

These questions and IP/ATM modem convergence have been on the mind of IEEE 802.14 and DOCSIS leadership for a long time with no easy answer in reach. This problem is further compounded since both the ATM forum and DAVIC are aligning or using a pointer in their specification to that ATM-based cable modem developed by IEEE 802.14.

On March 1999, and in an attempt to serve the industry better, IEEE 802 preliminary agreed to expand its work and converge future 802.14 to work with DOCSIS specifications. That is, the next generation of IEEE 802.14 is likely to become IP centric or at least much more IP friendly.

Convergence of DOCSIS and IEEE 802.14 is not simply aligning technical objects. There are more philosophical and political problems looming in this convergence scenario, especially if IEEE 802 becomes the body for further DOCSIS standardization. If such is the case, some of the crucial issues that must be dealt with are:

- Intellectual property policy
- MSO loss of control and the practice (or lack) of the ad-hoc decision making on their part
- IEEE slow process of standardization due to developing consensus
- MSO direct involvement in developing standards to protect their business interests

It is accurate to say that most of the crucial technical issues are debated in IEEE 802. All industry experts and intellectuals are drawn to the IEEE meetings to socialize and debate core technical issues. Once voted on, a specification is usually trustworthy and adopted by all involved. This, unfortunately, comes at a price of long-drawn-out debate that could impede time to market. Expediting time to market has been MSO's practice and business model.

A worst case scenario that may evolve will be similar to the present reality of creating forums, like ATM, ADSL, FR or Optical Interface forums. For example, IEEE 802.14 will develop core specifications, and a forum could be established by MSOs, CableLabs, and interested vendors where they will adopt the IEEE 802.14's fundamental recommendation, and accelerate the process by crafting interoperability specifications.

5.4.2 Abstract Cable Modem Operation

The technical challenge of developing a cable modem that will deliver integrated services over cable is significant. Sharing access among multiple users creates security and privacy problems. One user connected to a cable network can possibly receive transmissions intended for another or could maliciously make transmissions pretending to be another user. As such, the various components of the cable modem provide hooks for the management and security. The basic generic description of the cable modem below hides this complexity but provides an operational understanding of a cable modem.

The primary function of the cable modem is to transport high-speed digital data from the cable network to and from users on a cable network. Figure

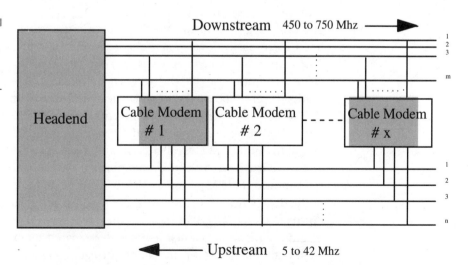

Figure 5.4
Cable Modem/Headend Physical Topology

5.4 depicts the physical landscape and interface negotiation between a head-end and the cable modem. Here "x" denotes the number of cable modems that are attached to this subnetwork. Depending on the traffic, 500 to 2000 homes would typically be served in this topology. At the channel level, the cable modem in the downstream direction must tune its receiver to a 6 MHz band within 450 to 750 MHz to receive data digital signals. In this example, "m" downstream channels are available in that cable subsystem. The QAM modulation scheme was selected by the industry as the modulation technique for the downstream direction.

In the upstream direction, the cable modem performs the transmitter function, transferring information to the headend using a 6 MHz band between 5 and 42 MHz. In this example, "n" downstream channels are used to transmit upstream data. The data is modulated and placed in the 6 MHz channel using the burst QPSK modulation technique. At this frequency range, the environment is very noisy because of interference from CB and Amateur radios, and from impulse and ingress noise from home appliances as described in section 5.3.9.1. For that reason QPSK was selected as the modulation scheme. QPSK is more robust in terms of its immunity to noise but at the cost of delivering data at a much lower rate.

Basic Operation. Basically, a cable modem must be able to tune onto any one of the downstream 6 MHz bands to receive data from the headend. At the transmitting end, the cable modem must also be able to transmit at any of the downstream channels from 1 to n. A dialog between the cable modem and headend is triggered when a station requests registration to join the cable network. There are certain associations between the downstream channels "m" and upstream channels "n". Depending upon the cable network span and traffic engineering, downstream channel 1 may be associated with one or more upstream channels.

During registration to join the network, the cable modem automatically starts listening to downstream channels seeking entry to register its device. When the cable modem receives a strong signal, it reads the subsystem frequency allocation layout and will send a request message to register to the headend using one of the assigned upstream channels. Once acknowledged by the headend, ranging, authentication, and initialization begin so as to legitimized the cable modem and provide the subscribed services. Plug and play is the philosophy used in designing these specifications.

Cable Modem Speed. Depending on the cable noise environment, typical bandwidths 10 Mb/s up to 43 Mb/s can be delivered to the cable modem in the downstream direction with the 64-QAM modulation technique. In the upstream direction, QPSK modulation can deliver up to 1.5 Mb/s. In a shared medium environment, such bandwidth is shared by many users who will be competing for access on the same upstream channel or channels. This competition function is handled by the MAC function in the modem. The MAC arbitrates access to this shared medium bus. The bandwidth allocated to a user in the upstream direction will invariably depend on the number of users sharing the bus. It will also depend, to a large extend, on the characteristics of the traffic being used by others who are sharing the bus. If congestion is encountered, a smart cable modem can re-tune its receiver and hop onto a different upstream channel when instructed to do so by the headend. This mechanism is referred to as a frequency agile capable cable modem.

5.4.3 Cable Modem Reference Architecture

The reference architecture is the building block and blue print needed to construct a device. The reference architecture for a cable modem is as shown in Fig. 5.5 and contains:

- Physical (PHY) layer
- Mac Layer
- Upper layers

Below is a brief description of the layers followed by the details necessary to have an operational understanding of a functional cable modem.

5.4.3.1 Physical Layer.   The physical interface for digital cable systems is the ordinary coax cable. This physical layer contains both the upstream and downstream channels. The characteristics of upstream channels on cable networks make upstream transmission more difficult than downstream trans-

Figure 5.5
Cable Modem/Head-end Layered Architecture

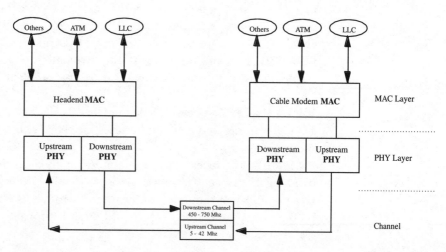

mission. This is due to the shared medium access collision and the multitudes of noise sources polluting that spectrum. This noise problem can be compensated for, to some extend, by using complex encoding technologies at the cost of reduced the data date. Hence, digital cable architecture may use only one downstream transmitter at the head and has several associated upstream transmitters. Transmitters are more expensive than receivers.

5.4.3.2 MAC Layer. The MAC layer is the most challenging to specify and has been and will continue to be a favored topic Ph.D. dissertations. The complexity comes about because of the shared medium coupled with the requirement to maintain quality of service for each user application.

A fundamental function of any MAC is to devise a mechanism that performs random access to the network, resolves contention, and arbitrates resources when more than one station wishes to transmit at the same time. The MAC is further burdened by the requirement of preserving quality of service for specific applications. If real time video or voice is being transmitted, then delay jitter must be minimized and a constant bit rate (CBR) bandwidth must be allocated. Unlike data packets, a real-time voice packet, if delayed only slightly, become useless.

The MAC for the cable modem offers even more challenging opportunities because it must operate under a more hostile environment that any MAC developed so far. Unlike MACs designed to operate in a LAN environment, the cable network MAC must operate in a public environment where quality of service and user expectations are of paramount importance. The cable modem MAC must deal with interactive and multimedia services with bandwidth hungry appetites requiring imposing exacting requirements.

Of the cable modems on the market today, most if not all address specific applications, particularly Internet surfing. This may solve initial user demands, but simply providing speedy cable modem access does not solve the multimedia service requirements. The concept of quality of service must be embedded in the development of the MAC protocol if it is to satisfy the future demands of end users. Market dynamics are very unpredictable and unfriendly to the future detriment of short-term planners.

5.4.3.3 Upper Layers. Cable modems should be designed to handle management entities and service interfaces, be they IP, native ATM or others.

5.4.3.3.1 IP Interface. Most cable modems today connect directly to a PC handling IP traffic. The physical layer of this connection is almost always Ethernet 10BaseT, although other interfaces such as USB or Firewire™ may become important in the future. Although it probably would be cheaper to implement the cable modem as an internal computer card, this approach would require different modem cards for different computers. It would also require opening the user's computer to install the service, something many cable operators would like to avoid.

5.4.3.3.2 ATM Native Interface.   The cable modem, being standardized by IEEE 802.14, is designed to handle native ATM services. That means ATM adaptation layers will be developed to handle ATM applications, including CBR, VBR and ABR services.

5.4.4 Cable Modem Fundamental Layers

The critical components of any cable modem, as shown in Fig. 5.6, can best be described using the two fundamental layers: the physical layer, and MAC layer. Both of these layers will be imbedded mostly in the hardware. Software in the cable modem will complement these layers to give the modem its service personality. Note (in the figure) that the MAC/ATM convergence sublayer (SAR) will not be present for the IP-based modems. These layers and sublayers define the dialog needed between the headend and cable modem. As each layer or sublayer is described, the overall operation of the cable modem will become clear.

Figure 5.6
Fundamental Layers
of a Cable Modem

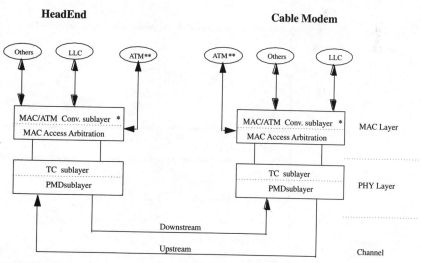

* MAC/ATM Convergence sunlayer
  is present for ATM-centric modem only
** ATM native service for ATM-centric modem only

5.4.4.1 Physical Layer. The physical layer, shown in Fig. 5.6, contains two sublayers: the Physical Medium Dependent (PMD) sublayer; and the Transmission Convergence (TC) sublayer. These sublayers take on the personality of the attached transmission link to perform the needed bit translation, synchronization, orientation, and modulation functions. Figure 5.7 shows the landscape of the sublayers and associated functionalities.

The PMDs are briefly described below so the reader has the needed background to assimilate the overall picture. The reader is encouraged to read recommendation ITU J.83 for the detailed technical specifications and critical timing requirements.

5.4.4.1.1 PMD Sublayers. The main functions of the PMD sublayer are to modulate/demodulate the RF carriers on the analog cable network into digital bit streams, as well as to perform synchronization code and run error checks. ITUJ.83 recommendation was adopted as the base in North America. The PMD, as shown in Fig. 5.7, is further subdivided into two sublayers:

- Downstream PMD
- Upstream PMD

Downstream PMD. The downstream PMD modulates/de-modulates the RF Carrier using QAM (Quadrature Amplitude Modulation) modulation techniques. QAM is used as a means of coding digital information over radio. ITU-J.83 specifies three types of downstream interfaces type A, B, and C. Type

Figure 5.7
Sublayers of the
Physical Layer (Cable
Modem Perspective)

. MPEG 188-Octet per ITU-T H.222.0 . Mpeg synchronization & recovery . Mpeg PID Muliplexing/ Demultiplexing . Delineation of PDUs & scrambling . Controls: Ranging; Power; Synchronization, etc		**TC**	**Physical**
DOWNSTREAM	**UPSTREAM**		**Layer**
. Scrambler . Reed-Solomon encoder . Interleaver . Differential encoder . RF. QAM modulator per ITU J.83	. Data to codeword comversion . Reed_ Solomon encoder . Scrambler . Preamble generator . Pulse Shaping . QPSK or 16-QA modulator	**PMD**	

B of ITU J.83 is the downstream as shown for this example. The cable modem supports both 64, and 256-QAM. At 256-QAM, the nominal symbol rate value is at 5.360537 Msym/sec (baud rate). At 64-QAM, the nominal symbol rate value is at 5.056941 Msym/sec. Nominal channel spacing is 6 MHz and center frequency is specified at 91 to 857 MHz.

Upstream PMD. The upstream PMD sublayer supports two modulations formats: QPSK and 16-QAM. The modulation rates of the modulator provide QPSK at 160, 320, 640, 1,280, and 2,560 ksym/sec, and for 16-QAM at 160, 320, 640, 1,280, and 2,560 ksym/sec. The upstream PMD supports a range of frequency from 5 to 42 at 6 MHz of the subsplit. The range for European cable plant is from 5 to 65 MHz and Japanese range is from 5 to 55 MHz.

Other functions performed in the upstream and downstream PMD layers are designed to transform FDM to TDMA. It will also improve the efficiency and robustness of the transmission by mitigating the effect of burst noise, using encoders to make phase rotation insensitive to QAM constellation, randomizing the transmitted data payload, correcting symbol errors within the information block (codeword), synchronization, and establishing a TDMA landscape. For the upstream, it performs pulse shaping, and a variable-length modulated burst with precise timing beginning at boundaries spaced at integer multiples of 6.25 ms. apart.

5.4.4.1.2 Upstream Frame Structure. Right after the PMD sublayers bit stream is processed, the frames begin to emerge and the information will look more comprehensive. The TC (Transmission Convergence) sublayer refines it further especially for the downstream, and processes it into formats readying it for further data processing. The frame format of the upstream is as described below.

Characteristics of the Upstream Frame. The headend generates a time reference identifying slot containers in the time domain. The slot containers enable a cable modem to transmit information to the headend. Based on that time reference, a minislot is then created whose size is in units of this time base ticks of 6.25 μsec. The duration (in time) of one minislot will equal the time required to transmit 6 octets (programmable) of data plus the time required to transmit the physical layer overhead and the guard time.

A MAC layer Protocol Data Unit (PDU) occupies a single minislot referred to in the industry as a MiniPDU. The upstream landscape can be thought of as a stream of minislots. Minislots are identified and hence are labeled using a free running counter assigned by the headend and incremented by the master clock tick. The headend determines the usage of each minislot in each of the upstream channels. This information is conveyed to the cable modems by broadcasting their usage using the downstream channel in the form of a map or image reflection.

Several minislots can be concatenated to form a packet as shown in Fig. 5.8. For an ATM based cable modem, consecutive minislots are used to form and transport an ATM cell to the headend. For an IP-based modem, minislots are allocated consecutively to form a variable-length packet for transmission.

There are various personalities defining each of the MiniPDUs. An information element in the MiniPDU (shown in the figure with header a, b, or c) defines the various functions allocated. A MiniPDU can be used as management messages for ranging or RF power adjustment (between the headend and cable modem), or the Mini PDU may be a request by the cable modem to vie for access to the shared medium, or the MiniPDU may be a portion of a payload of a data PDU. The number of minislots required to carry an ATM cell depends on the length of PHY overhead and guard time required by the upstream PHY (per the MiniPDU burst profile). For the ATM based cable modem, an integral number of minislots are allocated by the headend to transmit an ATM cell.

5.4.4.1.3 TC (Transmission Convergence) Sublayer. For the downstream, the TC sublayer refines the data further and the bits are assembled, and massaged to fit into a frame.

5.4.4.1.4 Downstream Frame Structure. For the downstream frame structure, the industry adopted ITU-T H.222.0. It is defined as the MPEG-2 (MPEG) packet format with 4-byte header followed by 184 bytes of payload (totaling 188 bytes). The header (PID field) identifies the payload as belonging to either DOCSIS' or IEEE 802.14-based MAC. Figure 5.9 illustrates the cable modem MAC frame, interleaved with other digital information payload(s). The interleaving rate must take into account the jitter that may influence service profiles. A constant rate of interleaving (1: n) is suggested (i.e., one Cable Modem Mac payload for very n digital video payload).

The digital video information payload in the downstream, MPEG-based, was not an accident. This was designed so it can be provisioned for future use when the cable network evolves into digital format with common receiv-

Figure 5.8

Minislots Landscape
for the Upstream

■■ ■■ ■■ ■■
Figure 5.9
Frame Format for the
Upstream

4 Bytes	184 Bytes
Header PID=0X1FFD= ATM-based Cable modem PID=0X1FFE= IP-based cable modem	Cable Modem MAC payload
Header PID= Digital Video	Digital Video payload
Header PID= Digital Video	Digital Video payload
Header PID= Digital Video	Digital Video payload
Header PID=0X1FFD= ATM-based Cable modem PID=0X1FFE= IP-based cable modem	Cable Modem MAC payload

ing hardware accommodating both video and data. This provides an opportunity for the possible future evolution to digital as described in section 5.3.5.

The first order of business of the TC is to establish synchronization and identify the MPEG frame boundaries. This is accomplished by the TC hardware (of the cable modem). When entering the Hunt State, the hardware in the TC, shifts, calculates, and seeks the correct CRC of the MPEG payload. Five consecutive correct parity checksums of the 188 bytes declares the MPEG packet "in frame". "Out of frame" is declared when 9 consecutive incorrect parity checksums are received.

Once the MPEG frame boundary is established, then the TC extracts the cable modem packet data from the MPEG payload. The format is shown in Fig. 5.10.

Beyond this point, the similarities between the ATM-based cable modem and IP-based cable modem more or less ends. Although philosophically and fundamentally they are similar, the data interpretation and manipulation differ in number of ways. Referring to Fig. 5.10, for an ATM-based cable modem,

Figure 5.10
Downstream Frame
Embedded in the
MPEG Frame

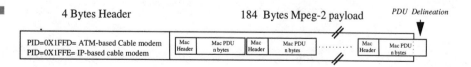

each PDU within the MPEG frame is fixed in length and, in fact, it is an ATM cell. For the IP-based cable modem, the PDU payload is variable in length and conforms to ISO8802-3 type PDU.

PDU delineation (where a PDU traverses from one MPEG frame to the next) is supported on both cable modems. For the ATM-based cable modem, the PDU boundaries (ATM cell) are marked using the procedure described in ITU-T Recommendation I.432. In summary, HEC (Header Error Control) checksum will be responsible for identifying cell boundary.

Once in the Hunt State, a number (7) of consecutive and correct HEC received (on the 5 ATM header bytes) marks the boundary of the cells that is in a frame. In this case, it declares the PDU "in frame". Out of frame is declared when 5 consecutive HEC error occurs. In this case, the TC will then go into Hunt State again to establish the cell boundary.

Mac PDU boundary for the IP-based cable modem, uses pointer_field. The pointer_field is the first byte of the MPEG payload (not shown in Fig. 5.10) and may be present to point to the start of the next Mac PDU. The header indicates if the pointer_field must be used. With this approach MAC PDU may begin anywhere within an MPEG packet, or a MAC PDU may span MPEG packets, and several MAC frames may exist within an MPEG packet.

Low Level Initialization.   Once the PDU framing and formats are established, the TC begins performing initial low level tasks such as: synchronization, ranging, and power adjustments. To accomplish this, two fundamental pieces of information are needed by each cable modem: global timing reference to all modems, and timing offset. Similarly, the attenuation from any cable modem is most likely different from another and the headend. Therefore, the modem must properly adjust its power level for its transmitter such that all stations' transmissions reach the headend at approximately the same received signal level.

Synchronization.   Once the cable modem successfully assembles the frames (by the TC as described above), it then must synchronize its clock with the headend clock. This is performed by the headend (periodically) sending a management message containing global timing reference. The management message contains the timestamp identifying when the headend transmitted this reference clock. The cable modem compares it with its own time and adjusts its local time accordingly. The cable modem periodically adjusts its local clock.

Ranging and Power Adjustment.   Once synchronization is established, the cable modem must then acquire the correct timing offset such that the cable modem's transmissions are aligned to the correct mini-slot boundary. In other words, it adjusts the cable modem timing offset such that it appears to be located right next to the headend (without delay).

First the cable modem must learn the map of the available upstream channels so it can send an Initial management message to the headend to perform the ranging. The Initial Maintenance region slot demarcation (subsequent to ranging) is large enough to account for the variation in delays between any two cable modems. When the Initial Maintenance transmits, opportunity occurs, and then the cable modem sends the Ranging Request message.

The headend responds and returns a Ranging Response message addressed to that particular cable modem. The response message contains the needed information on RF power level adjustment and offsets frequency as well as any timing offset corrections.

A dialog is then established again to fine tune and correct both the power and timing offset of the cable modem.

5.4.4.2 Overview of MAC. A single MAC Allocation and Management protocol operates over the collection of upstream and downstream channels. Its working environment includes the headend and all other connected modems. The headend services all of the upstream and downstream channels.

One can think of a MAC protocol as a collection of components each performing a certain number of functions. A cable modem MAC protocol can be broken into the following sets of components:

1. Acquisition process
2. message format
3. support for higher layer traffic classes
4. bandwidth request
5. bandwidth allocation
6. contention resolution mechanism

The message format element of the MAC defines the upstream and downstream message timing and describes their contents.

The MAC layer in the cable modem may contain sublayers. For an ATM-based modem, the MAC sublayer contains the SAR (Segmentation and Reassembly). SAR is used to assemble or disassemble ATM cells from the non-ATM service application (e.g., IP over ATM). Hence, it provides the interface for the upper service layers, be it IP-based or native ATM. This was illustrated in Fig. 5.6.

The main features of the MAC, however, are its ability to support the transfer of packets while maintaining the ability to provide Quality Of Service. The upstream channel is a precious resource hence collision and data flow must be managed very efficiently. The upstream channel is divided in time into basic units of minislots. There are several types of minislots. Their function is defined by the headend and conveyed to each cable modem by means of downstream control messages. Several minislots can be concatenated in order to form a single data PDU, or ATM cell. There is no fixed frame structure and there is a variable number minislots in any given time. Thus the upstream channel is viewed as a stream of minislots. The Mac layer also contains the controls and rules governing information processing and flow control. Management messages are defined to handle various tasks, but mainly the interaction between the cable modem and headend for modem initialization, authentication, configuration, and authorization. To the extent possible, plug-and-play is the philosophy adopted to perform these tasks.

Below are some (critical) examples of management routines and dialog between the headend and cable modem followed by description of the critical items such as bandwidth allocation, and contention resolution.

5.4.4.2.1 Channel Acquisition.   Channel acquisition is the process already described above. Once a cable modem accomplished its synchronization, framing, and establishes communication with the headend, it has completed channel acquisition. The headend acts as the traffic cop and could instruct the cable modem to change either the upstream or downstream channel(s). The cable modem must respond and re-initialize at the PMD and TC layers. Once accomplished, then channel acquisition is completed.

5.4.4.2.2 Registration.   During this process, several messages are exchanged between the cable modem MAC and headend to legitimize entry of the cable modem in joining the network so it can be declared operational.

Assuming channel acquisition, ranging and power leveling were performed, a cable modem must first register with the headend. This starts the MAC registration process. The cable modem is assigned a temporary service ID that has only local significance. This ID will be associated with the cable modem IEEE 802-48-bit Mac address. This address which is assigned during the cable modem manufacturing process. It is used to identify the modem to the various provisioning security servers during registration.

5.4.4.3 Security and Privacy in the HFC Network.   The security and privacy problems for HFC are different from the traditional point-to-point wireline networks. In the telco environment, the copper wires are dedicated to the user and connected directly to a line card at the central office. Eavesdropping on a telephone line cannot as easily be done. It certainly cannot be directly monitored by users in other homes. Registering a device illicitly (service theft) on a dedicated line is near impossible. The operator knows the identity of that line because it terminates physically at the site. In a cable network environment, the security problem is more difficult because many stations have physical access to the same wire.

Some in the cable industry understandably question the need for privacy and the added complexities needed to provide it. Cable telephony, for example, could fall under the same category as cellular, cordless phones or PCS. Eavesdropping on these systems can be done just as easily if not easier than in the cable modem. It is then argued that if a user chooses to use a cable phone, then privacy must not be of that great a concern. Not long ago, even copper wires were shared among several telephone users. Distinctive ringing was used to identify a particular called party.

5.4.4.3.1 Security Requirement at the MAC. Provisions were made on the cable modem to specify the access security mechanisms so as to make the security of shared media access networks comparable to that of not-shared media access networks (if not greater). The process is basically to exchange a secret key during registration in which a cable modem sends it's unique ID (IEEE 802 - 48-bit Mac address) to the headend and then proceeds with secret key exchange to register. The certification ID is simply used as authenticity (e.g., initial password) prior to secret key exchange that will follow. If the headend is not provisioned to accept this ID, then registration fails. A hacker using a legitimate ID (and illicitly obtained a secret key) will be able to register provided that the legitimate user had not registered first. During this process, the ID information is transmitted in the clear, a hacker might listen to a successful registration transaction and record the ID information.

5.4.4.3.2 Secret Key Exchange. The secret key exchange uses the Diffie-Hellman exchange during the registration process. The authentication procedure incorporating the secret key used to verify the identity of the station to the headend. A hacker who obtains a legitimate ID number of the device must also obtain the correct secret key.

5.4.4.3.3 Secret Key Exchange Using Diffie-Hellman Public Key. The Diffie-Hellman key exchange is used to establish a common secret key. IEEE 802.14 adopting the Diffie-Hellman key exchange procedure.

5.4.4.3.4 Maintaining Station Keys. The cable modem is usually equipped with more than one separate encryption/decryption secret key. They are exchanged during registration by means of a "cookie." Cookies are exchanged between the cable modem and headend during registration/authentication routine when entering the network or at any time network operator deems necessary. A 512-bit ephemeral Diffie-Hellman is used for main key exchange. Main key exchange produces a cookie.

This processor however does not differentiate between newly subscribed users with which it has no yet established a cookie. A hacker may very well be able to establish a network connection using a cloned MAC address during registration. This, however, is a futile exercise to the hacker since the legitimate user will be denied access to the network when attempting to register. This denial of service to legitimate user will prompt the operator to perform authentication by other means. For example personal intervention would thus reveal the attacker and remedy this predicament.

5.4.4.4 Fundamentals of Collision Resolution.

MAC operation in terms of flow control describes the entry mechanism and steady state operation of the station's behavior. In the shared medium environment, the communication is many to one. Subscribers will be competing on the shared medium bus to get the attention of the headend so it can be granted permission to start sending the data. MAC controls the behavior of users who want to access the network as well as honor the service contract promised by the application and network. Hence, the MAC arbitrates the communicating path and resolves any collision that occurs. The headend acts as a central office in that it controls and mediates all communications to and from connected cable modems. There are several MACs developed and specified in the public and private networks. The two contention/resolution mechanisms are the Time Division Multiplexing Access and collision resolution protocols such as Contention and Collision Resolution Mechanism (CDRM).

5.4.4.4.1 Time Division Multiplexing (TDMA).

In the TDMA approach, each connected device is allocated a timeslot in a specified timeframe. The frame contains fix number of time slots and each will be dedicated exclusively to one of the connected devices. When a device has data to send, it will simply use its dedicated time slot to send the information at its leisure. The obvious advantages of this mechanism are:

- No collision is experience in the shared medium and no contention resolution is needed.
- It is well-matched to constant bit rate traffic like voice or video telephony.
- It gives fair access to all connected subscribers.

The disadvantages of this TDMA mechanism are also obvious: an idle user's timeslot wastes network resources. In most multimedia applications, the traffic is bursty and unpredictable, whether the subscriber is on the Internet, or on the WWW, or sending email. In this case, the allocated time slot is used only occasionally. Providing full time access to stations in such a limited environment as the upstream channel is wasteful and likely prohibitively expensive.

5.4.4.4.2 Contention and Collision Resolution Mechanism. In this contention and collision approach, the mechanism requires devices to vie for the bus each time they have data to send. The MAC responsibility is to arbitrate the access, resolve contention and control the traffic flow. The shared bandwidth is only used as needed. Upstream efficiency is achieved. Collision is certain to occur, especially in a folded architecture. Devices must back off and retry until access is achieved. This mechanism is well known and serves the data transaction very well (for non-delay-sensitive traffic). The notion of QOS, fairness, and effective use of resources are not fully addressed in this mechanism.

5.4.4.5 Cable Modem MAC-Bandwidth Allocation. It is obvious from the above that a hybrid solution of both the TDMA and reservation/contention mechanism will best serve a multimedia application. Both the IP-based cable modem and ATM-based cable modem have adopted this technique, but differ in the collision resolution algorithm solution.

The headend algorithm is responsible for computing bandwidth allocation and granting requests. The flavor on the number of request slots/grand combination is vendor implementation specific.

The MAC protocol description will concentrate on the following issues:

- Upstream bandwidth control formats define the request minislot types and structure.

- Upstream PDU format describes the protocol data unit format for an ATM PDUs segmented into variable length fragments for an efficient transport of LLC traffic types.

Downstream format specifies the downstream data flow that can be seen as a stream of allocation units each 6 bytes long. As mentioned earlier, ATM cells can be sent by concatenating several basic minicells together. Each ATM cell can carry a number of information elements such as bandwidth information elements, grant information, allocation information (request minislot allocation), and feedback information (request minislot contention feedback).

5.4.4.6 Request for Upstream Bandwidth. The headend is responsible for allocating transmission resources in the upstream channel to cable modems which are queued for contention-based reservation. Each cable modem vies for access to obtain its share of transmission resources. One or more logical queues may contend for access from a single cable modem to serve the need of multimedia applications (several connections within a session with different QOS).

Each cable modem has various means of requesting bandwidth from the headend. Initially, it requests bandwidth through contention-based transmissions on the upstream channel. Once the headend grants the request, additional bandwidth may be requested by the cable modem by setting a bit (in the appropriate field) in the data PDU in transition. This method known as "piggybacking". It is most useful when contention access delay is high. A constant bit rate (CBR) permanent allocation can be requested by a cable modem for a logical queue, such that periodic grants at a desired frequency are allocated until the cable modem sends a CBR release message.

Request Minislots (RMSs) are allocated in the upstream channel by the headend to the cable modems for contention access. An RMS Grant message identifies a number of RMSs divided into groups for different distinct sets of MAC users which are at various stages of contention resolution.

Typically, the headend allocates more than one RMS to each cable modem, and an RMS may be allocated to multiple MAC Users. To reserve transmission resources in the upstream channel, a MAC user randomly selects an RMS from the group of RMSs available to the MAC user and then attempts to send a request message in the selected RMS. The request message, referred to as a Request MiniPDU (RPDU), identifies the MAC user and the size of its requested allocation. Since the MAC user does not necessarily have exclusive access to the RMS, collisions can occur. When there is a collision, a contention resolution algorithm is invoked to resolve the collision. The contention resolution algorithm for the ATM-based cable modem is based on a tree splitting algorithm and is specified with a flexible framework that permits a number of implementation variations. For IP-based cable modem, the contention resolution is based on the binary exponential backoff algorithm.

If the MAC needs to provide support for ATM, it also needs to differentiate between different classes of traffic supported by ATM such as Constant Bit Rate (CBR), Variable Bit Rate (VBR) and Available Bit Rate (ABR). Bandwidth allocation represents an essential part of the MAC and controls the granting of requests at the headend. Finally, the contention resolution mechanism, which is the most important aspect of the MAC, consists of a backoff phase and retransmission phase.

5.4.4.7 Contention Resolution. A collision resolution algorithm needs to be implemented because request packets and possibly data packets are transmitted in a contention fashion. A wide variety of algorithms can be used. Since cable modems cannot monitor collisions, feedback information about contented requests is provided by the headend. The algorithm must also take fairness into account whereby the delay it receives through the feedback information is compensated for. See ranging procedure. Below we describe two mechanisms for the contention resolution:

- Tree resolution and priority mechanism as adopted by IEEE 802.14 for the ATM-centric cable modem, and

- Binary exponential backoff as adopted by MCNS/SCTE for the IP-based cable modem.

5.4.4.7.1 Tree-Based Contention Resolution Algorithm. The principle of the tree-based contention resolution algorithms is that when a collision occurs, all the station involved in this collision split into n subsets. Each of them randomly selects a number between 1 and n (in a form of hierarchy). The idea is to allow different subsets to retransmit first, while the subsets from 2 to n wait for their turn. One can view the waiting subsets as a stack. The position in the stack represents the number of slots the station must wait before it can retransmit its request. If a second collision occurs, the first subset splits again. The subsets that are already waiting in the stack must be shifted up to $n - 1$ positions in the stack to leave room for the new stations that collided. If no collision occurs, the stations occupying the lowest stack level can transmit.

In its original definition, the algorithm assumed that stations receive feedback immediately. However, in the HFC system, the stations must wait at least until the beginning of the next frame before they receive feedback from the headend and thus are able to retransmit. The algorithm is modified as follows in order to accommodate this delay in feedback. For example, consider frame $j - 1$ containing $c(j - 1)$ collided minislots. All stations involved in the first collided slots are dispatched in the first n slots, those involved in the second collided slots, use the n next slots, and more generally, the stations involved in the ith collided slots select a subset between $(i*n + 1)$ and $((i + 1)*n)$ slots. If frame j contains p contention slots, the first p subsets are able to retransmit, the ith subset transmitting in the ith slot. The other subsets wait in the stack. If cj new collisions occur in frame j, the waiting subsets must be shifted by $n*c(j) - p$ positions in the stack to make room for the new subsets. Several studies in the literature that show optimal results with n equal to 3. New packets arrivals may be handled in two different ways. If the tree-based algorithm is non-blocking, new stations transmit without waiting in any slot selected randomly. On the other hand if the algorithm is blocking, newcomers are not allowed to use a slot reserved for collision resolution. In other words, they are directly put on top of the stack. When new stations are able to transmit, they randomly select a slot among the remaining available slots. Note that a variety of mechanisms could be used to limit the entry of new packets into the system. For example new packets could occupy different levels in a stack based on their arrival time with requests sent on a First Come First Serve (FCFS) fashion.

IEEE 802.14 also considered other allocation resolution algorithms such as the p-Persistence algorithm. The p-Persistence algorithm is an adaptation of a stabilized ALOHA protocol to frames with multiple contention slots. Newly active stations and stations resolving collisions have an equal probability of access p-Persistence to contention slot within a frame. The p-Persistence is determined by an estimate of the number of backlogged stations, computed by the headend and sent to the station in the downstream frames. After extensive simulation, it was concluded that the tree-based blocking algorithm performance is superior the p-Persistence algorithm.

The selection of this contention resolution scheme for the ATM-centric modem was based on the probability density function of the access delay measured from the time a packet is generated until it is received at the head-end. This measurement was thought to be very important as it relates to Cell Delay Variation (CDV) in ATM environments.

5.4.4.7.2 Binary Exponential Backoff Contention Resolution. This contention resolution was adopted by MCNS/SCTE for the IP-based cable modem. The headend controls assignments on the upstream channel through feedback and determines which mini-slots are subject to collisions. The method of contention resolution is based on a truncated binary exponential back-off with the headend controlling the initial and maximum back-off window.

When a cable modem enters the contention resolution process (due to collision), it sets its internal back-off window to the value conveyed that in effect. The cable modem randomly selects a number within its back-off window. This random value indicates the number of contention transmit opportunities that the cable modem defers before transmitting.

After a contention transmission, if the request is not granted, the cable modem increases its back-off window by a factor of two and again randomly selects a number within its new back-off window and repeats the deferring process. If the maximum number of retries is reached, then the PDU must be discarded.

5.4.5 Cable Modem Operation (Service Perspective)

We now described the cable modem beyond the abstract and provide a bird's-eye view of what an MSO hopes is likely behavior when deployed.

5.4.5.1 Review of the Cable Modem Operation. On power up, the cable modem will be in the unregistered mode. The PMD performs physical layer synchronization. This follows by the TC which performs synchronization as well as framing the information packet. The format for the upstream is MPEG-2 and for the upstream a MiniPDU is created from timeslots controlled and programmed by the head end. The cable modem then seeks and registers to join the network and exchange a security "cookie" and other parameters with the headend. Once ranged, a picture begins to emerge and operation then commences as shown in Fig. 5.11. Mini PDU's can be concatenated to form an ATM cell or the ATM-centric modem or IP packets for the IP-centric modem (as shown in Fig. 5.11). Delineation of packet/ATM cell is performed using available HUNT state machine techniques.

When a cable modem needs to send PDUs and has no allocation pending, it requests an allocation by sending a request PDU in a MiniPDU. If the contention is successful, the headend will allocate the requested upstream in grant information elements bandwidth and will then inform the cable modem in the downstream channel. The cable modem may then send the PDUs in the allocated slots. If contention occurred, the cable modem invokes either the tree resolution and priority mechanism (if an ATM-based modem) or the Binary Exponential backoff mechanism (if an IP-based modem).

Figure 5.11
Cable Modem Operation

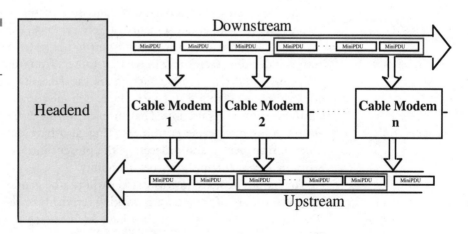

In the downstream direction, the cable modem receives a stream of ATM Cells or packet PDUs encapsulated within the MPEG packet. Management PDUs address a particular cable modem with an identifier that has only local significant. Each cable modem filters its incoming PDUs based on its identifier. IP connectivity (to obtain an IP address) for the IP-based modem is invoked by the cable modem using the DHCP mechanism (per RFC-1541).

5.4.5.2 Cable Modem Service Aspects. The MSOs have traditionally focused on quick returns to meet their short-term business goals. The MSOs' attitude toward customer service has been, admittedly, not especially commendable but they are becoming more receptive given the new dynamics of the market place. The AT&T/TCI merger will set a new paradigm for the industry. AT&T carefully guards its reputation in the customer services area.

The key differentiated services cable operators are likely to provide are these:

- High-Speed Data Services (Internet access)
- Voice over IP/Video over IP
- Broadcast, One-way Entertainment
- Telephony (Wireline and Wireless)
- Digital NVOD
- Work at home
- E-commerce/Shopping

Convergence of these services permeating across the millions of U.S. households will change the way people live, work, and play. Byproducts of this convergence will be a major source of growth and opportunity for the U.S. economy.

The major customer service differentiators of the MSO's are most likely to be Time to Market and Quality of Service.

Although time to market is important in the short term, it will not be the ultimate deciding factor to win over customers from the telephone company. It is only important if other factors are satisfactorily met, in particular customer service, reliability and cost. The telephone companies enjoy a very good customer service reputation. That in itself will be a marketing tool they will use in going after cable customers. They also have deep pockets to build a video infrastructure and to leverage the present infrastructure in billing, network management, and network reliability.

The battle for share in this market will not be gentle. Cable operators are, of course, aware of this and are working feverishly to regain customer loyalty. The road may be rough, but they have been known to survive.

5.4.6 High Speed Physical Layer

The IEEE 802.14 organization is cooperating with DOCSIS members to work on a new high speed physical layer for the upstream. Experts from both camps are finalizing the specification for high speed upstream physical layer. Hi_PHY is the term used to identify this working group.

The three proposals that were debated are:

1. Variable Constellation Multi Tone (VCMT)
2. Advanced FA-TDMA
3. S-CDMA

5.4.6.1 Variable Constellation Multi Tone (VCMT). The VCMT approach utilized a multicarrier modulation approach e.g., the 6 MHz would be divided further into discrete set of tones in the upstream direction. It is worth noting that the 6 MHz is only an arbitrary value the industry used because of the historical size of the video channels, but other values could be used. VCMT is based on transmitting multiple, harmonically-related tones in the assigned frequency channel, where the tones are independently modulated with a QAM constellation from QPSK up to and including 256-QAM. VCMT uses pilot tones at the beginning of each transmitted burst to help in the synchronization of the headend receiver and to facilitate dynamic adaptation to channel imperfections.

In this context, the frequency agility described in section 5.3.8 is coarse agility. For example, upon request from the headend, the cable modem hops to a different upstream channel in a 6 MHz increment. In this approach, the cable modem can tune to multiple sub-frequencies with very low resolution. The intention is to evade and selectively ignore that portion of the band when the noise becomes excessive or intolerable. This DMT (Discrete Multi-tone) technology was adapted for ADSL, but not necessarily because of the noise environment.

5.4.6.1.1 Advantages/Disadvantages

■ Evades noise by dynamically re-assigning sub-channels for transmission. This however may also be considered a disadvantage since the technology evades the problem rather than lives with it and take advantage of the precious upstream available resources.

■ Is more complex to develop with multicarrier controlling channels. Increased digital processing complexity.

■ Is Compatible with the DOCIS 1.0 or 802.14 MAC (considering the competing alternatives).

■ Provides robust performance in the presence of typical upstream cable channel conditions.

5.4.6.2 Advanced FA-TDMA. A TDMA channel access method is proposed which is the traditional mechanism used by DOCSIS 1.0 and IEEE 802.14. The FA-TDMA system uses a single carrier with adaptive and dynamic changing constellation size. Powerful newer and flexible Reed-Solomon FEC, with concatenated coding is used for inner Trellis Coded Modulation. The FEC, and the symbol rate range and modest interleaving provide robustness against impulse and burst and provide for a more robust system. Frequency agility, in this context refers to the cable modem ability to dynamically switch to a lower order constellation format if the channel is or becomes noisy. The constellation formats range from QPSK, 8QAM, 16QAM, 32QAM, 64QAM and 128QAM.

It should be noted that in a normal operation, the headend continuously monitors the channel noise and adjusts the constellation size accordingly.

5.4.6.2.1 FA-TDMA Advantages/Disadvantages

■ Has a single carrier approach which implies less development complexity.

■ Has a proven technology.

■ Is compatible with DOCSIS 1.0 and IEEE 802.14 MAC i.e., there is logical extension of the current TDMA system.

■ Co-exists in the same channel with the existing system.

5.4.6.3 S-CDMA Technology. Standard, IS-95 Code Division Multiple Access is a form of spread spectrum which works by coding and spreading the information to be sent over a wide band. While the idea of using conventional (asynchronous) CDMA for high-speed data has been proposed in the past, this technique has limitations, which are solved using a synchronous implementation. The problem with asynchronous CDMA is so-called self-generated noise: the users of an asynchronous system are not aligned with each other, thus losing orthogonality and creating a high degree of mutual interference. This in turn raises the noise floor and reduces capacity.

Synchronous CDMA (S-CDMA) minimizes interference by ensuring that the cable modem codes are synchronized with each other as they send information upstream. This prerequisite enables S-CDMA to support constant bit rate (CBR), variable bit rate (VBR) and available bit rate (ABR) bandwidth allocation. This feature is achieved by allocating different data streams (e.g., *nn* Kbps) for different types of services.

Bandwidth allocation is managed by the headend and bandwidth can be guaranteed on request to certain users for CBR applications. Other bursty traffic may vie for the remainder of the available payload.

5.4.6.3.1 S-CDMA Advantages/Disadvantages

- S-CDMA abates ingress interference and impulse noise in the up-stream path.

- S-CDMA provides for guaranteed data rates, enabling ABR, CBR and VBR traffic.

- Spreading is considered as an Alternative to Single-Carrier TDMA.

- Single-carrier is more robust against frequent, short, high power bursts (i.e., impulses).

- There is a very stringent synchronization requirement (very tight clock jitter requirement) leading to risk of deploying newer technology and added complexity.

- Uses a forward looking technology which may lead to even more robust physical layers in the future.

5.4.6.4 Hi-PHY Selected by DOCSIS/IEEE 802.14.   Preliminary analysis seems to suggest that the three proposals described above are not superior to the alternative. Each of the approaches has its superiority based on the noise assumptions, but behaves less so in the different noise characteristics as described in section 5.3.9.1.

In November 1998, DOCSIS and IEEE 802.14 experts met and debated the three proposals under study by the group. The group voted to adopt a combination of Advanced Frequency Agile (TDMA) technology and SCDMA (synchronous code division multiple access) technology (as optional) as the basis for the advanced physical layer standard.

In March 1999, an LB was approved by the IEEE 802.14 voting members. In that LB, the S-CDMA was assumed to be mandatory. Experts are yet to agree whether S-CDMA can handle the tight clock synchronization issues between cable modems and headend, and hence still are unable to synchronize the overall system.

5.4.7 Future DOCSIS/IEEE 802.14 Milestones

IEEE 802.14 approved an LB for high performance physical layer (Hi PHY). The draft was created by IEEE experts and vendors who are also working on

DOCSIS 1.2 specification. In light of that and for the issues detailed in section 5.4.1, IEEE 802.14 is contemplating the generation of a Project Authorization Request so that future IEEE development aligns with DOCSIS release 1.1

5.4.7.1 Future DOCSIS Timetable. DOCSIS future releases are expected to have the following milestones:

- DOCSIS 1.0 products from multiple suppliers are in the final stages of certification.

- DOCSIS 1.1 specification will be published in early 1999 and product availability in 3Q 1999. DOCSIS 1.1 specification will include QOS feature availability for real time application.

- The advanced physical layer created in IEEE 802.14 (Hi_PHY) to become part of DOCSIS release 1.2. DOCSIS 1.2 release is expected in 1Q 1999. Product availability and compliance to 1.2 should be ready by 1Q 2000. DOCSIS 1.2 release will also include all 1.1 feature releases.

CHAPTER  6

High-Speed Wireless Access

6.1 Introduction

In today's information society, individuals are on the move. The cellular telephone has changed the office model. Today's professional is no longer satisfied with a stationary desk linked by wires to a voice/fax circuit or to a Local Area Network. People expect the flexibility to work anywhere with a cellular phone in one hand and a portable computer in the other. The industry trend is to provide a rich set of computing and communication services that follow users in a transparent, integrated and convenient way.

This concept has been called **nomadicity**, a term used by the industry to define the nomadic user. The emerging picture goes far beyond the limits of the usual mobile and wireless technologies; nomadic computing is a concept that goes beyond the idea of being unplugged. It means a robust infrastructure supporting the nomad moving from place to place, whether on the road, in a hotel room or at home. The bonus in productivity that can be achieved is rewarded by a higher willingness-to-pay by the mobile user.

6.1.1 Technology Pull, Market Push

This vision of nomadicity has been recognized by the industry as ground breaking market opportunities. The basic idea is to enable a subscriber to call anyone, anywhere, from anywhere with a cell phone just slightly larger than the one you carry now. Satellites would blanket the heavens and transfer or switch calls until it beams down the information (video/voice/data) to its intended earth stations where the call, or more accurately the transaction, enters the terrestrial network for call completion.

Satellite communications deployment, therefore, is not so much intended to compete with or replace terrestrial networks, but rather to complement the existing network communications with capabilities best suited, at least initially, for multinational corporations. Satellite communications will extend terrestrial networks and help them solve the "last mile" problem in remote areas or in regions served poorly by telecommunications today, and would provide instant broadband connections to remote users previously only served by narrowband access. Eventually one will be able to buy IP service from a satellite service provider as easily and inexpensively as buying Internet access from an ISP today.

Satellites are expected to play a major role in nomadicity. Already geosynchronous satellites are widely used for video distribution both for backbone distribution and, more recently, for direct distribution to end users. Their use for voice communications, however, have been limited by the unacceptable delay introduced in propagating the radio signal to the satellite. As well, because the satellite is so high (35,786 km), relatively large disc antennas, rigidly pointed to the far-away bird, are needed to receive the weak signal: not a good solution for a mobile telephone. Low Earth Orbit satellites (LEOs) are a recent approach that solves these problems while introducing its own. LEOs fly just above the earth's atmosphere (1,457 km), thereby introducing low delay and attenuation. The area seen by the satellite is much smaller, meaning that its bandwidth is shared by a smaller group of users. LEOs thus are a better match for mobile telephony and is the satellite technology of choice for Internet access.

LEOs, however, have their drawbacks. First, since the satellites are not geosynchronous, a given satellite will be in sight only a few minutes after which the call must be handed off to another satellite. Second, since it is so close to earth, it take hundreds of satellites to ensure that a satellite will always be in site of the ground communication point. Third, satellites that low experience some atmospheric drag resulting in a life span of 10–12 years which means high replacement. Fourth, high-speed space junk is also in these low orbits which could easily cause damage the satellites. Fifth, because so many satellites are needed for this implementation, routing becomes complicated, increasing delay and reducing efficiency. Some companies are planing to use LEO and GEO satellites hybrid solution to reduce switching/routing complexities.

Other challenges and opportunities for LEOs are security, pricing, and multicast. Encryption will be very important because it would be very easy to ease-drop over the satellite. Pricing should be affordable and possibly compete with terrestrial high-speed access lines. The satellite system broadcast nature has an inherit advantage in multicast applications.

Wireless cable is yet another broadband wireless system the industry is seriously considering and in some cases presently deploying. Local Multipoint Distribution Service (LMDS) is an emerging wireless access that operates in the frequency bands 27.5–28.35 GHz, and 29.1–29.25 GHz. LMDS is somewhat similar in concept to cellular telephony, but at a much higher frequency. LMDS operates in a line of sight access approach, and depending upon the terrain, antenna height and transmit power, an antenna can serve as much as a 5 km cell radius. One can extend service to customers who are blocked from the line of sight of a cell transceiver by using passive reflectors. LMDS is yet another example of how to address the ''last Mile'' access dilemma for broadband services.

This chapter deals with wireless high-speed wireless access technologies. It describes the various infrastructures now under construction to satisfy the professional nomads and eventually all users.

6.1.2 Organization of this Chapter

This chapter briefly describes the fundamentals for deploying Low Earth Orbit (LEO) Satellites, Geo-stationary Earth Orbit (GEO), MEO, LMDS and DBS. It will then focus on three planned LEO deployments:

- SkyBridge
- Teledesic
- Iridium

This chapter will then consider the other promising wireless technologies:

- LMDS
- DBS (Digital Broadcasting System)

6.2 Satellite Constellation Fundamentals

Satellite-based systems offer unmatched potential in providing services to the largest possible number of users over the widest possible area. In addition to their broad geographical coverage, systems incorporating satellite transmission can be launched quickly, providing cost-effective coverage, especially in low- and moderate-density population areas. Satellites are characterized by the distance of their orbits as shown in Fig. 6.1. Satellites in each of these categories are circling or will be circling the heavens within the next few years.

Table 6.1 below illustrates the variables associated with the three basic satellite systems.

6.2.1 GEO Satellite

The GEO satellite is the furthest from Earth, located at 35,786 km or 22,300 miles in orbit. GEO satellite is geo-synchronous in that it rotates at the same speed as the Earth (circular plane with the same line-of-sight). At that elevation, only few GEOs are needed to scan the entire globe. Because of its high altitude, the resulting latency of about 250 ms (propagation delay only) make GEOs best suited for the delivery of TV broadcast, Video On Demand, and asymmetric data services (described in section 6.6). For interactive services, such as video conferencing or multimedia services, the service performance over GEO is very clumsy.

In 1993, NASA launched a GEO satellite called Advanced Communication Technology Satellite (ACTS). ACTS can transport bandwidth at hundreds of

Figure 6.1
Satellite Fundamentals

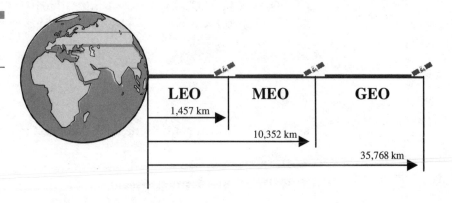

TABLE 6.1

Satellite
Comparison Chart

Orbit	LEO (SkyBridge)	MEO	GEO
Altitude	1,400 km	10,352 km	35,786 km
Max Distance	1,952 km	12,603 km	39,554 km
Round-Trip Delay	13 ms	84 ms	264 ms
Min Elevation	40 degree	30 degree	20 degree

megabits per second. Satellite access is a growing need in data communication, and several companies have started projects to move the Internet to the heavens. Internet in the heavens via GEO may not be as easy as claimed; there are several issues that need to be addressed. GEO is very inefficient when performing small transactions. Because of the long delay, a simple transaction of updating a customer's record (with multiple short transactions) could take as many as 8 to 10 seconds when using GEO.

The TCP/IP protocol's default "window size" will become a bottleneck in long delay links (high latency). Window size is the amount of information being stored at the source for retransmission in case of transmission errors. If the default buffer size in an application of TCP/IP is 64 kilobits, this means only 64 kilobits can be in transit and awaiting acknowledgment regardless of how fat the GEO pipe may be. It takes half a second (round trip delay) for the 64 kilobits to be acknowledged. So, the maximum data throughput rate is 128 kb/s. GEO can and should deliver about 2 Mbps of data. Fixing TCP is not an easy matter since it is an end-to-end protocol and that means modifying the protocol of every computer. Congestion controls with 500 ms delays (for GEO) may be hopelessly ineffective.

IETF is working on improving the performance of TCP over GEO. Connection spoofing is one method that could be used to make believe that the end-to-end communication uses a low latency link. This is an interim and a complicated solution, specially when data error is introduced to the system.

6.2.2 MEO Satellite

The second distance from the Earth being considered for satellites is the Medium Earth Orbit (MEO). MEO satellites are located from 6250 to 13,000 miles away from Earth. In many ways, MEOs are a compromise between

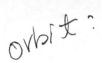

GEOs and LEOs having some of the faults and advantages of both. Its delay is one-fourth to one-sixth that of GEOs, making it suitable for voice communication. Like LEOs, the MEO satellites are not geosynchronous requiring multiple satellites to continuously cover a given point. However, the number of birds required is much smaller than LEOs and a given bird covers a location on earth a longer period of time, potentially eliminating the need for handoff for many applications. Nonetheless, MEOs are not as popular as LEOs, which provide even better latency and higher capacity.

6.2.3 LEO Satellite

Flying at a mere 500-1500 miles from Earth, LEOs achieve a latency of only 0.03 seconds round trip. This makes LEOs very effective for the delivery of time critical applications and hence multimedia interactive services. High-speed Internet accesses over LEO become very practical and effective. Applicable services include database access, audio, video conferencing, e-mail, new data broadcasting multi-media information services, telemedicine, direct-to-home video, electronic transaction processing, distance learning, and news gathering. Latency variation through intersatellite links and terrestrial components is an issue, but various well-known solutions by the industry are being considered and implemented.

LEOs are not without problems. Because of their proximity to earth, many more satellites will be needed in the sky to provide the needed coverage. This can be very costly, but launching can be incremental to cover certain profitable market segment.

A hybrid solution of LEO and GEO is under active study which, depending on the application, offers a better performance model. In this approach, the LEO and GEO satellites communicate directly through a satellite-to-satellite network, building a hierarchy similar to the hierarchy of class 4 and class 5 offices in telephone networks. LEO's shorter delays can be used for interactive real-time sensitive services while GEO's are used for bulk data transfer and broadcasting.

Several consortium where formed to develop and deploy LEO satellite systems. The three most wide known are the following:

- Iridium
- SkyBridge
- Teledesic

The first of these, Iridium, is already operational Given their robust infrastructure and potential to provide mobility in broadband services, LEOs promise to provide the right information in the hands of the right people for business success in the information age. Early in the 21st century, competitive LEO broadband satellite systems should be operational.

6.3 SkyBridge

The SkyBridge system called "Network in the sky" is expected to be operational in the year 2001. It will provide high-speed, broadband, interactive services to business and residential users around the world. In addition to Internet access, the SkyBridge system will provide bandwidth on demand for other types of high-speed data communications, at speeds up to 60 Mbps.

6.3.1 SkyBridge History

On February 28, 1997, Alcatel announced the filing with the FCC of an application to launch and operate a 64-satellite, LEO constellation, named "SkyBridge". SkyBridge is a Delaware Limited Partnership headquartered in Washington, D.C.

On April 22, 1998, a hearing was held before the Communications Subcommittee of the Senate Committee on Commerce, Science Transportation. The hearing centered on Section 706 of the landmark Telecommunications Act of 1996 on the deployment in the United States of infrastructure for high-bandwidth advanced communications services.

The testimony stated that SkyBridge is planning to deploy a global satellite system enabling operators to provide high-speed Internet, multimedia, video-conferencing and other interactive applications. As well, the testimony revealed that SkyBridge will "ensure that every citizen, wherever he or she is, can have a high-speed on-ramp to the information superhighway, integrating remote areas into the Nation mainstream."

6.3.2 SkyBridge Constellation

SkyBridge is a satellite-based system designed to provide global access to full range broadband services of interactive, multimedia communications including Internet access and high-speed data communications. SkyBridge consists of about 80 satellites, thus solving the local loop for all users while enabling interactive services to professional and residential users worldwide.

SkyBridge consists of 2 constellations of 40 satellites, each orbiting at an altitude of 1469 km as shown in Fig. 6.2. The constellation provides permanent worldwide coverage between latitudes $+68°$ and $-68°$. Each satellite illuminates an area of 3,000 km in radius. For each coverage area, traffic switching between satellites, transparent to the user, is managed by the local gateway. There is always at least one satellite visible within the coverage area of a gateway. However, most of the time, at least 2 to 4 satellites are visible and available to transmit traffic.

6.3.3 The SkyBridge Advantage

SkyBridge is a decentralized system which provides high-speed access at lower infrastructure costs for broadband service deployments. The decentralized architecture makes it easy to adapt services to local requirements. All switching and routing is performed in the ground station, making the satellite link operationally less complex and hence low-cost. A gateway is deployed to handle interconnections with local servers and with terrestrial telecommunications networks. The gateway also controls and manages all SkyBridge traffic within its respective coverage area and is able to offer customized service compatibility with existing applications and local content adapted to local market requirements and preferences. With low latency (about 20 ms), applications currently used or in development for existing broadband networks can be seamlessly transmitted via SkyBridge. All end-users get access to the same services such as high quality education, health care services, video

Figure 6.2
SkyBridge Constellation

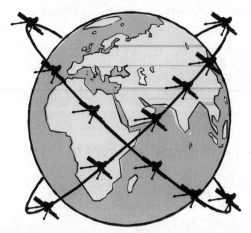

games, entertainment, cultural services, and video on demand, etc., regardless of their location. Other services will be added according to demand, such as voice, basic video conferencing, and data.

SkyBridge LEO is a "bent pipe" solution. That means that the satellites do not have to be very smart, and switching/routing is done in the ground gateways instead of switching them via the satellite network in the sky. This offers cost advantage and flexibility. Latency may become a problem if too many hops are required when routing the connection. The alternative to having LEO satellites perform switching between satellites introduces higher cost, while the performance gained may not be that important to most applications.

6.3.4 SkyBridge Architecture

SkyBridge will initially launch 40 satellites. A total of 80 satellites, with two planes, will be operational by the year 2001 (see Fig. 6.3). As new satellite links are rolled out, the system capacity increases, attaining over 20 million simultaneously connected end-users.

SkyBridge operates at the Ku-band instead of the Ka-band. This lower frequency makes it an inexpensive system in low earth orbit. The Ku-band is crowded with many GEOs, hence interference can occur when SkyBridge satellites are over the equator. SkyBridge solves that problem by shutting off

Figure 6.3
SkyBridge LEO Architecture

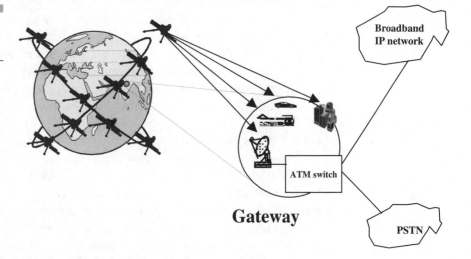

Gateway

its offending beams, and hence the ground terminal switches to another satellite. Ku-band is a proven technology and offers economic advantage, especially in the cost of end-user terminals.

SkyBridge users have an access capacity of nx20 Mbps on the downstream link and up to nx2 Mbps on the return link. It uses a combination of satellite links for local access and existing terrestrial broadband networks for long-distance connections.

SkyBridge is designed to make optimum use of existing and future terrestrial broadband networks. It offers interactive access at lower cost than land-line solutions for rural, suburban and sparsely populated areas worldwide. Urban areas, without sufficient broadband infrastructure, can also be accommodated to provide broadband services.

6.3.4.1 SkyBridge System Overview. A constellation of satellites provides the link between the system and end-users. The system architecture is based on three levels:

1. The first level consists of end-users that use satellite links as a means to access the system. A constellation of 80 satellites in low earth orbit provides the global, round-the-clock coverage required. The satellites transmit incoming and outgoing end-user data to a regional gateway ground station.

2. The satellite link is a simple, so-called ''bent'' function; all switching and routing occurs in the gateway, making for low-cost, low-risk space links. Through an ATM broadband switch, the gateway provides interconnection with the terrestrial broadband or narrowband networks.

3. The link between end-users and the system is asymmetrical, with data rates of up to nx20 Mbps to the user and up to nx2 Mbps on the return link. The asymmetrical design is optimized for Internet-type communications, characterized by random bursts of asymmetrical data transmission. Increments in data rates are in 16 kbps steps, thereby providing the user with ''bandwidth on demand.''

6.3.4.2 SkyBridge User Access. Each end-user terminal is connected via the constellation of satellites to a single local gateway that manages all the traffic within a radius of 350 km. From the gateway, user traffic is routed to local or remote servers, or to other user terminals. Such user-to-user links are transmitted via broadband or narrowband terrestrial networks, depending on the type of application and the needed data rate. If any gateway cannot be connected to high-performance broadband or narrowband terrestrial networks, its traffic will be routed via high data-rate links to a gateway, which does have access to the appropriate networks. These transit links are established via the SkyBridge constellation using the Ku-band

6.3.4.3 SkyBridge System Capacity.

With a constellation of 80 satellites, SkyBridge can handle over 20 million users worldwide depending on the services required. It can provide bandwidth on demand as required by data rate access. For downstream communications, from a server to an end-user, data rates range from 6 kbps to nx20 Mbps. For return link communications (from the end-user back to the server), data rates range from 16 kbps to nx2 Mb/s. Total worldwide SkyBridge system capacity amounts to over 200 Gbps. Each gateway can handle up to 350,000 users per visible satellite.

6.3.4.4 SkyBridge Satellites/Ground Components.

SkyBridge satellites have about 2,500 watts of power with an operational lifetime of eight years. Satellite design has been optimized for mass production. The spot-beams are generated by active antennas. Routing and switching functions are placed in the terrestrial gateways, which reduces space segment costs and risks, as well as providing flexibility in terms of server evolution. The decentralized architecture also offers flexibility in terms of adapting services to particular SkyBridge communications and includes:

- SkyBridge gateways
- User terminals
- Management of the network

SkyBridge Gateway. There will be about 200 gateways. Gateway design is derived from existing terrestrial broadband network architecture. Various gateway personalities are planned:

- Radio frequency subsystem, with up to four satellite-tracking antennas
- An access subsystem
- A switching and routing subsystem, providing interfaces with terrestrial IP networks such as Internet, broadband and narrowband switched networks, and leased lines
- A service access point with associated management functions
- A gateway management subsystem

SkyBridge User Terminal. Users will be equipped with low-cost terminals with a small, low-power RF unit. Two types of terminals are planned:

1. Personal terminals for individual subscribers will feature a small 45-cm in diameter at 2 W power.
2. Multi-user terminals for corporate and communal residential use will be able to serve several dozens of users with a 70 cm in diameter.

6.3.5 SkyBridge Costs and Partners

The budget for implementing SkyBridge is about $4.2 billion dollars. The costs include the following:

- Development of prototypes for the ground and space segments
- Manufacturing and launching the constellation
- Development and installation of the satellite control segment
- Launch and insurance

Alcatel is the primary partner. A number of other industrial partners have been announced:

- Loral
- Mitsubishi Electric
- Sharp
- Toshiba Corporation
- Spar
- Aerospatiale
- CNES
- SRIW
- COM
- DEV

6.4 Teledesic

Teledesic, founded in 1990, is a private company based in Kirkland, Washington, a suburb of Seattle. Using a constellation of LEO's, Teledesic is building a global, broadband network it refer to as "Internet-in-the-Sky". Teledesic is creating a worldwide network to provide affordable access to telecommunications services such as broadband Internet access, videoconferencing, high-quality voice and other digital data needs.

6.4.1 Teledesic History

The following is Teledesic LEO chronology:

■ 1994–Initial system design completed and FCC application filed

■ 1997–FCC license granted

■ 1998–Motorola, The Boeing Company and Matra Marconi Space join efforts to build the Teledesic system

■ 2003–Service targeted to begin

6.4.2 Teledesic Constellation

The Teledesic LEO Network will consist of 288 operational satellites divided into 12 planes as shown in Fig. 6.4. Each plane consists of 24 satellites. Satellites operate at a low altitude under 1,400 kilometers.

The combination of a high mask angle and low-Earth orbit result in a relatively small satellite footprint that enables spectrum re-use, but requires a large number of satellites to serve the entire Earth. Each satellite behaves as fast-packet-switching node network to link with other satellites in the same and adjacent orbital planes. This forms a non-hierarchical mesh that is tolerant to faults and local congestion. A large constellation of inter-linked switch nodes offer a number of advantages in terms of service quality, reliability and capacity. It adapts to topology changes and to congested or faulty nodes and links. Each satellite is able to concentrate a large amount of capacity on its relatively small coverage area. Overlapping coverage was intentionally designed to permit the rapid repair of the network when a satellite fails.

Figure 6.4
Teledesic Constellation

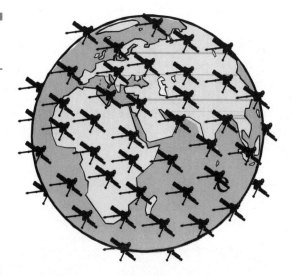

Teledesic will operate in the high-frequency Ka-band of the radio spectrum (28.6–29.1 GHz uplink and 18.8–19.3 GHz downlink).

6.4.3 Teledesic Architecture

Teledesic is a high-capacity broadband network that combines the global coverage and low latency of a low-Earth-orbit (LEO) constellation of satellites. Figure 6.5 shows the Teledesic network architecture. It can serve as the access between a user and a gateway in the terrestrial network and is designed to be compatible with today's applications. The network combines the advantages of a circuit-switched network (low delay "digital pipes") and a packet-switched network (handling of multi-rate and bursty data).

The satellite constellation covers nearly 100 percent of the Earth's population and 95 percent of its landmass. It is designed to handle millions of simultaneous users.

6.4.3.1 Teledesic Satellites/Ground Components. Teledesic ground segments consist of

1. Terminals
2. Network gateways
3. Network operations and control systems
4. Spaces segment (the satellite-based switch network that provides the communication links between terminals).

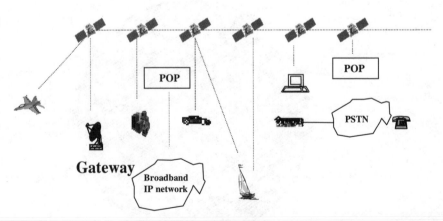

Figure 6.5
Teledesic Architecture

Terminals. Terminals are the interface between the satellite network and the terrestrial end-users and networks. They perform the translation between the Teledesic internal protocols and the standard terrestrial protocols. Terminals also communicate with the satellite network and support a wide range of data rates. The terminals also interface with a wide range of standard network protocols, including IP, ISDN, ATM and others. Most users will have two-way connections that provide up to 64 Mb/s on the downlink and up to 2 Mb/s on the uplink. A more robust Broadband terminal offers 64 Mbps of two-way capacity.

6.4.3.2 Teledesic User Access. The Teledesic LEO eliminates the long signal delay born by the communicating satellite as well as enabling the use of small, low-power terminals and antennas. The laptop-size terminals will mount flat on a rooftop and connect inside to a computer network or PC. The terminal-satellite communication links operate within the portion of the Ka frequency band. Downlinks operate between 18.8 GHz and 19.3 GHz, and uplinks operate between 28.6 GHz and 29.1 GHz. Operating frequencies at these ranges are degraded by rain and blocked by obstacles in the line-of-sight. To avoid obstacles and limit the portion of rain degradation requires that the satellite serving a terminal be at a high elevation angle above the horizon. The Teledesic's minimum elevation angle is 40° within the service area. This alleviates the problem and could achieve greater availability.

6.4.3.3 Teledesic System Capacity. The Teledesic system is designed to support millions of simultaneous users. Users can have access to two-way connections with up to 64 Mb/s on the downlink and up to 2 Mb/s on the uplink. Higher-speed terminals can deliver up to 64 Mb/s directionally. Frequencies are allocated dynamically and reused many times within each satellite footprint. The Teledesic Network supports bandwidth-on-demand, allowing a user to request and release bandwidth as needed in about 50 msec, resulting in an efficient statistical multiplexing.

This mechanism enables users to pay only for the capacity they actually use, and for the Network to support a much higher number of users.

6.4.4 Teledesic Costs and Partners

Design, production and deployment of the Teledesic system are estimated to cost $9 billion. End-user rates are likely to be set by service providers, but

Teledesic expects rates to be comparable to those of future urban wireline services for broadband access.

Principal shareholders/industrial partners are:

- Teledesic's primary investors
- McCaw
- Chairman Bill Gates
- Motorola
- Saudi Prince Alwaleed Bin Talal
- Boeing
- Matra Marconi Space

6.5 Iridium

Iridium World Communications, Ltd. is a Bermuda company established as a vehicle for public investment in the Iridium system. A private international consortium of telecommunications finances iridium and industrial companies and became operational in 1998. Motorola is the prime contractor and developer of the original system concept.

6.5.1 Iridium History

Iridium's first launch was on May 5, 1997 and its mission was to build the world's largest privately funded satellite constellation. Boeing, China Great Wall, and Khrunichev where chosen to launch Iridium satellites.

Iridium Chronology

- 1987–Motorola engineers research started Iridium system concept development
- 1990–Motorola file for application with the FCC
- 1991–Iridium, Inc. formed.
- 1992–Experimental license granted by FCC for Iridium system. Motorola becomes Prime Contractor.
- 1995–FCC grants operational license

6.5.2 Iridium Constellation

The Iridium satellite network is a constellation of 66 Low-Earth Orbiting LEO satellites, which are stationed in an orbit at 780 kilometers (435 miles) above the earth's surface. There are 6 orbital planes circulating Earth as shown in Fig. 6.6. Each plane contains 11 satellites with one spare. The six orbital planes are at an angle of 86.4 degrees to the equator. With such constellation setup, Iridium will cover 100 percent of the Earth's surface all of the time. Each satellite contains enough fuel to give it a life span of about eight years and weighs approximately 1,500 pounds.

6.5.3 Iridium Architecture

Iridium is structured as shown in Fig. 6.7. Its satellites are electronically interconnected to provide continuous worldwide connectivity. Communications are relayed via satellite and through terrestrial gateways where billing information and network management operations reside.

Inter-satellite and ground control links communications take place in the Ka-band frequencies. Telephone and messaging communications take place in the L-band frequencies.

6.5.3.1 Iridium System Overview. The receiving antenna is small enough to fit on a hand-held telephone or pager. An Iridium telephone is activated to the nearest satellite. The network determines account validity and

Figure 6.6
Iridium Constellation

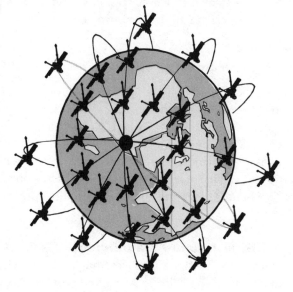

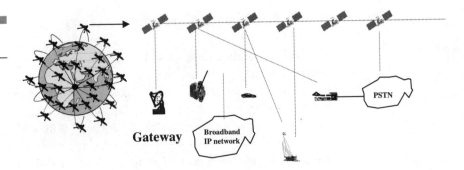

Figure 6.7
Iridium Architecture

the location of the user. The subscriber selects among cellular or satellite transmission alternatives. If the subscriber's local cellular system is unavailable, the call communicates with a satellite overhead. The call is then transferred from satellite to satellite through the network to its intended destination, either to another Iridium telephone or to ground station. The system gateways interconnect the satellite network with land-based fixed or wireless infrastructures worldwide.

The control system serves as the network management component for the Iridium. It performs satellite control and network management for Iridium Satellite and Network Operations Center in northern Virginia, U.S.A. Telemetry, Tracking, and Control Centers are also located in Hawaii and Canada, linking directly with the Satellite and Network Operations Center.

6.5.3.2 Iridium User Access. Iridium hand-held telephones are capable of delivering high-quality voice, facsimile, and data services anywhere on Earth. The handsets are available in two versions:

1. Single-mode handsets
2. Dual-mode handsets

The single-mode handset is designed to provide satellite communications around the world through the satellites. The satellite system tracks the location of the telephone and provides global transmission even if the subscriber's location is unknown.

The dual-mode handsets offer both worldwide satellite communications and terrestrial cellular telephone capabilities. They operate in Iridium-mode and or GSM-mode. This provides compatibility with existing GSM cellular service.

6.5.3.3 Iridium Ground Components. Gateways establish communications between Iridium appliances and any other telephone in the world. Interconnection to the existing wired and wireless networks, domestic and international, public and private, is performed through ground-based Iridium gateways which are located in key regions of the world.

These stations consist of equipment to communicate with the satellite constellation and with the international switching centers of the domestic fixed network. The functions include validation and registering of Iridium subscribers and providing access to existing ground-based telephone networks, so achieving the connectivity of the telecommunications networks is a worldwide venture.

6.5.3.4 Iridium International Gateways. Twelve gateways will be located throughout the world. The gateway's primary function is to switch calls between the satellite network and the landline network. This enables the Iridium user to call and receive a call from a landline or cellular phone anywhere in the world.

Iridium Italia and Iridium Communication Germany manage the European gateway that is installed at Telespazio's Space Centre in the province of L'Aquila, approximately 130 km from Rome. The Fucino station—the largest civilian telecommunications centre in the worl—was chosen for its geographic position, optimum for serving the European continent and outlying regions. The Fucino station is also to be home to a GBS (Gateway Business System) center, an information center for the processing and administration of the service's data (contacts, consumption and bills).

6.5.4 Iridium Operating Frequency

The Iridium system makes use of a combination of frequency division multiple access and time division multiple access signal multiplexing for best efficiency. The L-Band serves as the link between the satellite and user equipment. This user equipment has been authorized by the European Administrations and operates in the band 1621.35–1626.5 Mhz for both uplink and downlink. The Ka-Band (19.4–19.6 GHz for downlink; 29.1–29.3 GHz for uplink) serves as the link between the satellite and the gateways.

6.5.5 Iridium Costs and Partners

The $5 billion satellite system finally handled its first calls on November 1, 1998. Iridium, Inc. is an international consortium of telecommunications and industrial companies which funded the Iridium system. There are 19 investors worldwide who are partnered with Iridium:

- Motorola Inc. of the U.S. (primary)

- Korea Mobile Telecommunications Corporation of Korea

- Nippon Iridium Corporation of Japan

- Sprint of the U.S.

- Pacific Electric Wire

- Thai Satellite Telecommunications Co. Ltd. of Thailand

- Iridium Africa Corporation

- Iridium China (Hong Kong) Ltd.

- Iridium. Middle East Corporation

- Khrunichev State Research and Production Space Centre of the Russian Federation

- Lockheed Missiles and Space Co.

- Raytheon Company of the U.S.

- STET of Italy

- Cable Co. Ltd. of Taiwan

- VEBACOM GmbH of Germany

- Iridium Canada, Inc.

- Iridium India Telecom Ltd.

- Iridium SudAmerica Corporation

6.6 LMDS

Local Multipoint Distribution Service (LMDS) will no doubt play an important role in the grand unification network convergence. LMDS is a fixed wireless, two-way broadband technology designed to integrate video, voice and high-speed data.

Services over LMDS include:

- High-speed Internet access
- Real-time multimedia file transfer
- Remote access to corporate local area networks
- Interactive video
- Video-on-demand
- Video conferencing
- Telephony

LMDS operates in the frequency bands 27.5–28.35 GHz, and 29.1–29.25 GHz. LMDS was initially envisioned as an entertainment video distribution technology competing with CATV systems. Early experimental licenses demonstrated the feasibility a such a system. However, because foliage and rain attenuation is so high at this frequency and because terminals operating at this frequency are relatively expensive, current holders of LMDS licenses are now focusing on business access services.

Conceptually, LMDS is similar to the cellular telephony in that a service area would be divided into cells, with a transmitter serving each cell. It is not, however, targeted for mobile applications due its line of sight requirements. It differs from conventional microwave transmission systems due to its point-to-multipoint operation. Its characteristics are similar to that of a fiber-fed cable system, but operates in a line of sight (LOS) access approach limited to 5 km cell radius or less depending upon the terrain, the transmit power, and antenna height.

It should be noted that the Center for Wireless Telecommunications (CWT), a research facility of Virginia Tech in Blacksburg, Va., is organizing an ''LMDS Research Consortium.''

6.6.1 LMDS Architecture

LMDS uses low powered, high frequency (25–31 GHz) signals over a short distance. LMDS systems are cellular in that they operate at very high frequency signals over short line-of-sight distances. These cells are spaced about 4-5 kilometers (2.5–3 miles) apart. LMDS cell layout determines the cost of building transmitters and the number of households covered.

In a cell radius, LOS between the transmitter and receiver is required. LMDS signals are attenuated by water and have a path loss influenced by the amount of local rainfall, tree deflection and host of other barriers. To correct this, LMDS operators can reduce the cell size or increase the power of the transmissions when it rains. Trees, and branches can also cause signal loss, but overlapping cells and high roof mounted antennas generally overcome the problem.

Microwaves have long been used to transmit data. They are an excellent and a very cost effective point-to-point communication link.

Two-way wireless transmission links will connect small transceiver units located on the customer's rooftop (see Fig. 6.8) to a node sites. Coverage can be increased by overlapping hubs and by increasing hub antenna height. Depending on antenna height, terrain, weather and desired reliability, coverage between one to five kilometers in radius can be achieved. Transmission capacity depends on several factors:

Figure 6.8
LMDS Architecture

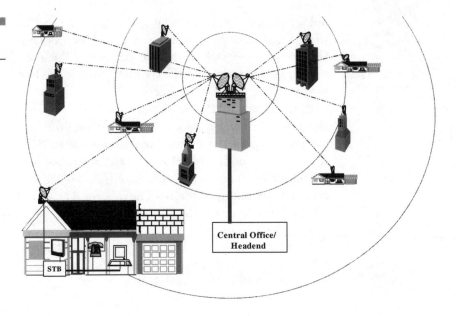

Central Office/
Headend

STB

- Modulation technology used
- Planned bandwidth offering
- Frequency re-use
- Hub sectoring equipment capabilities.

Since the signals have a limited reach, the transmission frequency band can be reused to increase the overall transmission capacity of the system. Dedicated fiber optic links will carry traffic from access nodes to the network operator's headend facility or central office. The headend can then be networked to a variety of service providers, such as IXCs, ISP POPs and video servers.

When engineering the system, the size of a cell has several critical components:

- Analog vs. digital signal
- Line-Of-Sight (LOS)
- Overlapping cells
- Environmental terrain (heavy rainfall)
- Antenna height (transmits well as receive)
- Landscape (foliage density)

One of the more critical components is LOS. LMDS services cannot be provided if a subscriber is in a shadow. Using reflectors or high power amplifiers to bounce signals, on most occasions, can restore light to the shadowy areas.

6.6.1.1 System Overview. LMDS is a broadband fixed wireless system that operates at 28 GHz and occupying 1300 MHz of the spectrum. LOS coverage over a 3-5 kilometer range can, in the best scenario in areas of concentrated traffic, provide voice/video and data for up to 80,000 customers. A 360-degree transmission pattern is sectorized into 4 quadrants of alternating polarity. Frequency reuse of the spectrum resources and an overlapping cell pattern drastically improve coverage for a targeted customer base.

The process of installing an LMDS system is a follows:

1. RF Engineering, with incline site analysis, LOS coordination and capacity planning

2. Right-of-way and municipal zoning approvals as well as lease negotiation to install the microwave dishes

3. Tower construction including zone approval and engineering design (tower elevation, etc.)

4. Installation, and testing of the networking to land-base systems

6.6.1.2 LMDS Modularity. An LMDS system is highly modular. The system is comprised of four elements.

1. The base equipment includes control and transport racks located at a central office for telephony or at a headend for video applications. Node equipment is located at the tower site, and includes solid-state transmitters, receivers and other elements.

2. The customer premise equipment includes a solid-state 28 GHz, roof-mounted, 12-inch dish and a network interface unit.

3. The antenna and network interface unit converts the wireless signal back into voice, video and data for distribution over existing in building wired plant.

4. A digital set-top box completes the end-user equipment for video services.

A single system will support a 5-km radius transmission pattern which is then segmented into quadrants with alternating polarities, allowing for overlap of signals to increase line-of-sight and allow for full reuse of the allocated frequency multiple times within a cell and in adjacent cells. LMDS systems are then repeated in a multiple cell configuration to cover an entire service area.

The capacity of such a system is enormous. For example, a single LMDS tower can deliver data and telephony lines to serve up to 80,000 customers, or each of 40,000 subscribers can be provided a voice/data line and 132 video channels. Because LMDS is broadband and digital, the service combinations are endless.

Physically, a system consists of two primary functional layers—transport and services. The transport layer is comprised of the customer premises roof-top-unit (RTU) and the Node electronics. The RTU solid-state transceiver is approximately 12 inches in diameter. The Node includes solid-state transmitters, receivers and other related elements located at the transmit site. The services layer is comprised of a network interface unit (NIU) at the customer premises and the Base electronics. The NIU provides industry standard interfaces to the customer and the Base provides control and transport functions remoted from the hub-site or central office/traffic aggregation site.

6.6.2 LMDS Enablers

A concept such as LMDS would not have been feasible a few years ago. Several recent technology advancements conspired to make LMDS a practical technology:

- Advances in Gallium Arsenide (GaAS) integrated circuits
- Digital signal processors
- Video compression techniques
- Advanced modulation systems with significant improvements in cost and performance

These, coupled with the decision by the FCC to make large chunks of high frequency spectrum available, have resulted in a delivery system capable of providing massive amounts of data throughput to customers cost effectively.

6.6.3 The LMDS Business Case

In building an LMDS business, one of the LMDS' most attractive features is its simplicity and the modularity of its building blocks. Competitive Local Exchange Carriers (CLECs) can deploy LMDS incrementally to suit the business or residential environment they are addressing. Best of all, the system can be installed in weeks instead of months. In such scenario, LMDS capital equipment can be the most cost-effective technology to deploy. The fixed expense associated with constructing LMDS cells is the cost of fiber to the hub and to the transmitters. The system is modular; hence only a minimum number of components would be deployed. More equipment is added as traffic increases. Unlike traditional landline networks, LMDS can be easily forklifted to a location where revenue generating business customers are located.

Competition to the local loop is yet to play a role, though the opportunity has been open to entrepreneurs since divestiture; the LECs have had a de-facto monopoly. The question was, how could a CLEC compete with thousands of miles of installed copper. LMDS, with low cost entry to the access market, will most likely change the present broadband access landscape. The CLECs are embracing this technology readily and penetrate the LECs initially, in the higher traffic markets of businesses, multi-dwelling units, and upscale residential area. The LECs are cognizant of this emerging technology, and in fact are investing in it heavily. One may conclude that LMDS had the effect of opening competition in local markets when others failed.

With 850 MHz bandwidth, using QPSK modulation, the LMDS pipe can deliver bandwidth capacity of about 180 T1 lines in per cell. With frequency re-use, hundreds more T1 lines can be created. Current LEC T1 tariffs are about $1000/month. These can be built into a Private Virtual Network for localized business for more revenue potential. These numbers are inspiring entrepreneurs to aggressively go after the LMDS market.

6.6.4 FCC Licensing

The FCC began the auction in December 1999 for the allocation of the LMDS 28 GHz frequency range which LMDS. The plan was to auction two licenses in each of 492 U.S. BTAs.

1. License "A" authorizes the use of 1150 MHz of non-contiguous spectrum in the 28 to 31 GHz bands.

2. Frequency block "B" will authorize service on 150 MHz of spectrum in the 31 GHz band.

Licensees can use any combination of system architectures to provide the services that suit their business case. Services can be a combination of two-way broadband broadcast video or a range of services including telephony, data and video.

Auctioning such a large chunk of the spectrum captured the attention of large and small corporations. One of the biggest players emerging in this market is NextLink. After its $695 million acquisition of WNP's LMDS licenses and buying out Nextel's share of a joint venture in LMDS, NextLink is now in a position to reach 95 percent of the top 30 markets in the United States with LMDS. Other major LMDS players are Teligent and WinStar, both of whom started to deploy LMDS systems in 1998.

LMDS is a ready solution to fulfill the growing need for high-speed access. It is capable of filling the void in the high-speed access market and may have cost advantages over legacy networks.

6.6.5 LMDS Standardization

On March 1999, IEEE 802.11 generated a Project Authorization Request, after a study group made the initial study to justify developing LMDS standards. The IEEE 802.144 PAR is initiated to develop standards for Broadband Wirelesss Access (BWA) Systems. The scope as stated in the PAR:

> This standard includes specifications for the air interface, including the physical layer and media access control layer, of fixed point-to-multipoint broadband wireless access systems providing mutiple services operating in the vicinity of 30 GHz. These specifications will be broadly applicable to system operating between 10 to 66 GHz.

The IEEE executive committee approved the proposed PHY. A new working group (elevated from the 802.11 study group) will soon become an official working group. It is expected that a good part of the technical materials discussed above will most likely be adopted by the new IEEE working BWA working group.

6.7 DBS

6.7.1 Introduction

Just when Cable companies began to relax over the much ballyhooed but never realized video challenge by the telephone companies, along comes the

Direct Broadcast Satellite (DBS). Overnight the Cable companies had a competitor in their back yards, offering 200 channels of high-quality digital signals. DBS' success, however, was anything but certain. Consumers had to be willing to pay several hundred dollars for a dish antenna on their house. The system served only TV per decoder and provided no local stations. But succeed it did, with the result that DBS is becoming one of the fast growth players in the video market. While Cable companies are fighting back, several have formed their own DBS consortium, Primestar.

6.7.2 DBS Architecture

With DBS, TV programming is beamed down directly from satellites to a stationary small satellite aimed at one position in the sky. The DBS satellites are positioned in geosynchronous orbits 35,768 km above the equator at various positions (about 25-35 degrees above the horizon) across the U.S. There are over 40 satellites operating for North American viewing audiences and over a hundred in the world. The TV signals are digitally compressed, allowing several programs to be broadcasted from a single satellite. Up to 200 digital channels can be broadcasted using the latest in video compression technologies.

6.7.2.1 DBS Uplink/Downlink. The uplink is mainly used for TV programming. DBS programming signals are gathered from the various sources (service providers) and delivered to the orbiting satellite via the uplink, as shown in Fig. 6.9, where they are handled and sent back via the downlink broad beam of radio waves.

DBS components include the following:

- Reception dish
- Decoder and
- Remote control.

Figure 6.9
DBS Architecture

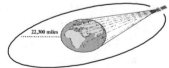

      Uplink   Downlink

The reception dish size varies, depending on services to be provided as well as the location a user is situated from the angle of dispersion of the downlink. Normally a dish antenna that is one meter (39″) or less in diameter is designed to receive direct broadcast satellite service, including direct-to-home satellite service. A reception dish may have to be mounted on a post to establish LOS. Some residents in the North region (direct-line-of-sight) of the Satellite may use the 18-inch dish.

The decoder decodes a TV single (channel) and then routes it to the household TV sets. Another decoder will be needed if a subscriber wants to view two different channels. There is no decoder DBS standardization so customers purchase their decoders from the service providers direct. Originally price range of DBS decoders from about $350 to $600 with added features. Customers with more than one decoder paid additional nominal fee. Recently, however, DBS decoder prices have dropped drastically in response to competition. One can purchase a decoder for as low as $200 with pre-paid programming subscription. Programming is similar to what the cable system offers with similar price range.

6.7.3 Compression Technique

A DBS channel transmits/receives a digital stream containing all the audio, video, and administrative control information (authentication, registration, etc.) The channel is compressed using the MPEG-2 standard. The DBS industry is yet to standardize systems, so service providers practice propriety solutions in order to differentiate their product. The decoders in subscriber's homes decode the digital bitstream converting it to its fundamental 6 MHz TV signal (video and audio components) which is then displayed on conventional TV sets.

Unlike Cable TV, DBS uses digital pipes to transport video channels. As a result, channel quality improves dramatically. Moreover, in a 6 Mhz TV analog signal, modern modulation techniques such as QPSK and digital compression technology can easily multiply the number of channels that can be carried in the same analog space. A 6 Mbps MPEG-2 digital video signal is far superior than it's analog cousin.

Another advantage to DBS is that when subscribers move, their cable network can move with them. Even a vacationing family can enjoy excellent programming if they install the dish on top of their RV.

DBS has some obvious disadvantages:

- Local programming is not available.
- DBS RF signals are more susceptible to rain. This can be addressed by directing more power to areas reveiving more rain.
- Digital encoding introduces artifacts. This is a subjective area, but some DBS users find them distracting.
- Power outages near the uplink affect all DBS TV viewers in the nation.
- Programs are sent at the same time to the entire nation. PST users are aggravated that prime programming appears very early since programming is East Coast originated.

6.7.4 DBS Services

There are several providers in the U.S. offering similar DBS services. Among the most popular:

- DIRECTV operates High Powered DBS satellites receivable with 18 inch dishes. DIRECTV has about 50 percent share of the DBS market.
- Primestar, a consortium of Cable TV companies, operates from conventional satellites using 27-36 inch dishes. Primestar has about a 28 percent share of the DBS market
- Direct TV and USSB (United States Satellite Broadcasting Company) merged and they are using the same satellite system operating at high power Ku-band.

6.7.5 FCC Regulation

As directed in the Telecommunications Act of 1996, the FCC adopted rules and restrictions regarding viewer's ability to receive DBS video-programming signals. Like LMDS, local governments could restrict antenna installation and maintenance. Potentially they could place absolute bans on antennas, require permits, or require planting trees to prevent them from being seen.

Congress, however, felt that local restrictions were preventing viewers from being able to choose DBS. Hence on August 5, 1996, the Commission adopted a new rule that is intended to eliminate unnecessary restrictions, installation costs or delay on antenna placement, while at the same time attempting to minimize interference with local governments. The new rule prohibits restrictions for those satellite dishes that are less than one meter (39″) in diameter. This new rule, in the spirit of TA '96, met it's objective of promote competition among video programming services providers. The new rule, in essence, means that subscribers should be able to install, use, and maintain an antenna on their property.

CHAPTER 7

Alternative Access Technologies: Power Line Carrier

7.1 Introduction

The previous chapters described the "mainstream" high-speed access technologies, which are becoming readily available and in some cases are being actively deployed. These are likely to be the means by which users gain access to future telecommunication, information and entertainment services. This chapter addresses a less traditional high-speed access technology that could provide an additional broadband information pipe into homes and businesses: powerline carrier by electric utilities.

One of the biggest challenges to a company becoming a facilities-based telecommunications provider is gaining access to right-of-way space. This is a cumbersome and bureaucratic process at best. This, coupled with the exploding demand for telecommunication services since deregulation, has been attracting fairly unusual players into this market. In the long distance arena, virtually anyone with right-of-way space has tried to leverage this into telecommunications revenue. For example Williams, a gas pipeline company, entered the telecommunications market by pulling optical fibers through their natural gas pipelines. Various railroad companies have buried fibers along their railroad beds (or leased this space for others to do so). In Georgia, the State government offered to let companies bury fibers along the state's interstate highways in exchange for giving the State government free access to some of these fibers.

With local competition, the same "land rush" is expected to occur in the access network. Already some churches have leased space in their steeples to hide cellular antennas. Land developers have partnered with telecommunications companies to roll the cost of burying fiber cables into the price of the lots. A group of companies which in particular has been eyeing this market is the electric utilities.

Power companies are facing difficult challenges. Environmental concerns are limiting growth in their core market and have increased the cost of their product. Nuclear power plants are being retired early, and regulators are resisting allowing the companies to recover their losses through increased rates. Utility deregulation is allowing competitors to sell electricity through the local utility's power grid. In response, power companies are looking for ways to grow their company outside their traditional services. An opportunity many utilities are finding attractive is telecommunications. Their enormous and ubiquitous power line infrastructure becomes a very attractive business case for the local distribution of deregulated communications services. Until recently, the electric utilities were forbidden by law from offering telecommunications services. This was changed by the 1996 Telecommunications Act, which allowed them to enter this market for the first time.

The technology to transmit data or voice over power lines is nothing new. For the past few decades, utility companies have used either carrier technology or Audio Frequency ripple technologies to transmit information over power lines. Usually this was used for the internal management of the power grid. Within the premises, this technology has been successfully marketed for household intercoms and lighting controls. With advancements in this technology and changes in regulatory structure, this technology might be usable for offering telecommunications services.[1]

---

[1] Alternatively, some utilities have proposed installing their own fiber or coax drops to homes ostensibly to be used for energy management, but then using the excess capacity to offer telecommunication services.

Power line communications technology basically adds frequencies above the 60 Hz electricity (50 Hz in Europe). In the past, audio frequency modems have been used on power lines for low-speed transmission supporting such applications as security, lighting control and other low bandwidth applications. However, electric wires are not limited to these frequencies. If higher frequencies can be used reliably, one can envision various applications including voice transmission, high speed Internet access, and perhaps even video communications occurring over power cables.

This technology is by no means mature enough today to enter the PSTN league. Additional research is needed in this technology before it is ready to function reliably in a shared environment outside the home.

This chapter describes the technologies for implementing high-speed access over power lines. The chapter will also examine some of the technical and marketing challenges the utilities face in deploying this technology.

7.2 Historical Perspectives

The AC powerline has long been recognized as a possible communications medium. As deregulation of both the telecommunications and electrical utility industries took hold, the utility industry, through their research arm, the Electric Power Research Institute (EPRI), began research into technologies that could transform their wireline landscape into a high-speed access network. Serious research to develop voice/data over power lines started in the early 1990's. The English utility PLC funded Dr. Brown to develop working models of his doctoral thesis at Lancaster University (England) on "Digital Transmission over Electric Power Lines" A prototype was developed and was tested in a neighborhood in England. Three years of trial and error yielding a dozen patents converged on the idea of using a multi-frequency approach. This new modem-like technology, placed at substations, allowed data to be transmitted over electrical power lines and into the home at speeds of about 1 Mb/s.

Powerline telecommunications technology and its market potential is still maturing. Standardization is in its infancy or yet to be addressed. There are several solutions that are under active experimentation by various vendors in the industry. Therefore, rather than focus on a specific solution, only general descriptions of this technology will be given in this chapter. Eventual standardization is critical to drive volume production and low unit costs that will be necessary for acceptance and deployment of this technology.

7.3 Power Lines as a High-speed Transmission Media

Transmitting information over power lines presents a host of challenges that must be overcome. Given the nature of the equipment connected to these lines, noise is a major problem. At the frequency range of interest, power motors and florescent lights play havoc and pollute the spectrum with unwanted noise. RF radiation is also a major concern. Cable design and installation practices were never engineered with RF ingress and egress in mind. Unlike telephone cables, the wires are not twisted and lengths of single wires are often seen at junction points.

The various step-down transformers used in the electric grid do not pass higher frequency signals. Hence, some means must be employed to shunt signals around these transformers. On the other hand, the segmentation due to these transformers enables controlling the number of homes that must share the limited power line bandwidth and can also aid in fault isolation.

There are numerous sources of signal reflections on power lines that impede transmission: impedance mismatches occur where appliances are attached; bridge taps exist due to branching in the wiring runs; and lack of transmission impedance terminations at the ends of wiring runs create reflections.

Power lines are a shared medium limiting the amount of bandwidth that can be offered to each user. The common medium must be used for both upstream and downstream communications, further limiting available bandwidth. Also this shared media means that all transmissions can be monitored at any point along the electrical branch. This will require some form of encryption for private communications.

Finally, in most countries the electrical transmission grid is much less reliable than the telephone network. This may need to be addressed before consumers will trust their lifeline telephone service to the electrical utilities. Note that in some countries such as Japan and Germany, the electrical utilities have a reputation for reliability approaching that of the telephone network. A network planner at NTT (the Japanese national telecommunications company) once mentioned to the author that for this reason, NTT fears the local telecommunications competition by the electric utilities in Japan more than they do their CATV companies.

7.4 A Power Line Transmission Architecture

The basic principle of providing access via power lines is illustrated in Fig. 7.1. RF signals are injected and stripped from the power line at power substations. A HE (Headend–like) node mediates access to all connected premises. The system is based on packet data principles. Voice services would typically be carried using voice-over-IP techniques for passing to the Internet or to be connected to the PSTN via appropriate gateways.

A hierarchical design is central to reliable powerline communications. Each level of the design must be optimized to overcome the unfriendly elements of the powerline environment. This results in significant use of error control coding to achieve acceptable error rate performance.

The topology of carrying data over power lines is shown in Fig. 7.2. There are five functional components to a powerline communication system:

Figure 7.1
Powerline Access
Connections

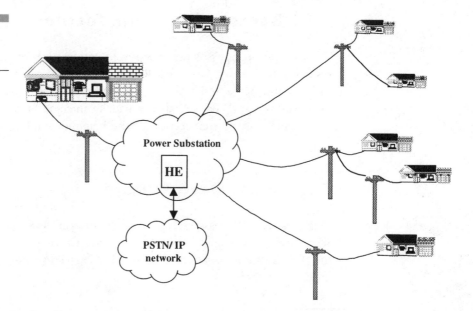

Figure 7.2
Powerline Access To-
pology

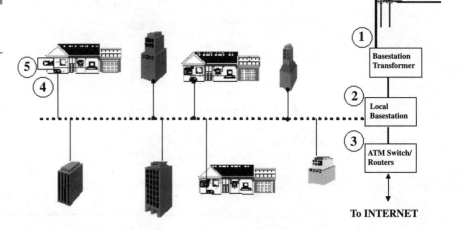

1. Basestation transformer
2. Local base station
3. Switching node
4. Fuse panel
5. Communication module

7.4.1 Basestation Transformer

The basestation transformer is a power station unit serving that business or resident community. Note that RF signals will not pass through power transformers and hence must be received and regenerated at transformer points in the power grid. RF signals are usually injected at the lower end of the transformer serving that section of the neighborhood.

7.4.2 Local Base Station

This local base station acts as concentrator and also a controller mediating access from customers. This node connects to the core data network where the data traffic is backhauled to the Internet and/or to the local telecommunications network.

7.4.3 Switching Node

The switching node connects the controller base station to the core data network. A fiber connection or other transmission medium connects to local telecommunication service providers and eventually to the Internet or PSTN.

7.4.4 Fuse Panel

At the home fuse panel, the telecommunications service is separated from the electrical supply at the connection point between the main service and the power meter.

7.4.5 Communication Module

The communication module is the unit terminating the service at the customer's PC. This can be implemented remote from the PC at a premises controller, as an external device next to the PC, or in an internal Network Interface Controller card within the PC.

7.5 Noise on Power Lines

Noise has played a major role in impeding development high-speed access over power lines. The operating frequencies used for power lines communication systems was prime for interference from electrical appliances. Working around this interference added complexity to powerline data transmission systems.

Arcs due to electric motors are a major source of noise for all transmission systems. Given that these motors are directly connected to the power line, significant background noise is created on electrical wiring.

Narrowband interference is the result of external, RF signals leaking into the power line. This interference creates random transmission interruptions and appears on a spectral display as a narrowband carrier less than 100 kHz in bandwidth, typically of short time duration. One finds time duration and carrier bandwidth variations as diverse as the types of signal sources. Common sources are open air RF transmissions such as CB, amateur radio, civil defense, aircraft guidance broadcasts, international short-wave, AM broadcasters, etc, and poorly shielded RF oscillators in the subscriber's household. Most electrical home appliances can be a source of narrowband interference. Factors contributing to coupling into the digital powerline system includes poor equipment grounding. Impedance mismatching and variations will also play a major role as appliances are plugged into electrical outlets.

7.6 Physical Layer Transmission

Previous attempts to use powerline for communications, such as home intercom products, used a modem approach with Frequency Shift Keying (FSK) modulation between 50 and 500 kHz. Amplitude Shift Keying (ASK) was also deployed on digital versions of FM and AM systems. At this range of frequencies, systems were plagued with noise and needed constant re-tuning as noise and other interference propagated power lines. The FCC now allows power communication systems broader use of frequencies except in the AM band of 535 and 1705 kHz.

7.6.1 Multifrequency Modulation

One method to address narrowband interference on power lines is to utilize multiple carriers to take advantage of the spectral "holes" around interferers. Potentially, a variation of the Discrete Multi-tone (DMT) technique described in chapter 4 for ADSL could be used. The system could automatically detect the presence of narrowband interferers and not use DMT tones associated with those frequencies. The powerline network operator could also disable tones associated with communication services to keep from interfering with those services.

7.6.2 Spread Spectrum Modulation

The key features necessary to provide immunity from powerline attenuation and noise are spread spectrum wideband modulation, fast synchronization, adaptive equalization, error control coding, and powerline-optimized network protocols. Direct sequence spread spectrum (DSSS) is a type of spread spectrum based on multiplying the incoming data with a much higher bit rate pseudorandom pattern. The resulting pattern is used to modulate the transmitted carrier. At the receiver, a correlator uses this same pseudorandom pattern to recover the original data. Without knowledge of this pseudorandom pattern, recovery of the data is difficult thereby providing a degree of privacy. With DSSS, the energy of the transmission signal is evenly distributed across the transmission bandwidth. This spreading of the data makes the signal resistant to noise, interference, and snooping. The redundancy inherent in the DSSS signal allows spectral portions of the signal to be lost (as due to narrowband interferers) yet the original data stream still recovered. Also, the spread signal is less likely to interfere with narrowband radio services and notch filters can be used to remove spectral energy interference to licensed services should this prove to be a problem.

Originally developed for military applications (due to its jamming resistance and difficulty of interception), spread-spectrum schemes are now commonly used in the United States for digital mobile phones operating in the PCS band, for wireless local area networks (LANs) and for cable modems.

In general, a spread spectrum system exhibits improved noise immunity over narrowband systems for powerline carrier transmission. This technique does not solve the difficult problem of signal synchronization on the powerline in the presence of constantly changing noise and frequency-dependent attenuation. However, specially designed physical layer spread spectrum technology can provide very rapid synchronization. Rapid synchronization is an important component of a fast, practical, and reliable powerline communications system. In the protocol, this allows data to be transmitted in short frames which are important for packet communication.

7.7 Media Access Control and the Data Link Layer

For upstream communications from the home, a means is necessary to ensure that multiple communication modules do not attempt to send data at the same time and also is necessary to mediate use of this limited resource between the multiple users. The responsibility for this resides with the Media Access Control (MAC) layer. MAC algorithms used on shared wires are generally based on either a carrier sense technique or token passing. Given the large distances between the users and the noisy environment, token passing appears better matched to the powerline medium. In token passing, nodes are only allowed to transmit when they hold an electronic token. In this way, there is no possibility of nodes starting to transmit in the midst of another node's transmission.

Several key data link layer features are required for reliable operation of large, multi-node powerline networks:

- Segmentation of larger packets into short frames
- Error correction and detection
- Effective adaptive equalization
- Reliable transfer of control

Transmission of short frames on the powerline is essential, since longer and contiguous information sent would almost certainty be corrupted in transmission. Both error detection and forward error correcting codes should be used. Forward error correction is needed to minimize the number of retransmissions, and error detection is used as a flag for a retransmission on a frame basis. For powerline carrier systems, the receiver acknowledges each frame before the transmission proceeds to the next frame (that is, a window size of one is used). Since powerline conditions can change over milliseconds periods, the receiver must be able to adapt quickly to these changing conditions. Using a low-level link protocol built upon short frames, the receiver can adapt on a frame basis and, through use of acknowledgments, data is not lost.

7.8 Applications of Powerline Data Communications in the United States

Once the data throughput of a powerline carrier system reaches about 100 kb/s, a number of interesting applications become possible. A number of these apply directly the utility's core business. With deregulation of both the telecommunications and electrical utility industries, utilities can then enhance this offering with a variety of telecommunications services and features:

- Remote meter reading (for electricity, gas, water, etc.)
- Load shedding

- Demand side management
- Remote user control of appliances
- Internet access
- Local telephone access
- Medical alert
- Burglar, fire, carbon monoxide, and natural gas security services

An obvious example is remote meter reading. Not only does this eliminate the expense of sending someone to the customer's premises monthly, the utility company can also use real-time information from the meter to provide value-added services. Some utilities have expressed interest in what has been called "demand side management". The electric company would continually send downstream the price of electricity which could vary with time of day and current demand. The customer would have the option of avoiding the use of high-energy appliances such as washing machines and clothes dryers and programming appliances (such as water heater and air conditioners) to cut back electricity use when electricity prices exceed certain levels. In a related service known as load shedding, the utility company would offer customers reduced electric rates if the company is allowed to remotely turn off high-power appliances in periods of excess demand. Customers could use this same mechanism to turn on air conditioners before leaving the office (or to turn them off if they forgot to before leaving home). In certain areas such a Florida, some homeowners use their homes only part of the year and discontinue utility service during the other portion. Powerline carriers could be used to accomplish this periodic disconnection/reconnection. Several utility companies (Duke Power perhaps the most aggressive among them) are already offering such services in various cities in the U.S.

Because of the packet approach used with powerline carriers, "always on" services can be offered. This would be particularly effective for Internet and for security services.

7.8.1 Data over Powerline Application in Europe

Phone-service trials have been conducted in several regions in England, Asia and Australia, and the technical results have been very encouraging. However, this has remained a technology looking for a solution. Technology used in the Public Switched Telephone Network is mature and competition has driven component costs, including line cards, very low. Hence the economics of competing for phone customers against the likes of British Telecom or other aggressive European carriers has not been attractive.

In the United States, a distribution transformer supplies power to an average of four to six residences. European utilities typically serves several hundred residences per power transformer. On one hand this is a disadvantage for U.S. utilities in that more equipment must be installed to deploy a powerline carrier system. With the concentrator costs spread over only four to six customers, this greatly raises the cost per user. On the other hand, since the transformer provides segmentation of the transmission line, the upper limit to the amount of bandwidth that can be offered to each user will be much less in the European case.

A particularly ripe opportunity for European electric utilities appears to be Internet access. Europe has been lagging the U.S. in Internet use in part due to the way local phone calls are billed. A typical U.S. customer pays a flat rate for both local calls and Internet usage and hence can (and sometimes does) leave Internet connections up for hours at a time. European telecommunication carriers impose per-minute fees for local calling, and ISPs typically also meter Internet access. This could provide an opportunity for power utilities to enter the telecommunications market with a package of services including flat rate, always on Internet access. However, this will have to await further progress in European telecommunications deregulation by which time ADSL may be generally available from European network operators.

7.8.2 The Future of Powerline Carrier

Does the powerline carrier have a realistic chance of being another communications pipe into the home? A look at equipment costs is not very encouraging. Digital powerline technology is very new and the cost per subscriber is over $1,600 today. This is not competitive with ADSL or cable modem both of which have per-subscriber costs well under $500. Both ADSL and cable modes are already commercially available and are being widely deployed; digital powerline carrier is still in the prototype stage. It may well be that utility companies are too late to enter this market. Moreover, they would be entering a market that is becoming fiercely competitive without providing operating skills in managing complex telecommunications networks or skills in marketing telecommunications services.

Nonetheless, the utility companies are highly motivated to expand beyond their core business and do bring certain advantages to the playing field. Investment costs of upgrading the electric infrastructure for telecommunications are low compared to burying new telecommunications lines. The utility companies have deployed over 50,000 miles of optical fiber, more than any other company except the phone companies. Most power companies have deployed SONET fiber optics rings for internal communications and to maintain their power grids. Hence they do have some operational experience with telecommunications networks. While burying these fibers in their right-of-ways, the incremental costs of installing a larger fiber count cable is small. They are therefore well positioned, as a first step in entering the telecommunications market, to lease dark fibers and excess capacity on these internal networks.

Rather than using the powerline carrier to offer telecommunications services themselves, the electric utilities could offer a last mile alternative to CLECs.

Utility companies are closely watching a remote meter reading pilot project involving sending NetWare LAN information over power lines. The demonstration is a result of an alliance between Novell and Utilicorp United. This alliance has two major goals:

1. Drive the development of Novell Embedded Systems Technology Powerline technology. The design requirement is to let NetWare run over power lines at speeds beginning as fast as two megabits per second and eventually as fast as 10 megabits per second.

2. Establish the Smart Energy Network Alliance, a group that would set *de facto* standards to facilitate the development and implementation of networking over power systems.

The meter reading application requires some home rewiring and installation of equipment in the home. Despite numerous trials in the past, this method of meter reading has never proved to be economical.

Novell and Utilicorp have maintained a low profile about the project since it was made public last fall. Many utility companies hope to revisit this market and plant the seeds for their future telecommunications services.

8

Home Networking

8.1 Introduction

The usefulness of the high-speed access technologies described in chapters 3 through 7 will be limited unless they are met with equally powerful home networks. These broadband home networks will allow multiple home terminals to share the access bandwidth and will also provide internal communications within the home. Indeed, broadband home networking can be considered an enabling technology for broadband access.

The subject of home networking has suddenly become very popular. From a rarely heard topic a little more than a year ago, home networking products are starting to pour onto the market. At the 1999 Consumer Electronics show in Las Vegas, more than two dozen companies were introducing home networking products. These companies are seeing a huge market opportunity. In a consumer study in late 1998, the Yankee Group found 30 percent of PC users were interested in the idea of a home LAN. This has come about because of the rapid growth in multiple PC households. Indeed, a recent market study revealed that most PC purchasers who already own a PC were getting an additional rather than a replacement PC. Dataquest estimates that 15 million homes have multiple PCs and that the market for home networking will reach $2 billion by 2002.

The use of a home network to today's family can be seen in the Ethnography study made by Intel Corporation for computing value to the home:

- Accessible from anywhere in the home
- Instantly available
- Involves multiple people in the same home
- Family decisions are made in short burst of time

The study also identified that in the modern American family, "the nuclear family online", the kitchen is the center of activity in the home.

These requirements lead to the conclusion that a home LAN is what consumers will be demanding in the near future.

Home LANs interconnect home PCs, peripherals and other consumer-electronics devices. The idea of a home LAN makes a lot of sense. Files can easily be copied from one PC to the other; printers, scanners and backup devices can be networked; high-speed network access can be shared: mom working at home connected by remote to the LAN at the office while dad executes an electronic stock trade and their son does his homework over the Web—all simultaneously over the same shared ADSL or cable modem. In fact, Internet access through a shared connection may well become the driver for home LANs. And then there is the home LAN application most frequently mentioned: networked games.

The above are more or less traditional LAN applications. Home networking products coming onto the market support these and much more:

- Providing the equivalent of multiple phone lines throughout the house
- Local calling between telephones in the house
- Music distribution within the house
- Room-to-room intercoms
- Video monitoring between rooms (acting as baby monitors; seeing who is ringing the doorbell, etc.)
- Remote lighting and appliances controls
- Distributing video signals to TVs from a central DBS receiver.

Adding to the home network a broadband access link to the Internet could allow any of these applications to be extended outside the home, e.g., checking up on the nanny while at work. Of course any application allowing remote access to the home network will need strong security mechanisms.

The terminals connected to home networks would certainly include PCs, computer peripherals and networking equipment. Also included would be other "information appliances" such as telephones and TVs. Some visionaries see a day when virtually every home appliance has a home network interface, although what the toaster would do with a network interface does stretch the imagination.

A major impediment to the development of home networks has been the poor state of inside home wiring. Today's inside telephone wiring is often little more than a random interconnection of phone jacks. Only the red/green pair is guaranteed to be connected throughout. CAT-5 cables are the exception rather than the rule. Home coax wiring is no better, especially that run by the do-it-yourselfer. Connections are often loose, lines are sometimes spliced into other lines and shielding may not even be connected. Electrical wiring is readily accessible in every room but suffers significant noise and radio interference problems.

Hence, part of the reason why home networks are not commonly deployed, at least in older homes, has been the need to install new wiring. All this is changing. New home networking technologies have been developed under the mantra "no new wires" allowing almost any type of existing wiring to be used—or no wiring at all.

8.2 Categories of Home Networks

Emerging home network products can be categorized by the four principal types of wiring they employ:

1. Telephone wiring
2. Coax
3. AC power line carrier
4. RF wireless

This following section examines the technologies being developed for each of these wiring approaches. Each present certain advantages and limitations with each racing to become the dominant home networking product. It is too early to tell which if any of these solutions will dominate. Cost, user friendliness, plug and play simplicity, software support and standardization will all play a major role in determining the eventual winner.

8.3 HomeRF LAN

Look Ma no wires! Wireless RF would seem an elegant solution for home networking. Today moving your LAN-connected notebook computer from the study to the family room requires unplugging cables then finding the RJ-45 jack in the next room that never seems to be where you wish to work—if there is even a jack installed in that room. Wouldn't it be nice to simply carry your notebook computer wherever you want to go—the breakfast table, patio, living room sofa, bed, wherever—and without connecting any wires, be able surf the Web, send email, print document, backup files? Inexpensive technology is coming on the market that will let you do just that.

There are several groups that are working actively to develop specification for home RF. Among the most credible is the HomeRF™ working group and Bluetooth. The HomeRF mission is ambitious:

> To enable the existence of a broad range of interoperable consumer devices by establishing an open industry specification for unlicensed RF digital communications for PCs and consumer devices anywhere, in and around the home.

8.3.1 HomeRF System Concept

Leading companies from the personal computer, consumer electronics, peripherals, communications, software, and semiconductor industries are developing a specification for wireless communications in the home called the Shared Wireless Access Protocol (SWAP). Founded in March 1998, the over 70 member companies have been actively collaborating in this venture. In January 1999, they released the SWAP Specification 1.0 having the following fundamentals:

- Defines a common interface supporting wireless voice and data networking in the home;
- Enables interoperability between consumer electronic devices available from a large number of manufacturers;

- Exploits the flexibility and mobility of a wireless solution. This flexibility is important in order to create a compelling and complete home network solution.

Future versions of SWAP are planned that will address wireless multimedia (SWAP-MM) and lower price point solutions (SWAP LITE).

The SWAP specification was designed for wireless home networks supporting voice and data communications between PCs, computer peripherals, PC-enhanced cordless phones, and new terminal devices such as portable display pads. Applications specifically addressed were the following:

- Accessing the Internet from anywhere in and around the home from portable display devices;
- Sharing an ISP connection between PCs and with other Internet-enhanced devices;
- Sharing files/modems/printers in multi-PC homes;
- Intelligently forwarding incoming telephone calls to multiple cordless handsets, FAX machines and voice mailboxes;
- Reviewing incoming voice, FAX and email messages from a small PC-enhanced cordless telephone handset;
- Activating home electronic systems by voice command into a PC-enhanced cordless handset; and
- Multi-player games and/or toys based on PC or Internet resources.

8.3.2 HomeRF Network Architecture

The HomeRF is designed to carry integrated voice and data traffic and interoperate with the PSTN and Internet. The HomeRF technology is an expansion of the today's cordless telephone and wireless LAN technologies developed in IEEE 802.11.

The SWAP specification supports multiple access techniques. For time-critical applications such as real time voice, TDMA (Time Division Multiple Access) is used. For bursty data traffic, CSMA/CA (Carrier Sense Multiple Access/Collision Avoidance) is employed. The system as shown in Fig. 8.1 depicts a set of information appliances that will become increasingly common. The network can accommodate up to a maximum of 127 devices with up to 6 full duplex conversations.

Wireless is an excellent platform to extend home networking to mobile devices around the home and into the yard. For example, a typical application

Figure 8.1

HomeRF Network To-pology

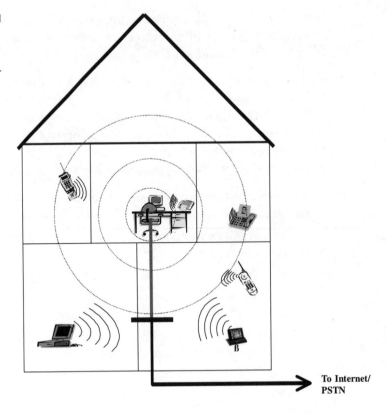

for device β, a wireless PAD, is to carry it to the garage or driveway to access the Internet for tips on how to fix the car. Wireless PADs like these, which are able to access the Internet network for the latest weather, traffic information and news updates, are expected to find a ready market.

8.3.3 Physical Plane

The HomeRF Working Group specified the physical layer as having the following characteristics:

- The system operates frequencies in the 2.4 GHz range
- Digital frequency hopping spread spectrum radio at 50 hops per second
- Data rate of 1 Mbps using 2FSK modulation or
- Data rate of 2 Mbps using 4FSK modulation
- Range of about 300-foot covering a typical home and yard

■ Data security using the Blowfish encryption algorithm (over 1 trillion codes)

■ Data compression using the LXRW3 algorithm

8.3.4 Control Plane

The system can operate in two modes:

1. Distributed control

2. Connection point control

The distributed control mode supports data applications. In this mode, all devices would access the LAN on an equal basis, the control distributed between the devices. This access would be based on CSMA/CA (Carrier Sense Multiple Access/Collision Avoidance).

In connection point control, a TDMA service is invoked for real-time applications such as voice and video. A centralized connection control mechanism is used to manage and coordinate information flow. The connection point provides connectivity to PSTN.

In summary, the system can provide the following service categories:

■ Voice devices would use TDMA service to communicate with a base station

■ Data would use CSMA/CA services to communicate with other data nodes

■ PCs with voice applications could use both TDMA and CSMA/CA services

8.3.5 HomeRF Advantages and Estimated Cost

Thirteen companies have committed to building products based on the specifications established by HomeRF working group. These include Compaq, HP, IBM, Intel, Microsoft, Motorola and Samsung. Such a broad consortium promises widespread product deployment and assures competition leading to price reductions.

Wireless home networks bring the following significant advantages:

■ No re-wiring

■ Simple installation

- Portability
- Location freedom
- When the customer moves, the ability to uninstall the equipment and to move it to the new residence

Wireless home networks do present certain concerns:

- Capacity loss due to neighbors buying similar equipment or other services using this same spectrum. This would be of particular concern in apartment complexes.
- Potential eavesdropping
- Limited total bandwidth. Some manufacturers are introducing products with capacities up to 10 Mb/s.

Several vendors have given price estimates for an RF home LAN in mass production of about $1,500 for a connection controller base unit, plus $80 to $150 depending on range and speed for the wireless unit needed for each terminal device.

HomeRF, however, predicts eventual inexpensive wireless connectivity to a wide variety of devices, costing about $30 from "antenna to bits" for a 300-foot radius in and around the home.

8.4 Powerline Ethernet

For a number of years, house wiring has been used as a communications bus for home automation. The most popular system has relied on the X-10™ specification from X-10, Inc. X-10 controllers can address simple messages such as "turn on" or "turn off" to modules plugged into electrical outlets throughout the house. Up to 256 modules can be addressed. X-10 controllers and receiver modules are now available from many manufacturers and distributors including HomeTech Solutions, Radio Shack, Leviton, Savoy Automation, Stanley and X-10 Inc.

While X-10 meets the needs of home automation, it is a low-speed specialized system not suitable for home networking. New technologies, however, are emerging that promise to provide powerful home networking capabilities over in-wall electrical wiring.

Several companies have jumped into the market for technologies and products for constructing home networks over home electrical wiring. Among the most visible are Adaptive Networks, Enikia, Intellon, and Intelogis. Unfortunately, proprietary solutions were adopted by these vendors on the physical and the MAC layer in anticipation of capturing an early lead in this huge potential home networking market. The attitude, in the absence of interoperability and standards, seems to be all or nothing.

This section will give an overview of the powerline technologies recognized by CEBus standard. CEBus is an open standard for home networking. It utilizes a separate physical layer specification allowing for communication over many media including over electrical wiring and others.

For powerline home networking, the CEBus standard technology specifies use of common electrical outlets and wiring yet allows ethernet-speed home networking using an off-the-shelf Ethernet controller system. This technology unlocks the information transmission capacity hidden in common electrical wiring despite the challenging environment that a home powerline network presents. The powerline approach to home networking satisfies two key requirements for bringing the Internet into every aspect of our lives:

1. High speed networking

2. Ubiquitous accessibility

The technology platform enables any home to be retrofitted for the high-speed digital communications of Internet, voice, audio, video conferencing, and home automation. This will spawn advanced home networking applications for sharing of computer resources and full automation of home electronics, home controls, home scheduling, communications, and news and entertainment.

8.4.1 Powerline Ethernet Applications

As in wireless, phone line and coax-based home networking systems, powerline-based home networks can support the full range of voice and data applications. Powerline techniques, however, bring unique advantages to home networking. Since virtually every home is equipped with multiple electrical outlets in every room, a powerline-based home network could give every device and appliance access to the home network, opening up numerous application possibilities. This will be particularly true as home appliances become Internet friendly, allowing these appliances to be monitored and controlled from anywhere, inside or outside the home. A premises controller could turn on or off a dishwasher or electric water heater depending on the cost per kW for that period. From the office or hotel room, a homeowner could program the VCR, turn on security lights or turn on the oven to start dinner. Several companies are already working on intelligent home appliances. Even a toaster or a refrigerator may one day have an IP address.

8.4.2 Powerline Ethernet System Concept

The home powerline networking fundamentals are conceptually similar to the one described for use on powerlines outside the home in chapter 7. The basic architecture is illustrated in Fig. 8.2.

Available technology has solved the problems of power interference and electronic eavesdropping that plagued its initial deployment. The home networking platform enables anything that plugs into an electric outlet to be connected together and, perhaps more importantly, connected to the Inter-

Figure 8.2
Powerline Ethernet
Topology

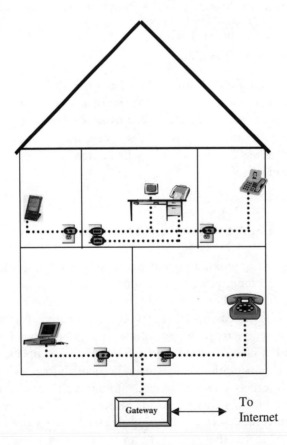

net. A Web PAD, for example, when plugged into an AC outlet, could immediately display real-time news from an Internet web site.

At each interface module an ethernet transceiver will transmit and receive data of up to 10 Mb/s. This module could be built into the appliance or connected to it via an ethernet, USB, or other types of cable.

8.4.3 Physical Layer Technology Alternatives

As mentioned earlier, there are several solutions to sending information over power lines. Below are various solutions adopted by the different vendors.

8.4.3.1 Legacy Powerline Communication.
Data over powerline was developed by X-10 two decades ago. It was targeted at home automation, providing simple control functions and a primitive data messaging system. An Amplitude Modulation (AM) technique was used to transmit binary data using the zero crossing point of the 60 Hz (50 Hz in Europe) AC powerline frequency of the positive and negative cycle. At this crossing, noise has the least effect on data transmission. Two zero crossing was required to transmit a binary symbol (0 or 1). An *ad hoc* message protocol is used with short commands for such things as dimming lights and setting time-of-day clocks. One can see that such a system, although cost effective, is not suitable for high-speed home networking.

8.4.3.2 Digital Powerline with FSK.
A simple yet robust two-way powerline communication technology was developed using Frequency Shift Keying (FSK). FSK modulation launches digital signals on AC lines using a range of multiple narrowband frequencies. The frequency range was selected above the range of most noise produced on AC lines, but low enough to minimize signal attenuation over long wires.

Each frequency symbolizes a 0 or 1 state. Information is therefore sent over the powerline using various selected frequencies to convey the content of an information packet.

Data transfer rates up to 1 Mb/s data rate is possible using this scheme.

8.4.3.3 Digital Powerline with Spread Spectrum.
Spread spectrum is a more complex and sophisticated communication technology that is beginning to penetrate the home networking industry. Due to advances in ASIC

and component technology, spread spectrum emerged as a practical and economic solution within the last few years for use in consumer and commercial applications.

Historically, spread spectrum was used by the military for the development of secure communication over radio frequency communications. Spread spectrum has been deployed for digital cellular application because of its bandwidth efficiency and, as in military applications, for its resistance to jamming and signal interception.

In principle, spread spectrum is a transmission technique that allows the frequency spectrum of a data signal to spread using a code uncorrelated to that signal. This, in effect, generates a multiple of the bandwidth for use. The spreading factor (bandwidth gain) is the ratio of transmission bandwidth to the information bandwidth. Fundamentally, the mechanisms to spread the signal are:

1. Direct sequence

2. Frequency hopping

3. Time hopping

4. CDMA

In direct sequence, the signal is multiplied by a Pseudo Noise (PNcode), which can be generated by a shift register. This PNcode has a noise-like property and therefore those trying to intercept are less likely to detect a data signal. The received signal is multiplied by the same PNcode, thereby recreating the original data signal.

For frequency hopping, the carrier frequency jumps among a set of frequencies in a pseudo-random sequence. This sequence must be known to the receiver to recover the incoming data.

Both techniques utilize a swept frequency pulse "chirp" to spread the signaling bandwidth to many times that of the information bandwidth. "Chirps" are of the same pattern and detectable by all nodes. This mechanism enables receivers to detect the presence of a signal immediately. This is important for implementing CSMA (Carrier Sense Multiple Access) LANs where stations need to determine if the channel is busy before transmitting.

With Chirps as a self-synchronizing carrier, it is it possible to implement a CSMA-based network. This is the technology adopted for the CEBus standard. The powerline carrier spreads its signal over a range from 100 Hz to 400 Hz during each bit in the packet. However, instead of using frequency hopping or direct sequence spreading, the CEBus powerline carrier sweeps through a range of frequencies as it is transmitted.

8.4.4 Control Layer Technology Alternatives

Again, as has been their habit, data-over-powerline vendors implemented various mechanisms to control data flow between nodes and appliances in the home network. Below is a generic description of the most popular ones.

8.4.4.1 CSMA. MAC protocols defines the rules that govern how and when a shared physical medium can be accessed by a participating node. The MAC for the CEBus standard specifies a traditional peer-to-peer communication protocol with Carrier Sense Multiple Access (CSMA) with carrier detect. The CEBus standard supports this protocol as it allows any node in the network to have access to the shared medium at anytime (if the medium is idle). If a collision occurs, the colliding node randomly waits before vying again for access on that shared medium. An exponential back-off mechanism is usually employed.

Compared with other techniques, CSMA/CA is complex thereby leading to greater component counts and hence cost. Each node must be aware of its environment and be fairly intelligent to perform its function.

8.4.4.2 Token Passing. The token passing scheme, versus peer-to-peer techniques, is less complex. The MAC protocol requires participating nodes to obtain the token before granting access to the shared bus. The protocol requires no collision-sensing logic. It is best suited for datagram application.

8.4.4.3 Client Server Topology. The MAC in the client-server approach allows most of the intelligence to be centralized, thus reducing costs and complexity at client nodes. The server functionality could be embedded in an existing device such as a PC. In this scheme, client nodes are unaware of their environment, but instead their access is centrally managed by the server. A priority scheme is invoked for synchronous voice applications. One obvious disadvantage of the client-server approach is system reliability, since a failure in the server disables the entire network.

8.4.5 Digital Powerline Advantages and Costs. Presently, the digital powerline is getting the least attention because there is no standards push behind it. CEBus specifies solutions, but as noted in the previous section, various solutions are under active development and interoperability will not be possible. Powerline product suppliers need to adopt a single standard before a strong market for their products will develop. Other alliances such as

HomeRF and PhoneLine Networking appear more likely to succeed because they are fully supported by heavy-weight PC and SW companies such as Microsoft, IBM, 3COM, Intel and others.

The Yankee Group predicts that when the technology is introduced in 1999, its price tag will likely to be on the high end of home networking alternatives. Spread spectrum technology is still expensive and lacks economies of scale in this application. Best price estimates for 1 Mb/s systems are $150 to $200 for a controller and under $100 for plug-in adapters.

8.5 Home Phoneline

Unshielded twisted pair copper wire has become the dominate medium for LANs in business offices due to its low costs and easy installation. Current systems operate at 10 Mb/s or 100 Mb/s over such wiring and standardization has already begun on 1 Gb/s version. However, this is not "regular" telephone wiring but Category 5 wiring with tighter twists and special dielectric constants specially designed for data applications. Although many newer homes are being wired with CAT-5 wires, most are not. Further, the 10BaseT LANs in business offices utilize two pairs of wire (for transmit and receive) dedicated for data in a point-to-point configuration. For many homes, there is only one pair of telephone wires, often daisy chain or random tree connected and already in use for telephone service. While it might be tempting to require rewiring the house with multi-pair point-to-point CAT-5 wires, this would not be practical, cost-effective or, as it turns out, even necessary. Technology now exists for providing high-speed networking over virtually any type of phone wire and does so without interfering with the underlying phone service.

The Home Phoneline Networking Alliance (HomePNA) was formed to develop specifications for interoperable home-networked devices over telephone wiring. The importance that the HomePNA is placing on standards is seen in their mission:

> The Home Phoneline Networking Alliance (HomePNA) is an association of industry-leading companies working together to ensure adoption of a single, unified phoneline networking standard and rapidly bring to market a range of interoperable home networking solutions.

The alliance recognizes that open standards will be essential for broad market acceptance of home networking products and is working closely with IEEE

802.3 and the ITU. The Home phone line technology is based on existing Ethernet technology but with a new physical layer. Essentially, the HomePNA is a 1 Mb/s ethernet running over phone lines. It is fully compatible with the ethernet MAC layer standard (IEEE 802.3 CSMA/CD) and meets applicable FCC regulatory requirements. This should greatly facilitate adapting existing ethernet software and hardware to operate over home networks.

By being designed to run over the existing telephone pair, this technology can be installed and operated at 1 Mb/s in 99.5 percent of homes, supporting computers and devices located up to 500 feet. A future version of the specification is expected to operate at 10 Mb/s, a good match for ADSL and cable modems. A design objective of the 10 Mb/s system is that it interoperate with the 1 Mb/s version on the same network. Another objective is that it coexist and interoperate with HomeRF, described in section 8.3.

Initially, the technology will be marketed only within the U.S. because of regulatory constraints in other countries and because the U.S. has the largest population of multi-PC homes. It is expected that this concept will expand to other countries as the market matures and regulatory constraints are lifted country by country.

8.5.1 Home Phoneline System Concept

Leading companies from the personal computer, consumer electronics, peripherals, communications, software, and semiconductor industries are developing the specification for Home Phoneline. The founding members of the Alliance are 3Com, AMD, AT&T, Compaq, Epigram, Hewlett-Packard, IBM, Intel, Lucent Technologies, Rockwell Semiconductor Systems, Tut Systems. Today, there are over 70 member companies.

This technology enables a 1 Mb/s LAN using existing in-home telephone wiring. It supports the complex, random-tree type of wiring typically found in the home as shown in Fig. 8.3. This networking solution uses only the single pair of an existing phone wires to make its connection and operates with any telephone or other appliance using the existing home phone line wiring infrastructure.

The initial specification, released in October 1998, adopted a 1 Mb/s LAN technology from Tut Systems. A proposed backward-compatible specification designed to achieve throughput of 10 Mb/s is being specified and contributions and calls-for-paper are ongoing. The Release 2.0 specifications (10 Mb/s rate) are expected in 4Q 1999.

No new wires and simple and friendly installation is the concept adopted by the alliance. It does not require any hubs, routers, splitters, filters or ter-

Figure 8.3
Phoneline Ethernet
Topology

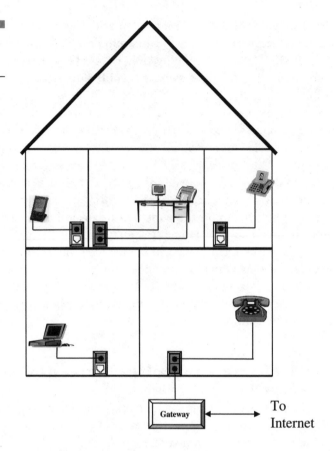

minations. Initial products will include both stand-alone adapters for use with devices having a 10BaseT interface and PC network interface cards (NICs) for direct connection of PCs. The initial specification allows up to 25 PCs, peripherals and network devices to be connected to the network.

To promote these solutions, HomePNA will provide field certification and interoperability test suites and will serve as a forum for technological and consumer issues. Certified products will be eligible to use the HomePNA brand logo and marketing programs.

8.5.2 Phoneline Access Requirements

Specifications for Home Phoneline were designed around the following objectives and conditions:

- Must operate simultaneously with telephone service and ADSL.
- Must comply with FCC Part 68.
- The network must be secure and the data must remain private; it must not be accessible to neighbors or anyone outside the home.
- Must operate in an environment of changing transmission characteristics. For example, going off-hook can change the data transmission characteristics and impedance of the phone wiring. Line impedance and noise will change as appliances are plugged into the network.
- Must operate with varying levels of signal noise as electric appliances are turned on adding to noise levels on the phone wires.
- Must operate with the varying signal attenuation that occurs within the random tree network topology of home wiring.
- Must operate in random and unspecified wiring topologies.

8.5.3 Phoneline Network Architecture

The overall Phoneline architecture is shown in Fig. 8.3. Potentially, every RJ-11 phone jack in the home can become a port on the home network as well as a phone extension. Phoneline home networks can act as a high-speed backbone. The specification's intention is to create an Ethernet-compatible LAN over the random-tree telephone wiring structure found in nearly every home. The system does not require any hubs, routers, splitters, filters or terminations. 10Base-T interface modules and PC network interface cards (NICs) will be initially used for direct connection of PCs to the in-home telephone jack.

8.5.4 Physical Layer

One of the most straightforward methods for operating multiple services over a shared medium is Frequency Division Multiplexing (FDM). With FDM, each communications service is assigned a unique frequency spectrum serving that particular service. Each interface device is equipped with a frequency-selective filter. In that way, devices using one type of service can exchange information without interference from other services.

The following frequencies are already in use on telephone lines:

- Standard voice occupies the range from 20 Hz to 3.4 kHz in the U.S. (slightly higher internationally)
- ADSL services occupy the frequency range from 25 kHz to 1.1 MHz

Phoneline networking was therefore assigned the frequency range above 2 MHz. The 1 Mb/s specification occupies the passband frequency range between 5.5MHz and 9.5 MHz. These passband filters attenuate frequencies below 5.5MHz very rapidly, hence there is no interference with DSL services or traditional phones.

Full rate ADSL utilize splitters outside the home with only the voice service passed on the inside wires servicing the telephones. However, the newer ADSL lite operates with no splitters, but instead passes the data signal onto the inside wiring. For this reason Phoneline networking was designed to operate above the ADSL frequency band. This means that consumers with ADSL service can access the Internet while using the phone and while using the home network to send a file to their printer.

8.5.4.1 Operation at 1 Mb/s. To achieve 1 Mb/s operation, encoding techniques are employed that place multiple data bits into each pulse. Patented line coding mechanisms were integrated into the receiver port to adapt to varying noise conditions. The transmitting circuit can vary its level of output signal strength based on the local conditions, and both the transmitting and receiving circuits continually monitor line conditions and adjust their settings appropriately.

Shared access to the phone line is implemented using IEEE 802.3 CSMA/CD (Carrier Sense Multiple Access/Collision Detect) methods.

8.5.4.2 Future Higher Speed Physical Layer. The HomePNA is targeting a 10 Mb/s specification based on field tests of working implementations. The impetus for developing a 10 Mb/s home LAN has to do in part with high-speed ADSL and cable modems. If ADSL can deliver 6 Mb/s, then it makes sense that the home wiring should keep pace.

The Release 2.0 specification planned for release in 1999 provides speeds of 10 Mb/s while remaining backward compatible with the 1 Mb/s Release 1.0 specification. This presents a difficult design challenge. Although a 10 Mb/s ethernet technology has been available for several years for operation in the corporate environment, it uses CAT-5 twisted pair cabling specially designed for data transmission. Homes in North America often have the sloppy wiring of the do-it-yourselfer and a wide range of wire types. While this presents a formidable task to designers, the availability of low-cost, high-performance signal processing makes solutions possible over the vast majority of in-place phone wiring.

Higher capacity can be achieved over most existing phone wiring using spectrally efficient modulation techniques in the 2-30 MHz band. Such schemes can encode up to 8 bits per symbol achieving a data rate reaching 100 Mb/s.

8.5.5 Phoneline: Services, Advantages/ Disadvantages and Cost

Below is a snapshot of applications being considered for the Phoneline networking environment:

Internet Sharing: Simultaneous access to the Internet from multiple users and PCs will become *de rigueur* in few years.

Peripheral Sharing: While multiple PCs in the home may be desirable, requiring multiple printers, scanners, etc. may not be.

File and Application Sharing: Phoneline networking can enable multiple users to share files and applications. This could be especially useful to SOHO users.

Entertainment: Multi-player network games perhaps enhanced with video and audio conferencing, either within the home or over the Internet.

Voice and Video-over-IP: With the maturing of Voice and Video-over-IP, all connected devices can establish voice calls or video calls over the home network without making use of the underlying analog telephone service on the phone wire. This can be done to other devices on the home network for intra-premises calling or through a home gateway to the analog telephone service. If ADSL or cable modem service is available, these IP voice calls can be made through the Internet to other users whose CPE also supports Voice-over-IP and to standard telephone users through IP/PSTN gateways provided by the network operators.

8.5.5.1 Advantages. The advantages of using phoneline for home networking are summarized below:

- The emphasis of the HomePNA on standards and interoperability should lead to greater user acceptance and lower costs.

- Data communications over phone wires is a cost-effective and proven technology for users requiring no costly rewiring of the home.

- Home phoneline networking (in the Release 2.0 specifications) will operate at the speeds of ADSL or cable modems.

8.5.5.2 Disadvantages

- Compared to HomeRF, phoneline requires tethered connections.

- Phone jacks are not ubiquitous in the home as power outlets. Additional phone jacks and wiring may be necessary especially in older homes.

8.5.5.3 Cost The cost of connecting most existing devices to the home network is expected to be less than $100 per device. Home Phoneline network interface cards (NICs) are expected to cost substantially less. Eventually this technology may be integrated into PCs, telephones, printers, or terminal equipment.

With its first release, the HomePNA's technology operates at distances of at least 500 feet between nodes. Based on typical home wiring analysis, this represents a home of up to 10,000 square feet.

8.6 Coax-based Home Networks

With over 70 percent of U.S. homes subscribing to CATV, coax cable is commonly available in most homes. Coax offers enormous bandwidth, at least 500 MHz and often 750 MHz or 1 GHz. As such, coax cable could be the basis of extremely powerful home networks. Operating a home network over coax holds the potential for tight integration of video and communication applications and of fulfilling the promise of the convergence of TVs and personal computers.

8.6.1 Coax-Based Home Network Architecture

A typical coax-based home network is shown in Fig. 8.4.

The central controller or switch manages channels on the internal network. The switch inserts channels received from VCRs, cameras, DBS receivers, etc., and modulates them onto unused or undesired channels of the signal from the cable company. All nodes on the network receive this composite signal. In this way an environment is created where there is no reason to dedicate a DBS receiver, DVD deck, etc. to a particular TV set. A VCR playing can be seen from any TV set in the house. Similarly, video cameras installed for security, baby monitoring, etc., can monitored from any TV set (or PC with video adapter card). The switch also provides the important function of ensuring that video and data do not travel outside the house.

For data applications, the switch also manages and creates the equivalent of a cable modem network within the house. This would provide interconnections between PCs and with computer peripherals or, through cable modem or ADSL interfaces, to the Internet.

Figure 8.4
Coax-based Home
Network

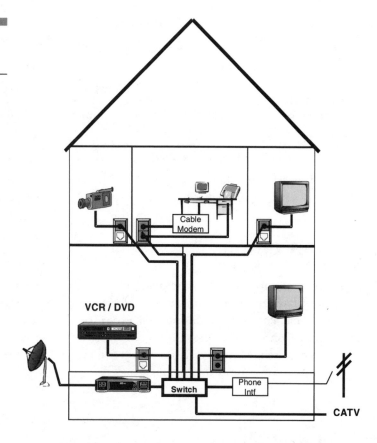

A cable-based home network could also allow interworking between PCs and television sets. For example, using a remote keyboard and the TV set as a display, one could surf the Web, run PC games, etc. from any room with a TV set.

IP voice can be carried over this data network for room-to-room calling. Gateways could be provided for connections to telephone lines to local exchange carrier or carried over cable modem or ADSL to the Internet.

8.6.2 Advantages and Disadvantages

Compared to other home networking alternatives, use of coax provides certain advantages:

- Enormous bandwidth up to 1 GHz
- Integration with video entertainment applications
- Relatively good shielding from RF ingress and egress
- Potential of tight computer, telephony and video integration

Some of the disadvantages of coax for home networking are these:

- Relatively few coax jacks are installed in homes. Hence additional wiring may need to be installed.
- Few standards have been developed for coax-based home networking systems beyond what is dictated by the CATV specifications.

8.6.3 Costs

Products providing this complete functionality over a coax-based home network have recently been announced. Initial prices seem very attractive. For example, Peracom has announced a coax-based home network product called HomeConnex. The suggested retail price of a complete system including video switch, PC interface, interfaces for TVs and video sources, and RF keyboard and mouse is under $300.

8.7 And the Winner Is . . .

At this point it is too early to predict which networking technology or medium will dominate the home networking market. Each provides unique advantages and best fits certain applications. Coax-based systems are the only good fit for entertainment video applications. Powerline systems may provide the only realistic interface for home automation applications. Given its dominance in the business data networking environment, phoneline-based systems may best fit data and voice applications. And, of course where portability and mobility are needed, wireless is the only choice.

The best guess at this point is that the dominant application for any particular user with determine the type of home network selected. Of course combinations of these networks are also likely to occur with gateways in between.

9

Services and Advanced Applications

9.1 Introduction

As the Internet grows at its phenomenal rate, so do opportunities for advanced communication applications. It is these applications that ultimately will drive the deployment of broadband access networks.

Solutions are becoming available with both business and "lifestyle" needs in mind. The success of the Internet should not be all that surprising considering we now live in a society that demands more to make our lives easier, more practical and better connected. This trend, coupled with changing social attitudes and the home PC revolution, were the catalyst that made it happen. People have grown tired of using their PCs merely for games or as a storehouse for useless information. Communication tools are being designed to fit home, office, travel and leisure, and advanced technology must be available to support this range of applications. Let's face it, being wired to the net is the "in thing" of the 1990s. The term *virtual* accentuates the present reality and its realization is the product of the 1990's.

Each day the Internet expands to include millions of Americans. What was once a technological tool utilized by a handful of scientists is now accessible to anyone. For instance, students can tap into information to complete papers or perhaps embark on virtual field trips. Launching a new business is not nearly the monumental task it was as entrepreneurs get the information they need to market their wares in any market. With the advent of on-line pharmacies, elderly citizens can now obtain the medications they need to lead productive lives with the click of a mouse.

Creative investments made right now will play a key role in tomorrow's advancements to the benefit of all United States citizens allowing them to become more competitive in the world. To remain on the cutting edge, we must promote and enhance partnering between government, school systems, families and industry. Here, high-speed access can help achieve this partnership as we enter the new millennium.

9.1.1 Organization of this Chapter

In this chapter we describe generic aspects of services and applications under investigation by the research community. Services are briefly described to show the fundamentals of the advanced applications being pursued.

Sections in the chapter cover the following:

- Service aspects as described by the ITU and other standard bodies
- A snap shot of the nuclear family application for the next millenium
- Legacy applications
- Present and future activities in application development

9.2 Services Application Definition

We are all aware of the services and applications the entertainment industry has been defining. It is very important that we distinguish the term "service" and "application." ITU and other standard bodies have defined services very narrowly. Services are classified as Constant Bit Rate, Variable Bit Rate, Available Bit Rate, etc. Applications, on the other hand, are defined as sessions used by a subscriber to perform a task. With that in mind, then services and applications can be generically defined.

Service is what the network must provide to the end-user so that the application can be run successfully and correctly.

Application is a session running on a PC or other home electronic appliances. Examples of applications are a video conference call, a voice call, downloading a video clip and voice over the Internet.

A network supporting Constant Bit Rate services can handle voice application successfully. Voice applications are not well-matched to the Internet because the Internet does not today provide a Constant Bit Rate, or similar service.

9.3 Services Aspects

The ITU recommendation I.211 describes all aspects of broadband services. It is a generic model of services one might anticipate. The I.211 recommendation lays a foundation for the broadband services the industry is developing for operation over the Internet and other networks.

Various business and entertainment elements have been folded into broadband services during the last few years as network operators continue to seek a profitable broadband business case.

This section describes services and their mapping to applications as defined by ITU. The service concept is then described in the context of service categories and QOS (Quality of Service) as applied to ATM.

9.3.1 Service Classifications

Two main service categories have been defined by ITU. This classification does not take into consideration the location or the implementations of the functions either in the network or in the terminals and hence Internet developers and entrepreneurs are able to use the generic definition and develop applications based on service classifications as defined by ITU. ITU defines two service classifications:

1. Interactive services

2. Distribution services

9.3.1.1 Interactive Services. For interactive services, three main subcategories are defined:

1. Conversational services
2. Messaging services
3. Retrieval services

9.3.1.1.1 Conversational Services. Conversational services in general provide the means for bi-directional communication with real-time (no store-and-forward) end-to-end information transfer from user to user or between user and host (e.g. for data processing). The flow of the user information may be bi-directional symmetric, bi-directional asymmetric and in some specific cases (e.g. such as video surveillance), the flow of information may be unidirectional. The information is generated by the sending user or users and is dedicated to one or more of the communication partners at the receiving site. Examples of broadband conversational services are video-telephony, videoconference and high-speed data transmission.

9.3.1.1.2 Messaging Services. Messaging services offer user-to-user communication between individual users via storage units with store-and-forward, mailbox and/or message handling (e.g. information editing, processing and conversion) functions. Examples of broadband messaging services are message handling services and mail services for moving pictures (films), high-resolution images and audio information.

9.3.1.1.3 Retrieval Services. The user of retrieval services can retrieve information stored in information centers provided for public use. This information will be sent to the user on demand only. The information can be retrieved on an individual basis. Moreover, the time at which an information sequence is to start is under the control of the user. Examples are film, high-resolution image, audio information, and archival information.

9.3.1.2 Distribution Services. For distribution services, two main subcategories are defined:

1. Distribution services with user control
2. Distribution services without user control

9.3.1.2.1 Distribution Services with User Control. Services of this class distribute information from a central source to a large number of users. However, the information is provided as a sequence of information entities (e.g. frames) with cyclical repetition. Thus, the user has the availability of individual access to the cyclical distributed information and can control the start and order of the presentation. Due to the cyclical repetition, the information entities selected by the user will always be presented from the beginning. Examples are Pay Per View, Video On Demand, Remote Education/Training and Tele-medicine.

9.3.1.2.2 Distribution Services Without User Control. Services in this subcategory provide a continuous flow of information which is distributed from a central source to an unlimited number of authorized receivers connected to the network. The user can access this flow of information without the ability to determine at which instant the distribution of a string of information will be started. The user cannot control the start and order of the presentation of the broadcast information. Depending on the point of time of the user's access, the information will not be presented from the beginning. Some examples of these services include broadcast services for television, audio programs and electronic newspapers.

9.3.2 ATM Service Architecture

The above services described by ITU are not necessarily ATM specific. However, ITU selected ATM as the target solution for broadband ISDN services. Hence, we will briefly describe services in the context of ATM being the underlining technology. This applies to ADSL, LMDS, APON, and ATM-centric cable modems.

9.3.2.1 Service Categories. Service categories are the foundations upon which a designer develops applications. It is the expectation the designer has of the network to accurately develop an application. There are four service categories defined by the ITU and adopted by all national standards organizations:

- Constant Bit Rate (CBR)
- Real-Time Variable Bit Rate (rt-VBR)
- Non-Real-Time Variable Bit Rate (nrt-VBR)
- Unspecified Bit Rate (UBR)
- Available Bit Rate (ABR)

These services above demand certain QOS and traffic characteristics from the network. The network performs services such as routing, admission control, and resource allocation for each of these service categories. For real time traffic, there are two categories—CBR and rt-VBR. The spacing between ATM cells must be controlled and minimized by the network in order to provide real time service. Near real time (nrt-VBR), UBR, and ABR traffic behaves in a non real-time fashion and it can vary, depending on the category the user requests from the network. Therefore, a service category is built from parameters the network must provide. A set of QOS parameters shown below defines a service category. They are defined as these:

- Peak-to-peak Cell Delay Variation (CDV)
- Maximum Cell Transfer Delay (Max CTD)
- Mean Cell Transfer Delay (Mean CTD)
- Cell Loss Ratio (CLR)

9.3.2.1.1 Constant Bit Rate (CBR) Service. This service is intended for real-time applications. This service is an applicable for voice and video applications. The ATM cells carrying voice or video information must maintain very short and bounded jitter (Cell Delay Variation) by the network. Cells that are delayed will reduce the value to the application. Some examples of applications for CBR are:

- Interactive Video   (videoconferencing)
- Interactive Audio   (telephone)
- Video Distribution   (television, distributed classroom)
- Audio Distribution   (radio, audio feed)
- Video Retrieval   (video on demand)
- Audio Retrieval   (audio library)

9.3.2.1.2 Real-Time Variable Bit Rate (rt-VBR) Service. Real time VBR service is similar to the CBR service and intended for real-time applications, i.e. those requiring tight delay and delay variation. This service is also applicable for voice and video applications. Unlike CBR service, the sources are expected to transmit at a bursty rate varying with time. Statistical multiplexing gain is assumed for this application. Cells that are delayed by the network beyond the specified parameter will reduce the value to the application.

An example of a real-time VBR service is a CBR application for which the end-system can benefit from statistical multiplexing by sending it at a variable rate, and can tolerate or recover from a small but non-zero random loss ratio (VBR voice is a possible application).

9.3.2.1.3 Non-Real-Time Variable Bit Rate (nrt-VBR) Service Definition.
This non-real-time VBR service is intended for non-real-time applications that have bursty traffic behavior. The parameters that define this service are Peak Cell Rate, Sustainable Cell Rate. This service may be applicable to large file transfer applications. Cells that do not conform to the specified traffic will reduce the value to the application and experience cell loss ratio. Example applications for non-real-time VBR include:

- Response time critical transaction processing
- Airline reservations
- Banking transactions
- Process monitoring
- Frame relay interworking

9.3.2.1.4 Unspecified Bit Rate (UBR) Service Definition.
The UBR service is intended for non-real-time applications; i.e., very bursty traffic and not requiring tightly bounded delay and delay variation. Examples of such applications are computer communications applications, such as file transfer and email messages. UBR sources are expected to be very bursty and therefore the service supports a statistical multiplexing gain among sources.

Like email applications today, UBR service does not specify service guarantees. No numerical commitments are made with respect to the cell loss ratio experienced by a UBR connection. Example applications for UBR include:

- Interactive Text/Data/Image transfer
- Banking transaction or credit card verification
- Text/Data/Image Messaging:
  - Email, telex or fax transaction
  - Text/Data/Image Distribution
  - News feed, weather station/ pictures
  - Text/Data/Image Retrieval
  - File transfer, library browsing
  - Telecommuting

9.3.2.1.5 Available Bit Rate (ABR) Service Definition.
Depending on the network traffic resources, the ABR service characteristics provided by the network change during the life of the connection. A flow control mechanism is

specified to support several types of feedback to control the source rate of information flow in response to changing network available resources. An end-system that adapts its traffic to the feedback (i.e., for reducing its traffic) will encounter a low cell loss ratio and also obtain a fair share of the available bandwidth network resources.

ABR service is well suited for support of the web browsing applications. An ABR connection is established by the end-system and specifies to the network the maximum required bandwidth (in peak Cell Rate) and a minimum usable bandwidth (in Minimum Cell Rate). The Minimum Cell Rate may be specified as zero. The bandwidth available from the network may vary and will be based on information gathered by a remote maintenance cell scouting the network from end-to-end.

Examples of ABR applications include:

- Any UBR application for which the end-system requires a guaranteed QOS
- Critical data transfer (e.g. defense information)
- Super computer applications
- Data communications applications requiring better delay behavior such as remote procedure call, distributed file service or computer process swap/paging

9.3.2.2 Internet Service Category Dilemma. The Internet, like cable TV and telephone networks, was designed for service-specific applications. It is for mainly connectionless services over UBR. Sophisticated browsers and their ease of use elevated the Internet to a state where some referred to it as the information superhighway. This image of the superhighway is different than the vision promoted earlier by the standards and telephone companies. The real superhighway integrates all communications, voice data and video. Some developed applications over the Internet and integrated the services but the quality remains and will remain poor unless major changes, mainly the concept of QOS categories, are adopted in the Internet infrastructure. Otherwise voice over the Internet, or more precisely Voice over IP will remain to be no more than Ham-radio quality (if that).

The Internet is plagued with access delay and it is often slow when downloading files or video clips. This slow speed access is turning off a lot of potential users to a point that it may threaten its growth. ADSL and cable modems and others will play a major and positive role to remedy this frustrating experience. Assuming the servers and Internet backbone are also equipped with high-speed technologies, the deployment of ADSL and cable modems will be like changing the highway speed limit from 55 mi/hr to over

11,000 mi/hr. This is similar to driving from coast-to-coast in less than 20 minutes, or downloading a 16-M bit file in about 3 seconds. This definitely addresses the access speed problem on the Internet. Unfortunately, this is not the whole story. The concept of QOS in the Internet network is practically non-existent today. There is nothing inherent in the Internet that prevents developing QOS. Presently it is simply a missing feature. Therefore, high-speed access alone will not solve the multiservice problems. QOS and associated service classifications, as defined by ITU, is analogous to a modern highway discipline: The road must be well maintained, with priority or a special lane given to passing cars, yielding to an emergency vehicle, or with special lanes designed for HOV. And, of course, a highway patrol to monitor traffic. This sort of discipline is necessary if the Internet is to behave as the information superhighway. One cannot travel coast-to-coast in 20 minutes if the highway is disorganized, i.e., trucks are blocking the road or there is no passing or wide enough lanes. In the interim, with a cost of bandwidth dropping and assuming a 100-lane highway, one will be able to fully utilize his or her high-speed access. Sooner or later, however, as more and more users access the network, it will eventually become congested. Internet2 community embarked in the development of a QOS aware network this year referred to it as Q-bone. The Internet Engineering Task Force is also developing related QOS specifications

9.4 Application Domain

In this section, various applications are described. It is assumed that the Internet network will eventually be QOS-aware. Several federal agencies are funding extensive research to the industry on how best to implement QOS that can accommodate the stringent requirements of NASA, NIH, and others, in terms of performance. The three key applications classifications are described below:

1. Legacy applications
2. Ongoing development
3. Application under extensive research

9.4.1 Key Legacy Applications

When developing and standardizing a high-speed access cable, all aspects of the services described in section 9.3 should be considered. A more realistic approach in developing services will emerge first which makes business sense. The key to survival in a multi-service, multi-provider market is differentiation. From today's viewpoint, it appears that all players/operators, either alone or in partnerships, will eventually offer the same services. Today's "most wanted" list of applications are:

- High-speed Internet access
- Digital VOD or NVOD
- Digital broadcasting
- Telephony over cable (wireline and wireless)
- Work at home
- Telemedicine
- Distant learning
- Video telephony
- Home shopping-"Virtual Mall"
- Digital audio/CD-ROM on demand
- Local information services
- Electronic games
- Telemetry

9.4.1.1 High-Speed Internet Access. High-speed Internet access is the key application that operators are banking on to build a rational business case. If there were a killer app, Internet high-speed access fits the definition accurately. The growth of the Internet has been documented in every journal and news media and hence it has the attention of every business executive in the communication industry, as well as all governmental agencies, including the White House.

9.4.1.2 VOD or NVOD. Video On Demand and near video on demand was originally thought to be "killer applications." VOD application is defined as the ability to check-out any movie electronically, watch it at home and have full VCR remote control capability. NVOD application is similar to VOD without the VCR control capability. In NVOD, only selected movies are played and repeated periodically (every five minutes or so) to accommodate the user's timetable. DAVIC developed a complete set of specifications addressing the VOD/NVOD applications. Most marketing studies concluded that VOD market and its periphery applications will be embraced on a large scale. Several factors were responsible for delaying the introduction of VOD to the market. They were priorities in developing strategies within the network operators to deal with the Telecommunication Act of 1996 (Local loop unbundling confusion); access technology choices; the Internet frenzy; RBOCs occupation with long distance market; and the risk involved in building such a huge and expensive infrastructure. As more and more ADSL and cable modems are deployed in the network for Internet access, VOD/NVOD will surface again and may take a central role. The revenue associated with the VOD market is important in justifying a business case.

9.4.1.3 Digital Broadcasting. Digital broadcasting will be of particular interest to the cable operators. A digitally compressed video signal can increase the HFC network capacity 10 fold. This allows cable operators the extra capacity to deliver NVOD movies and even provide broadcast premium programming on demand, for example HBO.

9.4.1.4 Video Telephony Over Cable. This new revenue application is of particular interest to the cable operators. Telephone service does not require a lot of bandwidth from cable networks, but it does impose a very short delay and delay variation requirements on the cable system. If long delay is encountered, then an echo will result disrupting the conversation. An expensive echo canceling technique can correct this problem. Telephony over cable is difficult to integrate, either in the cable modem, ADSL or other high-speed access described in this book. The problem the cable operators face is one of reliability and privacy. Cable operators are hinting toward implementing Voice over IP per DOCSIS™ Rev 1.1 specification (to be released in 1999).

9.4.1.5 Telemedicine. Telemedicine is the ability to provide high quality video telephony service interactively for the purpose of transmitting digitized medical data, X-ray images, and others. The communication will primarily be between hospitals, labs, pharmacies, etc. Other applications presently emerging that will help in a routine yearly physical consult with medical experts, and video care to the home. Special appliances are already on the market performing these and similar functions.

9.4.1.6 Home Office/Telecommuting. Home offices are linking the worker with the computer at the business home office and provide the user with the processing speed and processing power. It creates the environment of a virtual office. Telecommuting is growing at 12 percent CAGR.

9.4.1.7 Distant Learning. Distant learning is the ability to bring educational experiences of the campus environment into the home interactively. In its present form, it is not fully interactive. Internet2 is actively working on a virtual classroom that is fully interactive.

9.4.1.8 Video Telephony. Video telephony is the use of TV or PC screen to see persons or other images on the screen while talking. Video telephony (the death of the blind date) can be further segregated into the following:

- Communication with relatives "video visit"
- Alarm security monitoring
- Merchandising product

9.4.1.9 E-commerce. The virtual mall is where companies can market their products and services effectively. The application varies and may include interactive communication between a buyer and a seller. High quality video may be required for high-end items.

E-commerce is expected to become a $200 billion dollar market. Amazon.com phenomenal logarithmic growth became an icon. The company opened for business in July 1995, and now it is worth $22.1 billion (as of January 1999). On the sales menu were about 4.7 million titles for the sale of Books, CD, videos and games. The company is still loosing money but the sales growth is astonishing. Sales for 1995 jumped from $511 K to 3Q ending with $85 million.

9.4.1.10 Digital Audio/"CD-ROM On Demand." Users currently pay from $10 to $30 to purchase a CD-ROM title. Network providers can deploy CD-ROM servers to allow users to access high quality stereo "titles on demand." There will be no need for a local CD-ROM drive, and users could sample titles before purchasing them. Software developers would enjoy lower distribution and packaging costs, along with increased exposure and automated updating (especially valuable with reference-based titles).

9.4.1.11 Local Information Services. These services include multi-media-based classified ads, yellow and white pages, restaurant guides, "auto traders," local library, and matchmaking services, etc. Such advertising will become a viable source of revenue. The Internet is evolving into an interactive multimedia platform. This will, no doubt, nurture and accelerate direct marketing to the home.

9.4.1.12 Electronic Games. Some cable operators, and service providers like AOL already provide electronic game machines and deliver the software games through downloading to game machines. With directional capabilities, games can become interactive and play between different users on the network. This application applies to adults as well as children who are eager to find a partner to play chess or popular card games.

9.4.1.13 Telemetry. Telemetry application will play a major role when high-speed access, technologies are deployed. This application does not need high-speed access but will compliment the home service and help in developing the business case.

There are several potent sub-applications:

- Remote sensing and remote monitoring of gas, electric, and water utility

- Allowing utility companies to monitor usage frequently and to remotely initiate and disconnect service during peek hours at a cost saving to customers

- Provide burglar and fire alarm services and video monitoring

9.4.1.14 Phone Doubler™. Another interesting feature that is becoming popular is the "Phone Doubler" feature. This feature applies to the conventional connectivity to PSTN via the legacy modem (28.8 or 56 Kbps). While the end-user is logged-on to the Internet, incoming calls from the PSTN are forwarded to the user notifying him/her of an incoming call by a keypad graphic on the monitor. Upon clicking on the keypad icon, the phone conversation is carried by a PC microphone and speakers if voice over IP application is equipped. This happens without terminating the Internet session.

9.4.2 Ongoing Development

In the previous sections, we described the legacy key applications. In this section, we will examine the applications that are under active development mainly by the Internet2 community. The intention is to give the reader a snapshot of the most popular applications under development rather than extensive technical details that are yet to mature.

9.4.2.1 Emerging Applications. The emerging applications scenarios envisioned by NII (National Information Infrastructure) could best be described by reviewing the 10 applications crafted by the Cross Industry Working Team (XIWT). These applications depicts a snap shot and a way of life a nuclear family will function, be entertained, and do business in the next millenium.

XIWT was inspired by President Clinton and Vice President Al Gore's vision of a National Information Infrastructure. XIWT crafted a Functional Services application model and a Reference Architecture model. The Functional Services model describes the NII's building blocks, and components; particularly an "enabling services layer" that must exist to facilitate the rapid development and deployment of applications and the integration of all components. These enabling services include common capabilities needed by most of the applications and specialized services for specific application domains such as health care, finance, or manufacturing.

9.4.2.2 Applications Per NII/XIWT Definition. Ten applications were defined by XIWT. Some of the applications shown below either exist today or most likely will emerge in the near future. They are intended to illustrate how people, with different background, educational, and personal needs could benefit from the NII. These scenarios are just a few of the applications targeted for use in the NII. There is a vastly broader range of possibilities, most of which are difficult to imagine today. The people of the United States, and of the world, to which they are linked, will imagine and try out ideas for economic and human development, collaboration, and creativity that would seem wildly improbable if they could be articulated today. The 10 applications defined by XIWT are:

1. **Homing:** The multimedia application illustrated a fully interactive dialog with home, office, and school.

2. **Businesses Unit:** This Multimedia application described an interactive video email and video communication for business applications.

3. **Home Entertainment, Information, and Shopping:** This application describes a user surfing the network for leisure video activities (VOD) and shopping via the "virtual mall."

4. **Intelligent Transportation Systems:** This application described an Intelligent highway system with video surveillance cameras and roadway sensors to collect information to reach a destination using satellite positioning services and personal portable information devices with video display capabilities.

5. **Senior Citizen Use:** This application illustrates a video visit scenario via a large-screen, high-resolution computer display, and other access where users become information providers and show each other prerecorded video information.

6. **Starting a Business:** This application shows how small business can be established via video conferencing and accessing government databases, which provide both information and services on an immediate basis and allow online interstate registrations, online legal

services, and worldwide protection mechanisms for intellectual property.

7. **Middle School from Home:** This application illustrates a computerization and data networking in schools, with high-speed connection into the NII and audiovisual and networking "virtual classrooms" composed of students and teachers at different locations.

8. **Telemedicine:** This application utilizes a high-speed networking for videoconferencing, large file transfers, and real-time computer visualization of data and networked supercomputing resources for medical and educational applications.

9. **Government Services:** This application illustrates how government electronic service entities provide personal services using fast databases. These governmental information databases are or will be available to the public with easy-to-use navigational capabilities.

10. **Law Enforcement:** This application shows how a wireless network, both terrestrial cellular in populated areas and by low-earth-orbiting satellite with emergency video and image transmission will be used by law enforcement to apprehend suspects.

9.4.2.3 Applications per NCNI (Internet2). Over the last decade, North Carolina has been striving to create a framework for developing its academic and industrial information technology base. North Carolina Network Initiative is one of the handful of regional organizations around the country under the Internet2 umbrella. Others include California Research and Education Network 2, the Metropolitan REN serving Chicago area, the Houston Area Computational Science Consortium, and New York State Educational and Research Institution.

NCNI is a next-generation information technology, networking and Internet program with a focus on research, applications development and infrastructure of the next-generation technologies.

Clearly, one of the primary goals of the program was to target emerging technologies and applications and position this knowledge to facilitate a competitive advantage for the universities and businesses participating in NCNI. The universities currently include Duke University, North Carolina Sate University (NCSU), UNC at Chapel Hill, MCNC, and Wake Forest Road. Industrial partners include Cisco, IBM, Nortel and Time Warner Communications. Recently Alcatel and Interpath joined as partners to the consortium.

The N.C. Research and Education Network (NC-REN) was established in 1985. In its infancy, the Network focused on bringing interactive video and high speed data networking to few universities of North Carolina, and now the project has blossomed to a incorporate 34 private colleges as well as the state's four medical schools into the web.

Below are typical advance services that are under development by NCNI partners. Most of the applications that are under development by the various universities around the country are similar in scope and focus. Henceforth only NCNI application initiatives will be examined.

9.4.2.4 NCNI Applications. Applications that under active development by NCNI are summarized below. These applications will soon become part the Internet2 release.

9.4.2.4.1 OC3mon Analyses. The NCGN is fully instrumented with a passive network analysis system known as OC3mon. OC3mon uses optical splitters to send a portion of the signal to a specially configured PC, which records and condenses IP traffic data at speeds up to OC3c. The collected data is processed and transferred to another system for storage and analysis.

9.4.2.4.2 SoX Video. The Southern CrossRoads Networked Video Initiative is a regional Internet2 project to develop and implement standards-based video networks and middleware to support advanced multimedia communications throughout the member region and Internet2.

9.4.2.4.3 CS Nanomanipulator. The UNC nanoManipulator project is a virtual reality interface to a scanned probe microscope (SPM) constructed by UNC-CH computer scientists. The application allows chemists, biologists, and physicists to "see" the surface of a material sample at atomic (nanometer) scale and "feel" the properties of the surface through the use of force-feedback. This configuration requires only modest bandwidth (<1 Mbps), but demands very low and predictable latency (on the order of 20 ms) for the tracking and force-feedback necessary to allow teleoperation of the SPM in sample-modification mode.

9.4.2.4.4 Radiation Oncology. This application technology, which is designed for clinical implementation of planning and delivery of three-dimensional conformal radiation therapy, is ready for clinical trials, and a consortium of Carolina's academic radiation oncology departments has been formed to undertake cooperative clinical trials using three-dimensional treatment planning software.

9.4.2.4.5 Video-Over IP. One of the most powerful emerging application under intense scrutiny and interest is Video-over-IP, as it supports a variety of video applications, ranging from distance learning and collaboration to conferencing and video-on-demand. High quality interactive video conferencing will greatly increase the productivity and competitiveness of business as the network leverages expertise, resources without regard for time and distance. Video-over-IP is expected to have an impact that will redefine applications, require new connectivity, and bandwidth, and create new markets and businesses.

A Video-over-IP program has been initiated by the NC GigaNet and involves a variety of players from industry and academe. Below is a summary of the project under development.

9.4.2.4.6 MBone Virtual Classroom—NCSU. VBEE (Video-Based Engineering Education) which, including enrollments through the National Technological University (NTU), serves more than 1,000 students per year. The VBEE program provides an opportunity for practicing engineers to earn a Master's degree in engineering via distance education.

NC State has developed a "virtual classroom" model using a combination of Internet-based desktop conferencing (MBone) with multi-way audio, video and shared whiteboard for lectures and office hours, and web and "course lockers" (shared file space) for distribution of course materials and information. Workstations tied in to the NC State student-computing network have been set up at each site so that the remote students have access to exactly the same computing environment and on-line resources that are available to on-campus students. Circuits and logic labs identical to those on campus have been set up at the remote sites.

Seven engineering courses (three at the Master's level and four at the sophomore level) have been delivered from NC State to students at UNC-Asheville. On-site NC State faculty teach the introductory engineering courses, while discipline specific courses (e.g., circuits and digital logic) are provided via distance education.

With the MBone virtual classroom, courses are delivered live to these students in the evenings. In an evaluation of the first MBone course offering, the students reported that they greatly preferred the live delivery to videotape because of the opportunity for interaction that it afforded.

The key to success with the MBone virtual classroom model is that it supports the various types of interactions needed between the students and the teacher as naturally and as effectively as possible. Traditional videoconferencing provides high quality video, but typically involves expensive facilities, which require scheduling well in advance.

9.4.2.4.7 Teleconferencing. UNC-CH expects deployment of desktop teleconferencing to nearly every desktop on campus within two to three years. The campus data network can handle speeds up to MPEG2. The fully developed H.323 teleconferencing standard is being targeted as the standard for campus, enabling multi-vendor compatibility at several key price points. UNC-CH has been testing H.323 systems for over a year, and expects to implement directory services, multipoint capabilities, and H.323 to H.320 gateways in academic year 1998–99. Demand for high quality conferencing (better than H.323) continues to grow as well and is being met with digital television services.

9.4.2.4.8 Digital Television. UNC-CH is following the FCC "guided" transition to digital television by implementing an SMPTE 259M transport and switching system on campus that supports professional broadcast quality services. This service is aimed primarily at campus video production centers and researchers with critical needs for quality. The network will be migrated forward to HDTV when those systems become available. UNC-CH is one of the first campuses to make widespread use of Dense Wave Division Multiplexing to achieve multi-gigabit fiber transport of uncompressed digital television signals.

9.4.2.4.9 Video On Demand (VOD). UNC-CH has been working with packet-based on demand video systems for four years. Systems from Starlight Networks, Precept, Real and others have allowed delivery of everything from 28.8 video to full MPEG2 streams directly to the desktop via the campus data network. This work was originally undertaken to assist the campus in planning its data network to accommodate streaming video, but numerous pilot projects have uncovered many issues critical to the successful deployment and management of production VOD systems. UNC-CH plans to implement production VOD services in academic year 1998–99.

9.4.2.4.10 High Quality Interactive Video-Over IP Project. Video conferencing and high performance IP networks have been a hallmark in North Carolina, and the project will build upon that experience base. NCREN has been providing high quality (NTSC) interactive video conferencing and classes since 1985. Hundreds of programs and users are involved in interactive video programming each month, and there is a rapidly growing need to expand these capabilities substantially as the universities begin to seriously address distance learning, remote education and collaboration.

Project VIPER (voice-over IP, evaluation and research) is championed by Duke University. The initial research between some IVY campuses concluded that members submit their NPA for further analysis of the demography make-up. The submitted datasets were evaluated with SAS software and it was determined that there was enough traffic between the schools to warrant further research. Several vendors have been contacted and have expressed an interest in participating in the project and a NCNG voice-over IP working group has been formed.

Initial research efforts will concentrate on quality of service issues, the differentials between production Internet and Internet2 traffic, signalling protocols and billing/adminstrative functions. There is a particular interest in the integration of SS-7 (Signalling System 7) into the research voice network.

9.4.3 Applications Under Research

There are various emerging network services and advanced applications now under active research by the Internet2 research community as well as the White House sponsored Next Generation Internet Consortium (NGI).

One of the primary goals of the NGI is to enable experimentation with the next generation of networking technologies. For example, high quality video-conferencing is a wave of the future that includes real-time services. The impact of this application should directly increase the number of Internet users in the next few years. Advanced technologies are emerging at an extraordinary pace and the introduction of new commercial services is keeping tempo. The focus is on the NGI to provide innovative applications that support scientific research, national security, distance learning, health care and environmental monitoring.

The principal agencies involved in this initiative are the National Science Foundation, the Defense Advanced Research Projects Agency, the Department of Energy, NASA, and the National Institutes of Health. Below are just a few of the potential applications:

9.4.3.1 Distance Education. Universities are now experimenting with technologies such as two-way video to remote sites; VCR-like replay of past classes; modeling and simulation; collaborative environments; and online access to instructional software. Distance education will improve the ability of universities to serve working Americans who want new skills, but who cannot attend a class at a fixed time during the week.

9.4.3.2 Health Care. Doctors at university medical centers will use large archives of radiology images to identify the patterns and features associated with particular diseases. With remote access to supercomputers, they will also be able to improve the accuracy of mammographies by detecting subtle changes in three-dimensional images.

9.4.3.3 Energy Research. Scientists and engineers across the country will be able to work with each other and access remote scientific facilities, as if they were in the same building. "Collaboratories" that combine video-conferencing, shared virtual work spaces, networked scientific facilities, and databases will increase the efficiency and effectiveness of our national research enterprise.

9.4.3.4 National Security. A top priority for the Defense Department is "dominant battlefield awareness," which will give the United States military a significant advantage in any armed conflict. This requires an ability to collect information from large numbers of high-resolution sensors, automatic processing of the data to support terrain and target recognition, and real-time distribution of that data to the warfighter. This will require orders of magnitude more bandwidth than is currently commercially available.

9.4.3.5 Biomedical Research. Researchers will be able to solve problems in large-scale DNA sequencing and gene identification that were previously impossible, opening the door to breakthroughs in curing human genetic diseases.

9.4.3.6 Environmental Monitoring. Researchers are constructing a "virtual world" to model the Chesapeake Bay ecosystem, which serves as a nursery area for many commercially important species.

9.4.3.7 Manufacturing Engineering. Virtual reality and modeling and simulation can dramatically reduce the time required to develop new products.

Other challenging application that will funded include:

NSF funding

- Radiology Consultation Workstation
- Distributed Positron Emission Tomography (PET) Imaging
- Real-Time Telemedicine
- High Resolution Imaging Telemedicine
- Remote Control Telemedicine
- Medical Image Reference Libraries

NIST Applications

- Telerobotic Operation of Scanning Tunneling Microscopes
- Characterization, Remote Access, and Simulation of Hexapod Machines

NOAA Applications

- Advanced Weather Forecasting
- Chesapeake Bay Virtual Environment
- Distributed Modeling Laboratory for Mesoscale Meterorological Studies
- Real Time Environmental Data via the NGI

9.4.3.8 Low Orbiting Satellite. NSF is negotiating with a Japanese vendor to launch low orbiting broadband satellites to study the stratosphere so we can more accurately predict the weather. The existing high orbit satellite will exchange BB information with the low orbiting satellites. NSF believes that this technology enabler will open up a whole new science ranging from weather forecasting to atmospheric study in radiation.

9.4.3.9 Visible Human. NIH is enthusiastically working on a new project referred to it as "visible human." Two people (male and female) donated their bodies to the NIH. The cadavers were frozen and sliced into wafers at 4-mm increment. The microscopic information from the wafers was collected (terabits), crunched and recreated the human male/female image/anatomy. It gave all the intricate details of the physiology, nerve systems and brain structure, etc. With VR and 3D, surgeons, doctors and medical experts around the country are finding such information indispensable as it will lead to better diagnostics and cut down on surgical probing. The Internet will use multicasting to distribute the needed profile (e.g., of an organ) to surgeons and doctors

9.4.3.10 Gemini Project. The Gemini project funded mostly by the NSF, is taking shape. The project includes Australia, the U.K., the Americas, and others. Gemini will make it possible for astronauts around the world to observe telescopes remotely using high-speed Internet networking. To do this, one needs BW and near real-time operation (no compression to preserve quality). It is expected to be turned on in 1999.

9.4.3.11 Tele-Immersion.

Since its formation in 1990, Advanced Network & Services, Inc. has been dedicated to the advancement of education by expediting the utilization of computer network applications and technology by means of:

- Successfully demonstrating and disseminating the applications and technology
- Catalyzing adventurous experimentation and the creation of critical infrastructure and applications
- Extending the evolving best practices through studies and experimentation
- Participating in the creation of relevant standards and public discussions of this subject

Tele-immersion has the potential to significantly change educational, scientific and manufacturing paradigms. It will enable users in different locations to collaborate in a shared, virtual environment as if they were in the same physical room. For instance, participants would interact with a virtual group at a conference table, estimating what would be conceivable in a physical room.

It could be possible that the participants share and manipulate data, simulations and models of molecular, physical, or economic constructs. In addition, they could take part in the simulation, design review, or evaluation process. Consider for example, mechanical engineering students or industrial engineers collaborating to create a new bridge or robot arm through tele-immersion. Group members would be able to interact with other group members while sharing the virtual object being constructed.

Tele-immersion applications require advances in Internet infrastructure because of their high bandwidth, low latency, and time-dependent synchronous communications characteristics. Its potential will never be appreciated without high-speed networks which consists of advanced protocols (RSVP and Multicast, e.g.).

A well-coordinated research and development effort is needed on a number of fronts. Tele-immersion applications will require significant extensions to current Cave Automatic Virtual Environment (CAVE) technology. The areas which require improvement are tracking and rendering human interfaces that enhance the shared presence and manipulation experience, as well as shared work tools for communications and collaboration. The integration of real images into virtual environments to enable the simulation of realistic shared presence will be very important.

10
Access Performance Aspects

10.1 Overview

Today's network and infrastructure is unable to cope with the demand put by new broadband services and bandwidth hungry applications. The Telecommunication Act of 1996 coupled with the explosion of the interactive services are all the incentive network operators need to want to compete in the upcoming multi-billion dollar market. The two competing giants are the RBOCs and MSOs. The RBOCs (the $100 billion dollar access market) are almost five times the size of their $22 billion dollar cable competitors. They are determined to spend billions of dollars to deploy these new broadband networks based upon infrastructures such as HFC or Fiber to the Curb (FTTC) architectures.

Both the RBOCs and the MSOs have unique infrastructures in place in both landline base and switching equipment to manage their traditional services. Leveraging on the landline infrastructure to carry broadband and interactive services is presently the most technically effective and rational solution and meets the time-to-market criteria. For the

RBOCs, copper wires to the home is ubiquitous and hence the ADSL modem makes business sense. For the MSOs, coax cable lines is the physical medium RF landline infrastructure and therefore the most optimum solution is to deploy cable modems. There is every reason to believe that the consumer will enjoy this upcoming battle and reap the benefit of a true spirited competition in action. We can expect a cable or a telco operator in the near future, luring us by offering attractive packages, and throwing in POTS access charges for a mere few dollars. This may be wishful thinking, but we at least hope it will be similar to the present situation that makes customers feel wanted when negotiating with Sprint, AT&T, MCI, and soon the RBOCs, for long distant service.

Network performance will first be outlined as a prerequisite to describing the present status of NII network and other critical issues facing it. It makes no difference if a consumer installs an 8 Mbps ADSL modem if the connected IP network, for example, cannot deliver this data rate to the desktop, due to network congestion, lack of component performance and/or if the website server can only handle a T1 rate to all connected customers. In fact, in the short and medium term, an ADSL subscriber will unlikely notice any measurable difference in terms of speed if he or she is connected to a typical commercial website. The subscriber is unlikely to boost his or her pipe to much higher speed in the near future.

The word performance, understandably, has different meaning to different people. The generic definition of performance is *"the manner in which a network behaves"*. In this chapter we focus only on performance aspects that concern the industry in the access arena as well as technology comparison aspects. Since this book deals exclusively with access, our focus will be on network performance aspects that impact access in general and technology access comparison in particular.

The comparison will address the various technologies that are under active deployments: mainly ADSL and cable modem.

10.1.1 Basic Definition of Performance Terms

Below is a very generic definition of performance terms agreed to by various standards organization and the industry in general. The definitions are intended to simplify the complex nature of the subject at hand.

Availability: Is the probability that a system is operational at the time it is needed.

Component availability: Is a measure of availability of a single network component.

Network availability: Is a measure of availability of a network as seen by end-users.

Reliability: Is the degree that one can depend on something to perform as expected. In other words, it is the probability it will perform to expectation. Software reliability is the probability of a failure free execution for some period of time.

Serviceability: Is the ability to install product and to quickly identify and resolve problems.

Fault Tolerant/FT (hot standby): Is an effective means for increasing availability. For FT to work in the Internet environment, the network component failure must be unambiguous. Distributed/robust network protocol works against FT philosophy. Internet network node motto is "It takes a licking but keeps on ticking" (e.g., if a node is malfunctioning or imperfect it stays on-line until it is completely dead).

10.1.2 Organization of This Chapter

This chapter is organized by the following topics:

- Network performance aspects and access impact
- Services under consideration by the RBOCs and MSOs
- Brief overview of ADSL technology
- Brief overview of Cable modem
- ADSL cable modem comparison (analyze the strength/weaknesses of each solution)
- Who are the likely winners?

10.2 Network Performance Aspects

Increasingly, the emerging Internet-based information infrastructure is being used by public and private sector entities for "mission critical" applications. As the vision of the information age continues to unfold, today's application will transition to tomorrow's widely used application. Users and service providers will have limited alternatives to fall back on for commerce, healthcare, government services, and other important applications. No one can predict all the applications that will become dependent on information networks, or their specific dependability requirements. Yet it is desirable that the emerging information infrastructure be sufficiently dependable to support a wide range of application requirements in an efficient and effective manner. To that end, the following critical issues and questions will be explored:

- What are the performance requirements?
- Who will create the specifications for end-to-end dependability?
- What are the roles of the industry and government in defining and ensuring a dependable network of networks?
- How will dependability be measured, and how will such measurement results be shared with providers and users?

The network focus will be on IP-based performance aspects as perceived by NII (National Information Infrastructure) proponents. In that context, Internet, Internet2, and NGI is perceived to be the network that converges all services including voice, video, and data communication

10.3 Performance Requirements and Infrastructure Evolution

The information infrastructure has evolved both structurally and in terms of usage. The changing face of telecommunications has advanced from linear public switched telephone networks (PSTN) to a "network of networks" (NON) infrastructure. This evolving infrastructure is distinguished by a fast growing mix of legacy and new competitive telecommunications service providers, diverse in "last mile" access networks of copper over ADSL, cable modem, and fixed wireless and others described in previous chapters. The Internet, in particular, has greatly accelerated the interconnection of companies in a business environment, thereby accelerating electronic commerce as well as the application of information technology to all areas of the economy.

Networks are shifting from being voice-to-data centric to data-to-voice centric network. Voice becomes just one application among the many. In the interim, PSTN-IP gateways will flourish as new network elements. Telecommunications, or today more accurately communication, will enable convergence of telecommunications with cable/broadcasting, information, and entertainment industry. The rationale for the new emerging networks are:

- Lower cost technology
- Lower barriers to entry
- Growth in use of computers and data communications
- Growth in wireless mobility
- Ubiquitous on-line access
- Explosion of Internet
- Deregulation

10.3.1 How Many Nines are Enough?

The NON infrastructure raises concerns of uncertainty, diminished reliability, and inability to meet the high expectations of PSTN customers who have become accustomed to 99.999 percent switch reliability. How many nines are needed for a NON infrastructure? Integrity of an end-to-end connection in such infrastructure depends on the integrity of the weakest link. "Internet Operators Group: Industry Initiative for a Robust Internet" highlighted the tension between cooperation for robustness and compromise on reliability. The common belief on the part of the telco who are trying to build data networks is that equipment needs to be "five nines" (99.999%) reliable in order to be functional. Others believe that this stringent requirement would be an inhibiting factor for building and deploying data networks.

Given today's market dynamics, the costs of obtaining network trustworthiness have been questioned. The market is not driving more robustness, because consumers seem to prefer functionality to reliability. Users may not be fully aware of performance needs simply because they already have a five nine's reliability connection. Low demand for this public good underlies the need for a non-market driven solution, or a combination of approaches. In the meantime, customers are also demanding end-to-end Internet service reliability and performance, while focusing on the value for money spent. It will be interesting to see if the telco operators maintain such a posture in the future where they will have to compete for customers against other providers. Traditional phone companies, with Bellcore support, insisted on ultra reliable networks. Since they operated in a more or less captive market, the added cost was simply authorized by the local utility commissions.

This, however, was mainly driven by the FCC categorizing reliability issues into fundamentally three components:

1. E911. A-must-have for the consumer or I will see you in court.
2. Commercial viability (three minutes per year downtime)
3. Consumer confidence

It must be noted that the FCC is not chartered to mandate network dependability. The 1996 Act was intentionally written vaguely so as to enable useful competitive forces to thrive between the Internet and telco network.

Network dependability, however, was initially chartered by the FCC in 1992 when the SS7 network went bust. It established the NRC (National Research Council) in which senior officers and industry experts issued the 1000 page report. In case of a disaster, like the SS7 outage, there are two mechanisms in place that would address a problem. Depending on the nature of the outage, NRC would deal with reliability on network operations, while NSTAC (National Security Telecommunication Advisory Committee) would deal directly with security and threats.

In that spirit, the FCC also asked the telco service providers to develop a telephony outage database. This database (at Bellcore) is used by all to share outage information for the purpose of improving the overall network performance. The Internet community does not have such an operation. Competition is very stiff prohibiting ISPs from sharing valuable outage databases. For example, why would AOL declare their network outage, or its duration, to the competition? In fact, this declining-good-will of not sharing the information needed to improve the situation will worsen as more sophisticated applications requiring QOS and other call treatment evolve in the network. It will take a long time to remedy this situation and build trust among ISPs. Some ISPs, in fact, were also quoted as saying ''I cannot afford to provide end-to-end reliability.''

10.3.1.1 Distributed Architecture and Availability. In a fully distributed network, such as the Internet, network elements need not be fully redundant at the component level. The network hierarchy of the telco network (i.e., of class 4 and class 5 switches etc.) demand component reliability to maintain dependable operating functionality. This confusion in the industry is not well understood. The telco people are well aware of the mesh Internet architecture. The question is, will such architecture survive the vast Internet growth of fueling this furnace and be scalable at the same time.

For an end-to-end reliability, five nines for a network component (such as switch, or a router) is an overkill. TCP/IP is well suited to handle distributed architecture and in fact it was conceived from the outset. A graduate student had routinely turned off the light when leaving his Lab at night (which also turned off the router), but no one noticed this automatic rerouting until measurements were made later in the year when traffic was rerouted periodically and was traced back to that router. The question of network performance is much more fundamental than that. For example, what happen to the quality of voice, or more specifically latency, in a meshed network when rerouting? What about interdependency of elements of the information infrastructure, and the needed cooperation among them to assure proper operation of the overall system? There are a host of performance issues which the industry must address. Below are the most critical.

10.3.2 Network Performance Rationale

At the macro level, reliable network infrastructure is a strategic necessity for a competitive economy. The activities to improve elements of the network components and interdependencies of these components makes analysis and mitigation strategies difficult. There are many examples of this interdependency resulting in failure across the systems. This recently occurred in the San Francisco power blackout that started with a ground wire that was mistakenly left in place on the power grid. The result was failures for almost half a day affecting electronic mail and computer systems, air traffic control, electronic commerce systems, and the stock exchange, etc. Another area of concern for the attention of the State Department is the potential for widespread failure caused by terrorism through concerted attack. In its recently completed study on "Trust in Cyberspace", the National Research Council points out that the goal is an information infrastructure that can be trusted to do what is required despite disruption, errors, and attacks, and to not do what it shouldn't do.

Traditional approaches to improve the network reliability are not suited to the complex emerging information technology. System operators today police their systems for accidental and malicious malfunction. But the realm over which a system operator can work effectively is limited to the information and facilities owned by or controlled by operator. What is needed is to create a set of tools that allows separate organizations to work cooperatively to protect their systems.

The present rate of growth only hints at the ultimate size and value of the information infrastructure, but it will be unsustainable without better dependability and robustness. The emerging network, computer, and software intensive information technology infrastructure is efficient at meeting diverse communications requirements, but is complex and difficult to understand and manage with respect to dependability and robustness.

Assuring the dependability of the joint global information infrastructure in the presence of diverse threats will require concerted and cooperative action of multiple forms covering both technical and policy domains. The technical, economic and policy issues are multivariate across business boundaries and markets; thus there is real gain to common understanding and action. A mix of information exchange and cooperative research and development activities is required to deal with emerging issues and progress toward a satisfactorily robust and dependable infrastructure. Implementing solutions across the infrastructure will require cooperation and agreement of broad segments of the development, operations, and user communities. Without a more solid foundation of common understanding and commitment, it is unlikely that we will sustain the present growth of the "infrastructure-enabled economy" based on advances of the information-enabled infrastructure.

Dr. Robert Kahn, President of the Corporation for National Research Initiatives (CNRI) and Chairman of the XIWT Executive Committee, in a recent workshop on the subject, emphasized the need of collaborative cross-sector activities in achieving this goal, given the highly distributed, complex nature of this NII infrastructure, and the common interest in its dependability and robustness. The goal to advance the reliability and the dependability of the national information infrastructure must be addressed using the following theme:

- Information exchange activities
- Consensus activities
- Collaborative R&D activities
- Government role

10.3.2.1 Information Exchange Activities. Information exchange is essential a variety of between various industrial sectors as well as between government, industry, and academia.

As the industry must try to achieve a more reliable and robust information infrastructure, a development mechanism must be in-place to share "best practices" in the development and operations of various aspects of the information infrastructure. Reporting and sharing of incident reports, bug reports, etc., would help developers and operators. Similarly, an "Indications and Warning Center", tracking anomaly activities in the information infrastructure and providing early warning to system operators, would be helpful. There is already activity in this area, through the NCC (a subset of the National Communication System), but linking to industry initiatives and broadening the activity to go beyond communications would have significant payoff.

The government shares its experiences in building information infrastructure, thereby providing exemplars of both methods for achieving robust information infrastructures as well as a forum for discussing some of the problems. The rationale is that, while industry participants may have proprietary concerns in sharing what they do to achieve a reliable system, government agencies may be able to share such information with a selected audience more freely. For example, why would AOL share its network failure (downtime) with the competition?

10.3.2.2 Consensus Activities. A number of areas should be defined to develop a consensus across a broad spectrum of the relevant community. This could help move towards a trustworthy information system.

10.3.2.3 Collaborative R&D Activities. Moving the community forward towards a more trustworthy information infrastructure will require considerable research and development. Furthermore, the nature of the problem implies the need for collaborative research to assure widespread adoption and deployment of those results requiring such adoption.

Collaborative R&D activities develop the needed new techniques for information infrastructure robustness in a complex heterogeneous system and then assure widespread adoption and deployment of those results requiring such adoption. Across the component areas of the information infrastructure, discussion is needed on a number of topics:

- System outages
- Performance slowdowns
- System component vulnerabilities
- Component security breaches
- Best practices, (policies and technology configuration)
- Security, fault detection and repair
- Systems oriented solutions

10.3.2.4 Government and Industry Role. Janet Reno, as a keynote speaker on a related subject, urged the industry to start working on addressing network attack, be it from insider's theft, organized crime, or terrorism. The Federal agencies are making proposals to fund research suggested by the industry. It is expected that partnerships be developed to eventually create specifications to be brought to standards bodies. To this end, the government will bring its experience to secure network integrity including R&D funds.

There are, however, several things that government should not do. These are among the most obvious and significant:

- Legislate prematurely
- Set standards or bureaucracies to enforce standards
- Build unique systems for government requirements

As a reminder, deregulation is going to be a long process, very complex to undo the regulation imposed during the last 76 years. So any new regulation must be weighed very carefully.

10.3.2.4.1 The Role of Business. Business greatly depends on the Internet for financial transactions and services. Robustness is important for effectiveness of these uses, but it is also associated with economic and security trade-offs.

In the banking industry, the Internet includes home banking, electronic payment and authentication, corporate systems of cash and treasury management, and trading. Application characteristics range from real-time trading and daily cash management to once a month electronic bill payment. Hence a real-time non-stop operation of the Internet is not crucial for operations other than trading. However, reliable, even if not continuous access to Internet communications is very important. The Internet had several desirable robustness attributes by virtue of being distributed, adaptive, flexible, and diverse in media and technology. However, it also faced multiple threats, such as eavesdropping, stealing valuable information, manipulating information, denying service, and intruder-caused failure. Predicting and countering Internet emergencies is problematic, because the definitions of such emergency, its damages, and repair costs are unclear.

In a recent study, Bellcore concluded that the potential for disrupting critical components has increased. However, the potential for a total network outage is low. Intrusion detection is usually not in real-time, however, intruder mindset has changed to include denial of service attacks. The range of targets is increasing and the impacts vary considerably. To resolve this, the industry needs to support enhancements to critical protocols, promote R&D for secure software development and promote development of security guidelines for service providers.

Business economics involves a trade-off between the necessary amount of preparedness and/or built-in robustness to cover all emergencies, and the affordable amount, which would handle most reasonable contingencies and emergencies. There is the need to balance security with system performance requirements. This balance includes such trade-offs as customer convenience versus security, trust versus technological and contractual safeguards, and management of risk and financial crime versus privacy rights/demands of customer.

10.3.3 Present Research Consortiums

The Internet is already a rapidly evolving international enterprise, and embodies a great deal of collaborative and international "management". The underlying communication infrastructure operates under the sponsorship of the ITU, as well as numerous national regulatory bodies. A new international body has been created to oversee domain names, and the Internet Engineering Task Force operates under a set of voluntary processes.

There are several ongoing efforts engaging industrial, academia, and government sectors in cooperative activities that address the complexity of achieving a robust information infrastructure. The effort will most likely define required courses of action and framework in areas important to achieving a robust NII.

A trustworthy system must be built from untrustworthy components. The technical challenge lies in building more effective system architectures, algorithms, and software. In the software area, the Center for Survivable Infrastructures (a joint undertaking of the University of Virginia, Carnegie Mellon University, and the University of California, San Diego) personifies such opportunity. The efforts include researching software architecture and simulation, information management and control, security, systems and analysis.

At Stanford University, a Consortium for Research on Information Security and Policy (CRISP) is a new undertaking created for the purpose of developing a better analytical and policy understanding of how to strengthen national and international information infrastructures. It is also researching how to mitigate the effects of malicious actions directed at those infrastructures. CRISP is a follow-on to a successful Stanford program that assisted in the work of the President's Commission on Critical Infrastructure Protection and other U.S. government organizations. CRISP researchers have also recently worked on other problems related to IT and national security, including export controls, cyberterrorism, and international law and information warfare.

10.3.4 International Participation

In this global economy era, information infrastructure performance and security is a manifestly international problem. Usage and therefore dependence is becoming global and many organizations are truly international. As a result, walls erected in the name of security are not practical: cyber attacks can move easily across borders, and adequate remedies will require a high degree of inter-state cooperation. CRISP has active programs with the potential for bringing together experts in pertinent fields from many countries. It will undertake to build an international constituency to address the problems of securing information infrastructures on a global basis.

The Corporation for National Research Initiatives (CNRI), in partnership with the Stanford University Consortium CRISP is establishing a cross-sector, cross-industry community of interest working collaboratively to improve the robustness and trustworthiness of the information infrastructure. The focus of the activity will be on the issues requiring the sharing of information across the various industry sectors and developing solutions to problems in building a robust "meta-system" incorporating the various components both vertically and horizontally.

In the next sections, an objective access technology evaluation is attempted. Background materials are first described to put the comparisons in perspective.

10.4 Likely Applications

It is no secret that the battle between the RBOCs and MSOs is brewing. The two different cultures are bidding for the same business. The RBOCs are conservative in their approach; long term planners; insisting on open interfaces; fully support and shape the standards; and demanding product reliability. Some say the MSOs are entrepreneurial, and more focused on quick returns to meet their short term business goal. To the MSOs, standards, open interfaces, and reliability are all well and good, but they are usually in the way. The MSOs' attitude on standards may be changing because they are now cognizant that the era of the multi-service, multi-provider is a reality and they must become more receptive to customer service, and to the new competitive market place.

The key to survival in this multi-provider market is differentiation. These are the services likely to be offered by both MSOs and RBOCs:

- High-speed data services (Internet access)
- Broadcast, one-way entertainment
- Telephony (wireline and wireless)
- Digital NVOD
- Work at home
- E-commerce

Convergence of these services permeating across the millions of U.S. households will change the way people live, work, and play. By-products of this convergence will be a major source of growth and opportunity for the U.S. economy.

The major differentiators, as mentioned earlier, are the following:

- Time-to-Market
- Quality of Service

Although time-to-market is important in the short term, it will not however, be the ultimate deciding factor. It is only important if other factors are satisfactorily met. Those factors are customer service and cost. If the RBOCs were able to lure a premium customer from a cable company, it becomes extremely difficult for that customer to switch back if the price and service is acceptable. The RBOCs enjoy a very good customer service reputation. That in itself will be a marketing tool they will use to fearlessly go after the cable customers who are primed to make the switch. When it comes to protecting their legacy market or to offering the new multimedia services, the RBOCs are more apprehensive of the IXCs like AT&T, MCI and Sprint. The recent AT&T, Time Warner, and TCI mergers, spells trouble for the RBOCs, who now have complete control of the access market. The IXCs have an excellent record in customer service, billing, and most of all, because of the long distance stiff competition, they are well experienced in marketing their products and services.

Having noted all of the above, the battle is by no means over for the MSOs. In fact, it is just starting. Cable operators are, of course, aware of this wake up call and are working feverishly to regain customer loyalty. The road may be rough, but they have been known to survive.

10.5 ADSL Technology Brief Overview

This section describes ADSL technology in enough detail to make the appropriate technical comparison with its cable modem rival.

ADSL was standardized in 1992–1993 by T1E1.4. The committee agreed to develop the technology based on the DMT (Discrete Multi–Tone) concept. Digital Subscriber Loop techniques allow, through the use of digital signal processing, to transport high-rate services over the existing copper-based infrastructure. The state-of-the-art allows the transport of signals of a few Mb/s over several kilometers of twisted pair cables while assuring a BER of the order of 10^{-7} or better.

The most important feature of ADSL is that it can provide high-speed digital services on the existing twisted pair copper network in overlay and without interfering with the traditional analog telephone service (plain old telephone service: POTS). ADSL thus allows subscribers to retain the (analog) services to which they have already subscribed. Moreover, due to its highly efficient line coding technique, ADSL supports new broadband services on a single twisted pair.

As a result, new services like high speed Internet and on–line access, telecommuting (working at home), VOD, etc., can be offered to every residential telephone subscriber. The technology is also largely independent of the characteristics of the twisted pair on which it is used, thereby avoiding cumbersome pair selection and enabling it to be applied universally, virtually regardless of the actual parameters of the local loop.

The asymmetric bandwidth characteristics offered by the ADSL technology (64–640 kb/s upstream, 500 kb/s–8 Mbit/s downstream) fit in with the requirements of client–server applications such as: WWW access; remote LAN access; VOD; etc., where the client typically receives much more information from the server than he or she is able to generate. A minimum bandwidth of 64–200 kb/s upstream guarantees excellent end–to–end performance, also for TCP-IP applications. These basic characteristics are reflected in two important advantages of the ADSL technology:

- No trench diggers are required for laying new cables, making it an optimal solution in advance of fiber deployment in the local loop.
- ADSL can be introduced on a per–user basis; this is important to the network operators for it means that their investment in ADSL is proportional to the user acceptance of high-speed multimedia services.

The mature ADSL product combines the benefits of the DMT and ATM technologies:

■ Full bandwidth flexibility: upstream and downstream bit rates can be chosen freely and continuously up to the maximum physical limits. At initialization, the system automatically calculates the maximum possible bit rate, with a predetermined margin. The service management system can then set the bit rate to the level determined by the customer service profile, thus maximizing noise margin and/or minimizing transmit power.

■ Full service flexibility: a random mix of services with various bit rates and various traffic requirements (guaranteed bandwidth, bursty services, e.g.) can be supported within the available bit rate limits.

10.5.1 ADSL: A Technology on Unconditional Twisted Pair

A copper twisted pair cable is the basic access connection of a telephone user to the telephone network. The frequency band used for the transport of the analog signal ranges from 300 Hz to 3.4 kHz. Thanks to the use of digital signal processing techniques, Digital Subscriber Loop (DSL) techniques are able to transport high bitrates over the existing copper twisted line simultaneously with the analog POTS. The basics of ADSL are also valid for other DSL systems that are currently available or planned, such as HDSL (High-speed DSL), SDSL (Single pair HDSL) and VDSL (Very high-speed DSL). HDSL offers 1.5 or 2 Mbit/s bi-directional data transport over 2 or 3 twisted pairs and is intended as a mere replacement for T1/E1 repeater lines in the distribution plant. Therefore, it is often referred to as "repeaterless T1/E1". SDSL is a single pair version of HDSL that targets to transport 1.5 Mbit/s bi-directional data over the full CSA (Carrier Serving Area) range. SDSL can also be considered as Symmetrical ADSL. The latter system (VDSL) is intended to provide very high bitrate services, >10 Mbit/s over distances not to exceed 1 km.

The ADSL transmission system offers an asymmetric capacity to the residential subscriber. In the direction towards the subscriber, the ADSL system provides a capacity up to 6.1 Mbit/s (and above up to 9 Mbit/s). The bandwidth in the opposite direction (upstream) is in the range from 16 to 640 kb/s. In general, the ADSL data rate is a function of reach and wire gauge as shown in Table 10.1.

TABLE 10.1

ADSL Line Capacity

Reach	Data Rate—U/D	Wire Size (AWG)
18,000 ft	1.7 Mbps/176 Kbps	24
13,500 ft	1.7 Mbps/176 Kbps	26
12,000 ft	6.8 Mbps/640 Kbps	24
9,000 ft	6.8 Mbps/640 Kbps	26

The transmission channel capacity depends highly on the twisted pair characteristics and suffers from a number of impairments. The frequency dependent attenuation and dispersion leads to inter-symbol interference. Moreover, as coupling exists between wire pairs in the same binder or adjacent binder groups, crosstalk limits the transmission capacity of the copper loop. Furthermore, some subscriber loops have open-circuited wire pairs tapped onto the main wire pair, called bridged taps. The existence of bridged taps in the loop plant differs from country to country and depends upon the cabling rules used in the past. Their presence causes reflections and affects the frequency response of the cable leading to pulse distortion and intersymbol interference. A loop can also be built up of wires with different diameters leading to reflections and distortion. Further, copper transmission suffers from impulse noise that is characterized by high amplitude bursts of noise with a duration of a few microseconds to hundreds of microseconds. It can be caused by a variety of sources such as central office switching transients, dial pulses and lightning. Also ingress noise from radio transmitters (e.g. AM radio stations) affects the transmission. Lastly, the impedance mismatch between the hybrid transformer, responsible for the split between transmitter and receiver, and the line impedance causes unwanted reflections. This problem can be resolved either by echo-cancellation or by separation of upstream and downstream transmission by means of Frequency Division Multiplexing (FDM). To encompass the cited imperfections of the copper twisted pair, a highly adaptive transmission system is needed.

The ANSI T1E1.4 standard committee has selected Discrete Multi-Tone (DMT) as the line code to be used in the ADSL transmission system. There are some in T1E1.4 who are trying to standardize yet other line codes. This is unfortunate because it would confuse the market and the consumer in particular. This attempt is contrary to the spirit of standardization especially when the economy of scale is expected to play a role in reducing its cost. Moreover, multiple line codes will limit the choice a consumer would have if he or she opted to switch service provider, assuming they own a specific coding of the ADSL modem.

DMT modulation consists of a number of sub-channels, referred to as tones, each of which is QAM-modulated on a separate carrier. The carrier frequencies are multiples of some basic frequency (4.3125 kHz). The available spectrum ranges from about 20 kHz to 1.104 MHz. The lowest carriers are not modulated to avoid interference with POTS. The bandwidth used for upstream transmission is considerably lower than that used in the opposite direction. The up- and downstream spectra can overlap if echo canceling is used. The alternative is the use of Frequency Division Multiplexing (FDM) in which case no tones are shared by up- and downstream transmission. Both alternatives are accepted by the ANSI T1E1.4 committee.

The transmit power spectrum is flat over all used tones. The number of bits assigned to a tone is determined during an initialization phase as a function of the transmission characteristics as well as the desired bit rate. During operation, adaptation of this bit assignment is possible to compensate for variations in line conditions. In order to improve the bit error rate in the ADSL system, Forward Error Correction (FEC) is applied. ANSI specifies the use of Reed-Solomon coding, combined with interleaving. The additional use of Trellis coding is optional, but may further reduce the BER or increase the SNR (Signal to Noise Ratio) margin of the system for a given BER.

In the ANSI standard, special attention has been given to a service specific interface, referred to as the "channelized interface". This interface is based on a specific framing structure that can provide a combination of ADSL bearer channels envisioned up to now. In the current version, seven bearer channels can be transported simultaneously over ADSL: up to 4 downstream simplex bearers and up to 3 duplex bearers. The three duplex bearers could alternatively be configured as independent unidirectional simplex bearers, and the rates of the bearers in the two directions (network towards customer and vice versa) do not need to match. The bearer channel data rates can be programmed with a granularity of 32 Kbps. Other data rates (non-integer multiples of 32 kb/s) can also be supported, but will be limited by the ADSL system's available capacity for synchronization.

The variable bitrate inherent to ADSL ensures, in principle, the flexibility to support new services with other bit rates than those envisaged presently and takes advantage of the decrease of the required bit rate. This holds for video services, specially with MPEG 2, images can be offered with reasonable quality at low bit rates from 1.5 to 6 Mbit/s. However, this inherent flexibility of the ADSL modem is limited by the definition of the framing structure combined with the definition of consistent physical interfaces. Therefore, it is advisable to offer a flexible transport mechanism that is future safe, adaptable to the varying demands in bandwidth and that can rely on a flexible switching technique. The ATM offers this flexibility and when the ADSL system provides an ATM interface, a gradual evolution can be planned for the provision of B-ISDN services towards the Residential Subscriber. B-ISDN envisages a full fiber based access network, which involves large investments from the telecom operator companies as the present access network is merely based on copper twisted pair. By providing ATM over ADSL, ATM-based services can be offered to the residential subscriber right away, while the access network is evolving towards a full fiber based network (FTTC, FTTB, FTTH: fiber to the curb, the building, or the home). Meanwhile for the subscribers and service providers, ATM cell transport over twisted pairs by ADSL technology remains transparent and emerging new services can be introduced rapidly and independently of the state of evolution towards a fiber-based access network. As the ADSL transmission system offers a large bandwidth to the residential subscriber, while still offering a reasonable upstream bandwidth capacity, it is specifically tailored to services that are asymmetric in terms of the bandwidth such as Internet services and VOD.

Internet Application.   Some identify the Internet to be the "killer application" that eluded the industry in the last few years. In any case, ADSL is well suited for Internet traffic, and specifically for multimedia services including voice, data, and video now everywhere on the World Wide Web. Downloading these services is tailor made for ADSL. The ABR service, specified in the ATM forum, will no doubt improve the traffic flow immensely.

VOD.   As video compression techniques improve, several VOD services could be offered simultaneously to the subscriber by means of the ADSL system. With an ATM-based ADSL system, the VOD service is independent of the state-of-video compression and full transparency is guaranteed between VOD server and VOD set-top unit to the residential subscriber. For the deployment of a VOD service, a mix of access network techniques (FITL, ADSL, coax) will be applied. When full ATM transparency is assured, there will be no need for interworking functions in the access network.

10.6 Cable Modem Technology Overview

The cable modem was described extensively in previous chapters. The overview below is intended to touch only briefly on topics that would be instrumental when comparing it with its ADSL rival.

10.6.1 Summary Description of the Cable Modem

A conforming cable modem has downstream data contained in one of the 6 MHz TV channels that occupy spectrum above 550 MHz. The upstream channel assigned band is between 5 and 45 MHz. The downstream uses 64 QAM modulation technique and can deliver over 30 Mbps data rate. Presently the upstream channel uses QPSK modulation technique and can deliver up to 2 Mb/s. Unlike ADSL, the cable modem must operate in a shared medium environment and hence a MAC is needed to mediate the upstream traffic. The MAC protocol is IP- or ATM-based in which data is received and delivered using ATM/IP packets as transport concept. Status and control information are looped back from the downstream to the upstream so each station determines its pecking order.

A variety of cable modems are now available in the market, and most use QAM for downstream and QPSK for upstream. Some cable modems divide the upstream into frequency channels and allocate a channel to each user. Others combine the two multiplexing methods. Unlike ADSL, cable modem rates do not depend upon coaxial cable distance, as amplifiers in the cable network boost signal power sufficiently to give every user enough signal power. These factors, however, are built in the MAC, and during initialization, time synchronization and power are aligned to the cable modem so it can function properly. In the upstream, cable modem capacity will depend on the ingress noise, and the number of simultaneous active users who are accessing the shared medium.

10.7 ADSL/Cable Modem Comparison

Comparing ADSL with cable modem is not an easy task. There are strengths and weaknesses in each of the technologies. The weaknesses of either will have more to do with the assumptions made, specifically, in service offering. Without a criteria yardstick, the comparison becomes subjective, and therefore unproductive. End-user expectations is the ultimate test. The consumer cares less about the underlying technology, be it ATM or Morris code. To the endures, the service and its ability to deliver the promised application is the only thing that matters. With that in mind, most agree that the services mentioned in chapter 9 are applicable and hence will be used as a guide to compare ADSL and the cable modem.

10.7.1 Capacity

Cable Modem. A cable modem user, in general, has little over 30 Mb/s for the downstream. This bandwidth is shared among 500–2,000 users. With proper traffic engineering, applications can be performed without service degradation. For the upstream path, the shared bandwidth is about 2 Mb/s. This again should not present noticeable service degradation for all asymmetric applications in a properly engineered network.

ADSL Modem. The ADSL modem operates in a point-to-point application. The downstream bandwidth is not shared, but the bit rate varies depending on the quality of the copper line and terminating distance. As shown in Table 12.1, the rate can be 1.5 Mb/s up to 6 or 9 Mb/s. Unless an ADSL subscriber resides in the rural area (farm), most users fall into the 6 Mb/s category. The ADSL modem can dynamically adapt it's bandwidth to suit its environment, even during the life of the call (being standardized). For the upstream, the un-shared bandwidth also ranges from 16 to 640 Kb/s. For all asymmetric applications, ADSL can perform without any service degradation.

For IP related traffic, most Internet WWW servers operate at 56 kbps and a few at the 1.5 Mb/s T1 speed. This is likely to stay with us for the short term. Therefore, the inherent speed for the cable modem or ADSL cannot take full advantage of its available rate. Users, however, will experience a noticeable performance in speed compared with the 28.8 kbps analog modems. For the short term, both the cable or the ADSL modem will operate adequately.

10.7.2 Throughput

Cable Modem. In the long term, the cable modem will exhibit service deterioration if a large number of users attempt simultaneous transmission. This may require re-engineering of the traffic more often than the norm. It will be based on the viewing habit, time of day, and new bandwidth hungry application that is yet to be defined but surely will come on the Internet. A conforming IEEE 802.14 cable modem, can hop onto other less congested channels, but in the end, the laws of physics prevail. Adding more upstream and downstream channels in the spectrum to meet the extra demand is technically feasible, but costly. The amplifiers in the HFC network need to be replaced or at least modified. The cable modem may also need replacement if its frequency agility feature is not robust enough.

ADSL. Assuming a 6 Mb/s rate, the ADSL modem is fully dedicated to a single user. The limited upstream bandwidth will become a problem if video telephony application becomes popular. Variable bit rate MPEG 2 and advances in video comparison technologies could save ADSL if the user is not demanding high quality video. The ADSL access node is also a multiplexer. As more and more traffic is added, re-engineering of the trunk capacity will be needed. A bigger capacity trunk, extending from the access node to the network, needs to be replaced to account for the traffic increase. Changing from DS1 to DS3 or to OC-3 is a way of life in the telco business environment. The procedure is routine.

Both the cable modem and ADSL can survive the unpredictable future asymmetric hungry applications with acceptable performance. For the cable modem, it would need more intensive upgrades engineering.

10.7.3 Scalability

Cable Modem. For cable modems, subscriptions can only be made after modernization of the entire cable network into HFC. The $225.00 upgrade cost is the incremental cost per home passed.

ADSL. The ADSL is more scalable. An operator can provide service to any copper-based customer regardless of his or her geographic location. Customer copper is simply rewired to an ADSL access box located at the central office. When enough people subscribe, then an access node would be economically justified and be deployed in that area. In this scenario, the ADSL has the advantage over cable modem.

10.7.4 Performance/Service Categories

Cable Modem. A conforming cable modem should be able to handle CBR, VBR and ABR services with a proper traffic engineering. It is not clear yet how ABR services will be performed in the cable modem. If ABR services are to be performed by the cable modem itself (as it should), then the resource management (RM) cells from all the active cable modems will unduly tax the precious upstream bandwidth resources. The upstream bandwidth is shared among all active users. It is expected that the headend will act as proxy to all the active ABR connections (see chapter 8).

ADSL. The ADSL can also handle CBR, VBR and ABR services. For ABR, the connection is point-to-point and hence it is not expected to have the cable modem limitation.

10.7.5 Security

Cable Modem. In a shared medium environment, the cable modem is more vulnerable to misuse, eavesdropping, and service theft. All signals go to all cable modems on a single coaxial line, creating serious prospects of intended or inadvertent wiretapping. Encryption and authentication will be important parts of both systems, but vital for cable modems.

ADSL. In an ADSL point-to-point architecture, eavesdropping will not be possible from a home premises. Copper lines are buried underground, and therefore, inherently secure. Wire tapping requires invading the line itself and knowing the modem settings established during initialization.

10.7.6 Cost

Cable Modem. The cable modem is expected to cost less than ADSL. Only one modem is needed to connect to a subscriber. The complexity of developing security modules on the MAC, and the extra hardware used in developing the MAC and RF tuners in the cable modem should be noted. Today's cable modem cost is from $300 to $500. Cable modems conforming to IEEE 802.14 are expected to be in that range.

ADSL. Two ADSL modems are required to connect a subscriber. It is expected that with the economy of scale (more than cable modem) and competition, ADSL costs will drop. The inherently lower network costs of cable modems compared to ADSL access systems may be neutralized by higher infrastructure costs incurred when upgrading to HFC. The telco will not have to bear a high modernization cost and can afford to offer an attractive package.

10.7.7 Voice Adaptation

Cable Modem. The cable modem can provide POTS. If POTS functionality is embedded into the MAC logic, then it becomes vulnerable to power outage. Cable phone with voice modulated over its physical layer is a more attractive approach. Voice-over IP will most likely be the long term solution as DOCSIS release 2 version becomes available.

ADSL. POTS over ADSL is provided over the physical layer. On power outage, the central office provides enough current to power the physical layer electronics and keep POTS operational. One can also implement voice-over-IP or voice-over-ATM supporting additional telephone applications.

10.7.8 Reliability

Cable Modem. Cutting a CATV line in the street or losing above ground cable in a windstorm will bring down all users on that line. A single noisy transmitter or defective cable modem on a shared bus will also bring down all users on that line. Amplifier failure can cause outage to the entire neighborhood. Each additional user creates noise in the upstream channel, and thereby reduces reliability. A well maintained HFC network would obviate these problems.

ADSL. A DSL modem operates on a point-to-point basis. Failure affects a single user only. ADSL itself suffers no degradation based on traffic or number of users accessing the network.

10.7.9 Internet Application Comparison Scenario

Present drivers for both ADSL and cable modem is high-speed Internet access. It is viewed as the market entry to jump start all other broadband services and application.

The White House is challenging the business community to structure the Internet so it can provide video multimedia access to the nation. The FCC vowed to support regulations in order to keep the Internet momentum in high gear. The Internet is both educational and entertainment, and will be instrumental in maintaining the U.S. competitiveness in this global economy.

Beyond the high-speed Internet access, both technologies address remote access for work at home and telecommuting, distance learning, and access for the millions of personal computers in place today and to be sold over the next 10 years. Below is a typical Internet access configuration for both the cable modem and ADSL.

10.7.9.1 Typical Internet Access for Cable Modem.

Figure 10.1 illustrates a typical Internet access configuration for the cable modem in an HFC network. The IP router is located at the headend. In the shared medium arrangement, the headend concentrates all IP traffic and sends it to the IP router. Proxy servers or cache memory are optional and can be located at the headend. They contain copies of popular WWW pages and texts. These copies, e.g., Netscape home page and CNN news-scripts, can be sent on demand to the requesting cable modem users with noticeable speed. This approach also mitigates congestion in the Internet network.

10.7.9.2 Typical Internet Access for ADSL Modem.

Figure 10.2 illustrates the alternative architectures for the ADSL modem. The access to the Internet network is using existing twisted pair copper telephone lines. The ADSL modems connect at both ends of a subscriber's telephone line that was originally used for POTS. The POTS splitter, not show in the diagram, transparently forwards the analog voice to the POTS CO in frequency below that of ADSL domain. The receiver (Home) ADSL modem, connects directly to the Ethernet port of a personal computer or a local Ethernet hub. The access switch serves to concentrate access lines into expensive router ports. A DS1(1.544 Mbps) line from the router to the Internet can support five or more continuous users or as many as 55 subscribers with 10 percent usage. A DS3 (45 Mbps) pipe could support up to 1500 subscribers.

Figure 10.1
Cable Modem/HFC Typical Internet Access

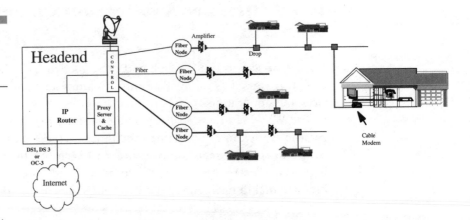

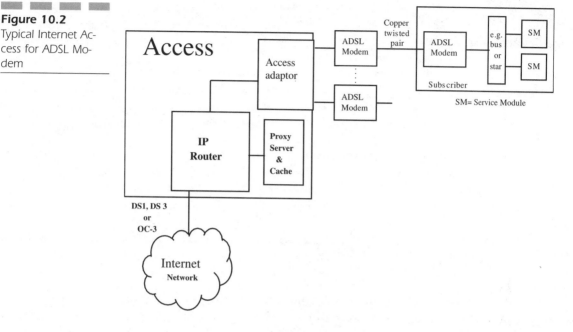

Figure 10.2
Typical Internet Access for ADSL Modem

It is most likely that the access adapter includes ATM switch fabric and from there connects to the various Internet service providers located in the Internet network. A proxy server or cache memory can reside in the access housing.

10.7.10 Cable Modem Market Size

There are about 180 million basic CATV lines in the world now. Table 10.2 shows the U.S statistics.

Most of the lines are old, one-way, and coaxial only. It was estimated that 6 million subscribers are currently served by HFC. All major U.S. CATV companies have upgrade programs underway. With the exception of the U.K. and Belgium (which have mostly completed upgraded programs), the picture looks similar worldwide.

10.7.11 ADSL Market Size

ITU estimated that there are about 700 million copper telephone lines in the world today, 70 percent of which connect to residences. The rest are business

TABLE 10.2

U.S. Cable Statistics

Home Passed	91 million homes	98% of TV households
Home served	56 million homes	61% of TV households
Growth of penetration	5% per year	
% of subscribers connected to 30–54 channels	60% of subscribers	
Rate of fiber deployment	100 miles/hour	
Average bill per month	$30	
Revenue of CATV business	$23 billion per year	
Revenue of Advertising	$3.5 billion per year	

Source: ADSL forum

and private lines. By the year 2001, the projection will exceed 900 million. In the U.S, as shown in Fig. 10.3, approximately 80 percent of these lines can operate with ADSL at 1.5 Mbps, and 50 percent can support rates of 6 Mbps or more (see Table 10.1).

Variable rate ADSL modems, with minimum speeds at 1.5 megabit, will enable connection to all users. When enough users subscribe to the service

Figure 10.3
Copper Line Distribution

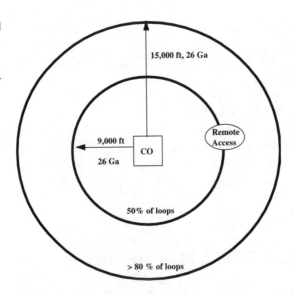

and a remote access node is economically justified, then the access could be re-located to that region. This will reduce copper distances and hence improve performance.

10.8 The Likely Winners

Based on the technical analysis above, cable modems and ADSL have comparable capabilities and can easily adapt into broadband/IP-based infrastructures. Serious deployment of both technologies, having similar capabilities, have begun. ADSL and cable modems deployments can rapidly grow from a few million to tens of millions.

By the sheer numbers of copper access lines, it is apparent that ADSL or xDSL will dominate the future market. Moreover, telephone companies are already connected to the entire customer base while CATV passes a smaller fraction. The economies of scale will play a major role in the ultimate outcome of who controls what share of the market. The ADSL Forum estimates that telephone companies will achieve 70 to 80 percent market share in the U.S. (a somewhat biased estimate).

When the untamed battle first started between the MSOs and telco, it was postulated that each will take 30 percent share of the other's market. The MSOs will provide cable phone and access charges to long distance carriers, while the telco will take 30 percent of the video market share. This theory is questionable. The cable operators can compete for the POTS services, particularly for a second line. The size of that market depends on the long term strategies of the cable companies to improve their customer service image, network management and reliability. Another problem the MSOs face is disfranchising. Eleven cable companies are serving the city of Philadelphia alone, so networking and billing become a real problem if all or some cable operators are to provide cable phones in that area.

The likely MSO strategy is to go after prime telco customers and offer attractive packages. The competition will be fierce. Cable operators must compete with the reputation of AT&T, MCI, and Sprint who have already started competing for the same share of this market. The RBOCs are more apprehensive of them than the cable operators. The most likely and profitable scenario for the cable operators is to make alliances with the newcomers or traditional telco operators and leverage both their strengths to compete effectively. TCI and two other cable companies joined with Sprint Corp. to bid on the wireless personal communication services spectrum. The friendliness of cable infrastructure to PCS, and Sprint's reputation in customer services and billing, made a good business match.

Telco executives feel their service reputation gives them a competitive edge over cable operators and will likely get 15 percent of the video market share. Bell Atlantic's recent poll of 500 of their customers found that 46 percent of cable subscribers said that they would switch their cable company if a similar service were offered by the phone company. Fifty six percent said they would switch if the service includes VOD.

CHAPTER 11

Standards

11.1 The Role of Standardization

Telecommunication standards will play an important role in this transitional period. Historically, the reason for standards was to enable the interconnection of national and international networks. Since then, other factors created the desire for harmonized networks and for cost reduction due to economies of scale. The standards, however, must be receptive to the market and attractive enough to be adopted by all (big and small organization or forums) so that products, services, and applications can be used world-wide.

Forums are becoming very popular in the 1990s. This can be traced back to the slow pace of ITU and that coupled with liberalization of the telecommunication market. Time-to-market is the key driver of the emerging forum. Forums are created by those eager to work together to insure interoperability of building next generation networks.

11.2 Access Related Standards

There are several standard organizations responsible for the development of access standards. Below is a summary of their activities and ongoing development.

11.2.1 ADSL/VDSL

11.2.1.1 Technical Subcommittee T1E1. T1E1 is an ANSI affiliated standard. T1E1 works with the mother T1 technical subcommittees, as well as with external standards setting bodies. There are six working group withion T1E1:

1. T1E1.1 Physical Interfaces and Analog Access
2. T1E1.2 Wideband Access
3. T1E1.4 DSL Access
4. T1E1.5 Power Systems
5. T1E1.7 Electrical Protection
6. T1E1.8 Physical Protection and Design

T1E1.4 develops standards and technical reports for systems and associated interfaces, for high-speed bi-directional digital transport via metallic facilities. The scope of work undertaken by the Digital Subscriber Line Transmission Working Group includes the development of standards and technical reports for transmission techniques, user interfaces, and interface functionality for transmission services providing access to telecommunications networks. T1E1.4 maintains a close and coordinated liaison, through T1E1, with other external standards bodies.

Work on standardization of ADSL started in T1E1.4 in the early 1990's with studies of the requirements for and the specification of ADSL systems. In the beginning of 1993, the Bellcore ADSL Olympics were held in the U.S. The purpose was to compare various technologies on prototypes. The DMT prototype performed better than the CAP system and a 64-QAM device. Based on these results, DMT was accepted as the (only) line code for ADSL. In 1995, the ADSL Issue 1 specification, T1.413, was released defining two categories of DMT-based ADSL modems (Categories I and II). ETSI contributed an annex to T1.413 to reflect European requirements. In 1996, there were heavy and time consuming debates in T1E1.4 on the introduction of CAP as a second line code. After rejection in the T1E1.4 working group, the creation of the CAP Ad-Hoc group to address an "Alternative ADSL specification" was accepted in the T1E1 plenary meeting. Meanwhile, the main group of T1E1.4 worked on a second version of T1.413 that was approved in 1998. Main additions to Issue 1 are support for ATM transport, rate adaptation at startup, the provisioning for rates of 8 Mbit/s and more, the description of reduced overhead and an enhanced embedded operations channel and the specification of a tighter psd mask to improve spectral compatibility with VDSL. In October 1997, the CAP/QAM proposal was released for letter ballot. At the time of this writing, it is unclear if or when this single carrier RADSL proposal will be approved.

Besides ANSI/T1E1.4 and ETSI/TM6, the ADSL Forum has been very active in the promotion of ADSL and the specification of many technical, non PMD-related, issues such as ATM transport, packet mode operation and network management. Since mid-1997, the ITU-T has started activities on ADSL in SG15, Q4, with G.DMT, G.Lite, G.HS (handshake), G.Test, G.CAP/QAM. Also within the ATM Forum, ADSL specific topics have been discussed.

11.2.1.2 VDSL. In T1E1.4, a VDSL standard project was defined in 1995. After early discussions on modulation techniques, a moratorium was established until sufficient progress was made on system requirements (bit rates, symmetry, noise, interfaces, operating conditions, etc.). In April 1996, the moratorium was lifted. Basically, four different transmission techniques have been proposed so far: SDMT, Zipper, CAP/QAM and MQAM (see preceding sections). Also in ETSI/TM6, a moratorium on modulation techniques was established. It will be lifted at the first ETSI meeting of 1998. At that moment, the system requirements document will be available (and frozen). DAVIC has standardized a short-range asymmetrical transmission system for use in a Fiber-To-The-Building or Fiber-To-The Curb topology. ITU-T has no activities yet on VDSL. The ADSL-Forum has focused till now on ADSL but will start an activity on network migration to VDSL soon. GX-FSAN is an organization of 10 major operators making recommendations on access systems, including VDSL.

11.2.1.3 ADSL Forum. The ADSL Forum was formed in late 1994 to help telephone companies and their suppliers realize the enormous market potential of ADSL. With nearly 300 members, the Forum consists of the key members of the communications, networking and computer industries. The ADSL Forum is an association of competing companies, which abide by various anti-trust rules.

The ADSL web site offers short explanations of all DSL technologies, the system environment for ADSL, and the basic features of the ADSL market.

At present, the Forum's formal technical work divides into the seven working groups:

1. ATM over ADSL

2. Packet over ADSL

3. CPE/CO configurations and interfaces

4. Operations

5. Network management

6. Testing and interoperability

7. Support to the VDSL study group

Each working group develops technical reports through a working technique called "Working Texts". As work is completed, the Working Text becomes a Technical Report, subject to membership approval.

The ADSL Forum has established formal liaisons to key standards bodies and working groups, including these:

- UAWG

- ATM Forum

- ANSI T1E1.4

- DAVIC

- IETF

- ITU

11.2.1.4 UAWG. The Universal ADSL Working Group (UAWG), was created by leading PC industry, networking, and telecommunications companies. UAWG's mission is to build on the present T1.413 standard and generate a create quick interoperability standard for Universal ADSL modem deployment. UAWG's main goal is to provide consumers with interoperability product and foresees Universal ADSL modems being a preferred PC modem technology by the year 2000. These are its major advantages:

- Universal, single standard for the PC and telecommunication industry

- Simple plug and play

- Performance—With a speed of up to 25 times faster than today's fastest modems.

- "Always On"—connectivity the Internet which facilitates consumer's daily life and enriched Web lifestyle

11.2.2 Cable Modem/HFC

Standardizing the cable modem for some cable modem companies is quite unique. In the past, the MSOs have formed partnerships with vendors and developed product that was unique, proprietary and service specific. Some MSOs believe that standards impede ingenuity. Although this is true to a certain extent, the advantages of developing a standards-based product far exceeds that of deploying a proprietary solution which sometimes may become the default standard. However, in a competitive industry where networks must be interoperable, a standard solution has been proven to be the best and most cost effective alternative. With a standard product, the economy of scale will further reduce the cost of operation, and open competition to the benefit of both operators and consumers.

With few exceptions, MSOs are not in the habit of attending standard organizations and, unfortunately, this continues to be true. CableLabs, which represents 85 percent of cable operators in North America, is doing an excellent job in the standard bodies and is providing valuable and critical inputs. However, direct participation from all MSO is still essential, not only to provide their local input, but also to protect their own interest in the technical or business environment. The technical issues related to HFC and cable modem are very complex, an honest technical debate is needed to flush out political decision or special agendas.

11.2.2.1 Organizations Affiliated with Cable Modem. There are several standards organizations, forums, and associations dealing directly or indirectly with standardizing or specifying cable modems over HFC. Among the most active are:

1. SCTE
2. MCNS
3. IEEE 802.14
4. CableLabs
5. ATM Forum
6. DAVIC

Direct and indirect liaison between these organizations, to great extent, is well established, and good communication flow is maintained. Despite some overlap in the charters, cooperation has been very good.

11.2.2.1.1 Society of Cable Telecommunications Engineers (SCTE). In August 1995, the Society of Cable Telecommunications Engineers was created to tackle interoperability and develop the appropriate interfaces. SCTE now hopes to use IEEE 802.14 advanced PHY when available. SCTE, however, is going beyond the cable modem standards and dealing with interoperability specification of HFC system in particular. SCTE is ANSI affiliated. Both MSOs and vendors are members of that organization.

The charter is to explore the need for SCTE involvement in the development of standards for digital video signal delivery through coordination of efforts with NCTA, the FCC and other related organizations.

11.2.2.1.2 SCTE Sub-Working Groups.   There are eight SWGs each dealing with specific topics. The SWG charters are as shown below.

Digital Video Subcommittee. The charter of this SWG is to create standards for developing for digital television services delivered by cable, including: video and audio services; data and transport applications; network architecture and management; transmission and distribution; encryption and access control.

Data Standards Subcommittee. The charter of this SWG is to advance the cable industry's interest in high-speed data delivery and develop standards for hardware interoperability.

Emergency Alert Systems Subcommittee. The charter of this SWG is to develop filings to the Federal Communications Commission Office of the Emergency Alert Systems. The subcommittee will inform the FCC about cable's role in this country's Emergency Alert System, focusing on the industry's capabilities, limitations, and cost effectiveness.

Maintenance Practices and Procedures Subcommittee. This SWG is chartered to develop recommended practices for the proper maintenance and operation of cable television systems.

Interface Practices and In-Home Cabling Subcommittee. This SWG approved 69 standards for specifications and test procedures by mid-year, and it had 12 standards in the proposed stage.

Design and Construction Subcommittee. This SWG is chartered to develop the Coaxial Basic Construction and the Fiber Optics Construction manuals. The SWG is also responsible for cost analysis of network upgrades and modernization of cable plant.

Material Management/Inventory Subcommittee. This SWG is working on automated material management and inventory management real-time computerized control. This SWG is also responsible for the bar code system for cable television product packaging and shipping as recommended in the MMI IP 001 standard.

11.2.2.1.3 Multimedia Cable Network System (MCNS). MCNS is an organization preparing a series of interface specifications for early design, development and deployment of data-over-cable systems. The specifications is open, non-proprietary and multi-vendor interoperable. MCNS contracted with Arthur D. Little, Inc. to develop these interface specifications. MCNS comprises four leading cable television operators:

1. Comcast Cable Communications Inc.
2. Cox Communications
3. Tele-Communications, Inc.
4. Time Warner Cable

Partners of MCNS include:

1. Rogers Cablesystems Ltd.
2. Continental Cablevision
3. Cable Television Laboratories

11.2.2.1.4 Participation. Interested vendors are free to participate and contribute without fee or any other restrictions. Participants, however, must execute the Information Access Agreement (essentially a non-disclosure and hold-harmless agreement).

11.2.2.1.5 MCNS Goal. The objective of MCNS is to have accelerated specifications for developing a cable modem. The goal is to eliminate intellectual property, if possible, and simplify and minimized the costs of developing a cable modem. This facilitates the manufacture of compatible equipment by multiple vendors. The protocols should be developed to support future upgrades and changes by negotiation at the session establishment of physical and higher-layer protocols. The specification must specify the protocols and algorithms to achieve seamless and reliable operation of the cable modems under all possible conditions, including robust fault recovery mechanisms and scalability of the algorithms.

11.2.3 Data Over Cable Interface Specification (DOCIS) Project

DOCIS project was prepared by Arthur D. Little, Inc. on behalf of cable industry system operators who provide service to a majority of cable subscribers in North America. The goal of the DOCIS project is to rapidly develop, on behalf of the North American cable industry, a set of communications and operations support interface specifications for cable modems and associated equipment over HFC. A reference model for a data over cable system has been established that includes the data communications elements as well as these needed operations and business support elements:

- Security
- Configuration
- Performance
- Network fault
- Accounting and management

11.2.3.1 Project Scope and Timeline. The scope of this project is described in the MCNS request for proposals. In summary, the project entails the development of interfaces between system elements organized into three phases:

Phase 1. Phase 1 deals with the data communications aspects of the data over cable system at the subscriber location and the headend. Timeline for Phase 1 specification release was July 1, 1996. Phase 1 interfaces include:

- The Cable Modem to CPE Interface (typically a personal computer).
- The Headend-based Cable Modem Termination System Interface. It would typically be connected to a backbone network of a server complex (e.g., local content servers, gateway server to the Internet, and servers for security, spectrum management, network management, accounting and configuration).

Phase 2. Phase 2 interfaces deal with the operations support system and telephone return interfaces. Timeline for Phase 2 specification release was September 1, 1996. These are the interfaces:

- The Data Over Cable System Operations Support System Interfaces are used to support: management of the network; performance; security; accounting; provisioning; spectrum management; customer service and network faults.
- Cable Modem Telco Return Interface is an interface between the cable modem and the public switched telephone network.

Phase 3. Phase 3 interfaces deal with the RF communications path over the HFC cable system between the cable modem and the cable modem termination system. Timeline for Phase 3 specification release was December 1, 1996. These are the interfaces:

■ The cable modem to RF Interface is a single-port RF interface between the cable modem and the cable system coaxial drop. The specification includes all the physical, link (MAC and LLC), and network level aspects of the communications interface. This includes RF levels and frequency, modulation, coding, multiplex, contention control and frequency agility.

■ The Cable Modem Termination System Downstream RF Side Interface and the Cable Modem Termination System Upstream RF Side Interface are two separate ports on the cable modem termination system equipment at the headend.

■ Cable Modem Termination System Security Management Interface.

11.2.3.2 Other Data Over Cable Related Documents

Data Over Cable Service Interface Specification Project

Multimedia Cable Network System Partners (MCNS Holdings L.P.)

Comcast Cable Communications, 1500 Market Street, Suite 3400, Philadelphia, PA, USA 19102.

Cox Communications, 1400 Lake Hern Drive, NE, Atlanta, GA, USA 30319.

TCI Technology Ventures, Inc. Terrace Tower II, 5619 DTC Parkway, Englewood, CO, USA 80111.

Time Warner Cable, 160 Inverness Drive West, Englewood, CO, USA 80112.

Cable Television Laboratories, Inc., 400 Centennial Parkway, Louisville, CO, USA 80027.

Continental Cablevision, Inc., The Pilot House Lewis Wharf, Boston, MA, USA 02110.

Rogers Cablesystems Limited, 853 York Mills Road, Don Mills, Ontario, Canada M3B 142.

Data Over Cable Service Interface Specification Project Manager and Consultants.

Arthur D. Little, Inc. Acorn Park, Cambridge, MA, USA 02140.

11.2.4 IEEE 802.14

IEEE 802.14 is a Working Group under the umbrella of IEEE 802 (International Electrical and Electronic Engineers). IEEE is an international organization which is affiliated with the US ANSI standard body, and also affiliated internationally through ISO/IIEC JTO.

IEEE 802.14 was a spin-off of 802.6 which, at the time was working on DQDB WAN standard. In the November 1993 meeting, a study group was formed to begin work on cable TV-oriented protocols in a more formal way. IEEE 802.catv study group was hence created and worked to develop the Project Authorization Request (PAR) as required by IEEE when creating a standard working group.

On July 14, 1994, the IEEE Project 802 Executive committee unanimously approved the formation of a new Working Group to produce standards for two-way communication over Cable TV and similar systems.

Attendance IEEE 802.14 is open to all. A meeting fee is imposed to anyone who attends any or all of the IEEE 802 Working Group meetings.

To create a standard, the IEEE requires these criteria:

- Broad market potential
- Compatibility with IEEE 802 architecture
- Distinct identity
- Technical feasibility
- Economic feasibility

11.2.4.1 IEEE 802.14 SWGs.   IEEE 802.14 members created three official subworking groups:

1. Physical Layer (PHY) Subworking Group (SWG)
2. MAC Subworking Group
3. Architecture Subworking Group

11.2.4.1.1 PHY Layer SWG.   The PHY SWG was delegated to develop the standard for the cable modem Physical Layer. Their charter is to develop the modulation coding scheme which must also be optimized in terms of bit rate, efficiency, bit error rate and availability to support the tree-branching multiple access MAC over long distance for a large number of users. The PHY must be architected to support improvement in transmission technologies for the forward and reverse channels. The advance PHY study group was also established to look into the long term PHY technology advances, specifically in the DMT modulation technology where the number of frequency bands used scales to the needed bandwidth.

11.2.4.1.2 MAC Layer SWG.   The MAC SWG was delegated to develop the standard for the cable modem MAC layer. The MAC layer is used to arbitrate transmissions on a shared medium accessed by multiple DTE using a set of rules that are followed by each node. The Cable TV Network based distributed environment with forward and reverse channels, long propagation delays and support of multiple types of services (data, voice, image and video) which make the design of the MAC layer protocol a great challenge.

11.2.4.1.3 Architecture SWG. The architecture SWG was tasked to oversee the linking, interworking, and functional interfaces between the MAC and PHY SWG. The SWG also developed the OSI reference model of the cable modem and its layering service architecture above the MAC and PHY. One of its main functions was also to insure compliance with the requirement document created by IEEE 802.14 WG.

IEEE 802.14 has created a web site (walkingdog.com) containing the relevant informaiton of the Working Group.

Official contact to obtain standard documents for the IEEE office is:
445 Hoes Lane
P.O.Box 1331
Piscataway, NJ 08855-1331
908-562-33800
908-562-1571 (fax)

11.2.5 CableLabs

CableLabs was created in 1988. It is an R&D consortium sponsored by the North American CATV operators. The organization indirectly represents 85 percent of CATV subscribers in the U.S., 70 percent of CATV subscribers in Canada, and 10 percent of CATV subscribers in Mexico. The staff and the facility are located in Boulder, Colorado.

11.2.5.1 CableLabs' Organization and Missions. CableLabs performs R&D projects for specific members or group of members, and also transfers relevant technology to its members. Presently CableLabs is collaborating with the computer industry to develop HFC digital cable networks. Another function is to provide publications to MSO members only. However, some publications are accessible if CableLabs considers their distribution to the community of vendors beneficial to the Cable business. Newsletters are published periodically. CableLabs organization is built around the following operational entities:

- Advanced TV: In charge of new TV techniques (video compression, HDTV).

- Operations Technologies Projects: In charge of all operations related issues in existing CATV networks (improving connectors, etc.).

- Network Architecture Design and Development: In charge of designing new architectures (PCS, SONET hubs, etc.).

- Clearing House: In charge of conferences, publications and external communication (Congressional and State Government, public relations, etc.).

11.2.5.2 CableLabs' Affiliation.   Like Bellcore, CableLabs is not affiliated with the American National Standards Institute (ANSI). Hence, its mission is not to generate standards, but vendors might use CableLabs to check compliance of technical approaches or standards with the CATV market.

11.2.6 ATM Forum

The ATM Forum is an international non-profit organization created with the objective of accelerating the use of ATM products and services through a rapid convergence of interoperability specifications. In addition, the ATM Forum promotes industry cooperation and awareness. Since its formation in 1991, the ATM Forum has created a very strong interest within the communications industry. Currently, the ATM Forum consists of over 700 member companies, and it remains open to any organization that is interested in accelerating the availability of ATM-based solutions.

The ATM Forum consists of a single worldwide Technical Committee, three Marketing Committees for North America, Europe and Asia-Pacific as well as the Enterprise Network Roundtable (users of ATM group) through which ATM end-users participate. The marketing group is very active in marketing research and recently commissioned a marketing study on "ATM Residential Broadband Market Requirements".

11.2.6.1 ATM Forum's Affiliation.   The ATM Forum is not affiliated with any standard organization. There are several MOUs established, such as ITU, IETF and others. The Forum works closely with the T1S1.5 (T1 committee/ANSI affiliated) to establish U.S. position and hence forward to ITU. Liaisons are established with almost all other technical organizations.

11.2.6.2 Residential Broadband Working Group.   The RBB Working Group was founded by the ATM Forum in February 1995. The RBB WG is chartered to bring ATM to the home, and defining a complete end-to-end ATM system specification both to and from the home to a variety of devices, e.g. set-top boxes and PCs. Its emphasis is on defining and specifying the home UNI and access network interface (ANI).

The RBB WG is planning to adopt IEEE 802.14 PHY and MAC standards when available. The reference model of the access portion was generically developed to accommodate HFC, FTTC, FTTH, etc. The RBB group is also working on specifying the requirement of the core network and working with ETSI and ITU to coordinate requirements. Work started on security requirement on end-to-end HFC. This work is well coordinated with IEEE 802.14. Its home appliance interface deals with wiring inside the home. There are three proposals now under consideration:

- PCIA
- DAVIC interface
- IEEE 1394

11.2.6.3 ATM Forum Other Working Groups. The ATM Forum consists of 14 technical working groups, a marketing working group and a user group.

1. SIG   P-T-P and P-T-M Signaling point-to-multipoint
2. SAA   Services aspects and application
3. TM   Traffic management (ABR/QOS & BW negotiation)
4. B-ICI   Interoffice interface; + PNNI/B-ICI IW
5. PNNI   Private NNI interface specifications
6. NM   Network Management (M4, M5 interface specifications)
7. RBB   Residential BroadBand services
8. MPOA   Multiprotocol Over ATM
9. PHY   Physical Layer
10. LANE   LAN Emulation
11. Testing   Testing (PICS)
12. Security Security
13. VTOA   Voice Telephony Over ATM
14. WATM   Wireless ATM

11.2.7 DAVIC

DAVIC (Digital Audio Visual Council) is a non-profit association established in Switzerland. DAVIC's goal is to be the home for all those who see the potential of digital technologies applied to audio and video. As such there is no "constituency" because the range of industries represented in DAVIC covers content, service and manufacturing companies. Total membership is 218 companies/organizations worldwide.

Membership in DAVIC is open to any corporation and individual firm, partnership, governmental body or international organization. Associate member status is available to those who do want to take an active role in the precise technical content of the specifications.

11.2.7.1 DAVIC's Goal. The purpose of DAVIC is to favor the success of emerging digital audio-visual applications and services, by the timely availability of internationally agreed specifications of open interfaces and protocols that maximize interoperability across countries and applications/services.

The primary instrument is production of technical specifications that maximize interoperability across countries and applications/services. DAVIC specifications are issued in versions: DAVIC 1.0; DAVIC 1.1; DAVIC 1.2; etc. DAVIC 1.0 version of specifications allows the deployment of systems that support initial applications such as:

- TV distribution
- near video on demand
- video on demand
- some basic forms of teleshopping

Future versions will extend on previous versions to provide more functionalities while keeping, as far as possible, backwards compatibility with previous versions.

11.2.7.2 DAVIC Technical Committee Organization. DAVIC established six technical committees (TC). They are:

1. Applications
2. Information representation
3. Sub-systems
4. Physical layer
5. Systems integration
6. Security

A TC produces successive revisions of the DAVIC specification part(s) assigned to it. A TC Chair will usually assign sufficient time to clarify an issue and will resort to membership voting when consensus cannot be reached.

Specifications are drawn from existing standards, if available. However, each TC makes its decision to balance other criteria such as technical merit, cost and wide usage.

11.2.7.3 DAVIC Work Plan/Timeline. The DAVIC work plan of relevance in the HFC environment are these:

- Modulation and associated technology to transmit digital audio-visual information over terrestrial UHF and VHF TV channels
- Modulation and associated technology to transmit digital audio-visual information over 28-40 GHz (LMDS)
- High-bandwidth symmetrical digital connectivity
- Telephony
- Audio conferencing
- Video telephony
- Integrated video-conferencing

11.2.8 Internet Related Standards

This section below provides a snapshot of the various projects and organizations working toward Internet-related standards. The government is expected to play a major role in shaping the Internet of the future.

Figure 11.1 below depicts a snap shot and association of the various projects, organizations, and research activities. The picture is not complete, but it should provide a good guide of today's realities.

11.2.8.1 Next Generation Internet (NGI) Initiatives. On October 10, 1996, in Knoxville, TN, President Clinton and Vice President Gore announced their "commitment to a new $100 million initiatives, for the first year, to improve and expand the Internet." A total of $300 million was committed for the three-year period. NGI goals are:

1. High-performance Network Fabric (providing 100 to 1000 times the access speed of today)
2. Advanced Network Service Technologies
3. Revolutionary Applications

The vision is to have the Internet provide a powerful and flexible environment for business, education, culture, and entertainment. Sight, sound, and

Figure 11.1
Snap Shot of Inter-
net-Related Standards

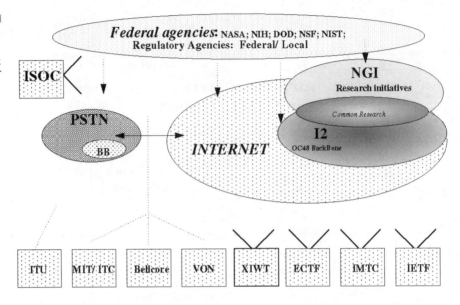

even touch will be integrated through powerful computers, displays, and networks. People will use this environment to shop, bank, study, entertain, work, and visit with each other. Whether at home, at the office, or on travel, the environment will be the same.

11.2.8.2 ITC.   MIT sponsored the ITC (Internet Telephony Interpretability Consortium). ITC's mission includes:

1. Original research related to Internet telephony;

2. Sharing of knowledge among interdisciplinary Consortium researchers and member companies.

The Consortium welcomes visitors from member companies to help fulfill both of these components of the mission. ITC currently envisions that visitors with appropriate background could contribute to research in the areas of:

■ Technical architecture

■ Economic, business strategy, industry structure, and policy issues

■ User feedback studies

11.2.8.3 Internet2 Project.   Internet2 is a collaborative effort of about 130 U.S. research universities. The program started in October 1996. The White House adopted the Internet2 mission as the goal for NGI, which is building a backbone, and necessary peripherals and applications to accommodate 100 to 1000 faster access. Internet2 will also leverage the research to be done on NGI to build the network and make sure interoperability is being addressed. Internet2's mission is to coordinate its development, deployment, and operation and accelerate new applications and network service.

11.2.8.4 *v*BNS. The NSF was one of the major players in the development and commercialization of the Internet. Currently, the NSF is focusing on the development of technology for high performance networking. Through the *v*BNS, in conjunction with MCI, and the Connections Program, technology is being developed to make the Internet faster and more efficient. The NSF is also involved in assisting the worldwide growth of the Internet to support the research and education community. It is funding the development of software tools for the discovery and retrieval of information from the distributed databases on the Internet and educational tools for classroom use of the Internet.

11.2.9 Abeline

The Abeline project is a competing solution to *v*BNS. Unlike *v*BNS, it is IP-centric. QWEST and NORTEL are championing this effort and if successful it will kill ATM as the transport for IP at the hub.

11.2.10 Consortium/Organizations and Standards

The fundamental workings of the standards are changing, and the market now more than ever is setting the agenda and future directions. ITU, is the next few years, will become a repository of standards and specifications that have been successful in building and passing the free market test.

11.2.11 XIWT

XIWT is an organization of major industry (telco/ data) with good connection to federal agencies. XIWT's goals are to foster the understanding, development and application of technologies that cross industry boundaries, facilitate the conversion of the NII vision into real-world implementations and establish a dialogue among member representatives in the private and public sectors.

XIWT will:

- Organize NII technology forums to discuss and disseminate XIWT results and sponsor periodic publications of White Papers on specific technology issues
- Develop common technological and architectural approaches that bridge industry gaps and match the evolution of information
- Plan information infrastructure pilot projects that leverage industry capabilities and state-of-the-art research
- Promote cross-fertilization of technology and stimulate advanced infrastructure development

11.2.12 ISOC

Internet Society (ISOC) is a group of voluntary membership organizations whose purpose is to promote global information exchange through Internet technology. There is no single person or agency in charge of the Internet—it is unofficially coordinated by ISOC. The volunteers make up the Internet Architecture Board, which meets regularly to approve standards and allocate resources. The Internet Engineering Task Force (IETF) consists of volunteers and determines solutions to near-term technical problems of the Internet.

11.2.13 VON

The VON (Voice Over the Network) is a coalition. Its mission is twofold:

1. To act as a key advocate in the fight against regulation. It maintains an ongoing relationship with the Federal Communications Commission (FCC).
2. To educate consumers and the media on Internet communications technologies.

Since March 1996, when the VON Coalition was founded by Jeff Pulver, they have successfully been able to prevent the FCC from banning Internet Telephony and regulating content on the Internet.

By forging relationships with various governmental agencies, they have also been able to bridge the gap between those who develop these technologies and those organizations which have tried to have it regulated. VON continues to provide "common ground" meetings by facilitating discussions concerning the future of these technologies. They have been involved in the interoperability standards discussions being held by the Internet Engineering Task Force. The VON Coalition will be holding seminars and conferences throughout the year. VON Coalition membership includes over 90 Internet telecommunications-related companies, plus more than 460 individuals in over 25 countries. Alcatel is not a member. There are no membership dues. Alcatel should attend the regular VON meetings

11.2.14 IETF

The Internet Engineering Task Force (IETF) is a large, open and international community of network designers, operators, vendors, and researchers concerned with the evolution of the Internet architecture and the smooth operation of the Internet. It is open to any interested individual. The actual technical work of the IETF is done in its working groups, which are organized by topic into several areas (e.g., routing, transport and security). Much of the work is handled via mailing lists. The IETF holds meetings three times per year.

The IETF working groups are managed by Area Directors, or ADs. The ADs are members of the Internet Engineering Steering Group, or IS. Providing architectural oversight is the Internet Architecture Board, or IAA; the IAA also adjudicates appeals on complaints that the IESG has failed. The IAB and IESG are chartered by the Internet Society (ISOC) for these purposes. The General Area Director also serves as the chair of the IESG and of the IETF, and is an ex-officio member of the IAB.

11.2.15 Bellcore

Bellcore has been researching voice-over Internet research for the last three years. In its new role it is expected to act as consultant and create consortiums to address critical issues related to the Internet and NII. Bellcore, with its telco experience, is also expected to play a major role in developing specifications and solicit funding from the industry to develop specifications for the industry. Thus far, Bellcore, has failed in that capacity because of a low participation rate and hence has a high cost and implementation risk.

11.2.16 ECTF

The ECTF (Enterprise Computer Telephony Forum) is an industry organization formed to foster an open, competitive market for Computer Telephony Integration (CTI) technology. By gathering a broad group of suppliers, developers, systems integrators, and users, the ECTF will work to achieve agreement on multi-vendor implementations of computer telephony technology based on *de facto* and *de jure* standards. The Forum is an open, mutual benefit, volunteer membership, non-profit corporation incorporated in the state of California. It is modeled after other industry forums such as the ATM Forum.

The basic objective of the ECTF is to support the rapid advancement of an efficient and compatible technology base that promotes a competitive CTI market. The ECTF will facilitate the implementation and acceptance of CTI solutions by bringing together suppliers, developers, systems integrators, and users to discuss, develop and test interoperability techniques for dealing with the diverse technical approaches to computer telephony integration. The ECTF will address both the technical and market education issues necessary to achieve interoperability among CTI standards and technologies across the heterogeneous client and client server enterprise.

11.2.17 IMTC

The International Multimedia Teleconferencing Consortium, Inc.'s (IMTC's) fundamental goal is to bring all organizations involved in the development of multimedia teleconferencing products and services together to help create and promote the adoption of the required standards. IMTC is a non-profit corporation composed of, and supported by, more than 140 organizations from North America, Europe, and Asia/Pacific. The IMTC's mission is to promote, encourage, and facilitate the development and implementation of interoperable multimedia teleconferencing solutions based on open international standards. In this regard, the IMTC is currently focused on multimedia teleconferencing standards adopted by the International Telecommunications Union (ITU), specifically the ITU-T T.120, H.320, H.323, and H.324 standards.

Key activities include sponsoring and conducting interoperability test sessions between suppliers of conferencing products and services based on these standards and educating the business and consumer communities on the status, value and benefits of the underlying technologies and resultant applications. IMTC members are encouraged to make submissions to standards bodies that enhance the interoperability and usability of multimedia teleconferencing products and services.

IMTC is open to any and all interested parties, including Internet application developers and service providers, teleconferencing hardware and software suppliers and service providers, telecommunications service providers, end-users, educational institutes, government agencies, and non-profit corporations.

IMTC maintains one permanent class of Members, called Voting Members. These Members participate in all the IMTC activities and functions, and have full voting rights on all issues. The cost of a voting membership is $5,000 a year.

12

Major Players

12.1 Introduction

This chapter is a brief summary of the major players in the telecommunications access market. These are the companies that will be deploying broadband access technologies to compete for future broadband interactive services.

12.2 RBOCs

12.2.1 SBC Communications

SBC Communications Inc., headquarters located in San Antonio, Texas, is determined to become a national and international telecommunications powerhouse. Having recently gobbled up Pacific Telesis and Southern New England Telephone Company (SNET), SBC is awaiting Federal approval for its $62 billion merger with Ameritech. In exchange for this approval, SBC has promised to invade the rest of the ILEC market as a CLEC with Boston, Miami and Seattle to be first. On the cellular side, SBC announced in January 1999 that it will acquire Comcast Cellular Corporation, the wireless subsidiary of Comcast Corporation, in a transaction valued at $1.674 billion. SBC markets wireless services under the Cellular One brand.

SBC is a global telecommunications leader with more than 36.9 million access lines and 6.5 million wireless customers across the United States and in 11 other countries including Mexico, France, Chile, South Africa, South Korea, the United Kingdom and Israel. Its 1998 earnings (before exceptional charges associated with the SNET merger) were $4.1 billion on revenues of $28.8 billion, a 19.3 percent earnings growth from the previous year. SBC's wireline and wireless operations continue to see strong growth with 1.5 million wireline additions in 1998 and 900,000 wireless customer additions.

At the beginning of 1999, SBC announced the country's largest rollout of Asymmetrical Digital Subscriber Line (ADSL) service. By the end of 1999, SBC intends to have equipped 526 central offices with ADSL reaching 8.2 million residential customers and 1.3 million businesses. On the video side, in 1998, SBC began a marketing and distribution agreement with DirecTV to market DBS services to apartment buildings.

12.2.2 Ameritech

Ameritech is a global leader in the telecommunications industry, with more than 37 million access lines and 6.9 million wireless customers across the United States and in 11 other countries. Internationally, Ameritech has focused primarily on eastern Europe with operations in Austria, Belgium, Croatia, the Czech Republic, Denmark, Germany, Hungary, Luxembourg, the Netherlands, Norway, Slovakia, Slovenia, and Switzerland. Current revenues are over $16 billion on assets over $25 billion.

Ameritech provides cellular, long distance (for its cellular customers), paging, cable TV, security monitoring, electronic commerce, managed services and wireless data communications. Ameritech provides security monitoring services to 1 million homes and businesses. Ameritech is unique among the RBOCs in offering security monitoring, as the MFJ forbade the RBOCs from offering security monitoring services with the exception of Ameritech who was "grandfathered" since it already provided these services.

While all of the RBOCs have made efforts to get into the CATV business, Ameritech is the only one aggressively going forward. Ameritech has received CATV franchises in 99 communities in the Cleveland, Detroit, Columbus, Ohio, and 20 in the Chicago metropolitan area. Currently 100,000 customers receive CATV services form Ameritech. In each of these areas Ameritech is a second operator competing with an incumbent Cable operator.

In June 1998, Ameritech announced that it had agreed to be merged with SBC.

12.2.3 Bell Atlantic Corporation

With more than 42 million telephone access lines and 8.6 million wireless customers worldwide, Bell Atlantic is one of the largest providers of advanced wireline voice and data services, a market leader in wireless services and the world's largest publisher of directory information. Bell Atlantic is one of the world's largest investors in the high-growth global communications markets, with operations and investments in 23 countries outside the United States.

Lately Bell Atlantic has been following a growth-through-acquisition strategy. In August 1997, Bell Atlantic absorbed NYNEX, thereby unifying a region covering the east coast megalopolis from Washington DC to Boston. Most long distance calls from this region terminate in this same region. Hence Bell Atlantic will be in a powerful position to exploit inter-LATA calling relief when it comes.

Bell Atlantic owns the PCS provider PrimeCo jointly with AirTouch Communications. In January 1999 Bell Atlantic offered $45 billion to buy AirTouch but lost the bidding war to Vodafone. Had Bell Atlantic won, and then coupled with its purchase of GTE, it would have created a wireless company rivaling AT&T Wireless.

In July 1998, Bell Atlantic announced its intention to absorb GTE. If approved the merged company will have combined revenues of $53 billion and a combined market capitalization of approximately $125 billion.

Bell Atlantic has made numerous attempts to get into the video entertainment market, attempting at times video-over-ADSL, HFC, and MMDS. Its current strategy through its Bell Atlantic Video subsidiary is to resell DIRECTV® and U.S. Satellite Broadcasting^{SM} service.

12.2.4 BellSouth

At divestiture, BellSouth Corporation was the largest of the regional Bell holding companies. Headquartered in Atlanta, Georgia with $37 billion in assets and 86,000 employees in 19 countries, it has over 23 million access lines in nine southeastern states and 4.5 million domestic cellular customers. BellSouth International, primarily focused on South America, ended 1998 with more than 7 million total customers, up 88 percent over the previous year. BellSouth earned a revenue stream of $23.1 billion in 1998 with telecommunications, wireless communications, cable and digital TV, directory advertising and publishing, and Internet and data services.

After trials in 1986 of video delivery using Fiber-to-the-Home, BellSouth later entered the video service market deploying HFC networks in Chamblee, Georgia (an Atlanta suburb) and Vestavia Hills, Alabama. BellSouth has now chosen wireless MMDS as its preferred video distribution technology having rolled out MMDS service in Atlanta, New Orleans and Orlando. For content BellSouth is a member of the Americast® consortium with Ameritech, GTE, SNET and The Walt Disney Company.

In domestic wireline, BellSouth has consistently followed a regional strategy and has eschewed the merger mania going on elsewhere. Driven by the strong economy in the southeastern U.S., BellSouth delivered a record increase of 1,066,000 lines in 1998, producing the highest growth rate in access lines for any calendar year in the company's history.

12.2.5 US West

US West, headquartered in Denver, Colorado, provides wireline and wireless telecommunications service in 14 western and mid-western states. With 47,000 employees, it is the smallest of the RBOCs earning $1,374 million on revenues of $11.4 billion and assets of $17,740 billion in 1997.

US West's video strategy has been somewhat troubled. In May 1993, US West formed a joint venture with Time Warner Entertainment to build and operate broadband cable networks outside its 14-state region. Then in September 1995, it sued Time Warner over its acquisition of Turner Broadcasting. In February 1996, US West announced it would merge Continental Cablevision into its US West Media Group, making it the nation's third largest cable company with interests in cable systems in 10 countries. Then in June 1998, US West chose to divest itself of the MediaOne Group altogether. At the end of 1998, the company began deploying video over "VDSL" (very high-speed digital subscriber line) to customers in Phoenix.

The large, low population density area of the United States in which US West operates is one of the most expensive to serve. In January 1999, US West announced its intention to sell approximately 500,000 of its most rural access lines–about three percent of its total 16.5 million lines. This is part of a strategy to achieve revenue growth through advanced, high margin services such as data communications. US West began offering its MegaBit ADSL service in 1998 with 20,000 ADSL customers by the end of that year.

12.3 MSOs

The cable company Multiple System Operators (MSOs) are a second group of competitors in the upcoming broadband access battle.

A cable company which operates in more than one system (strictly defined geographical area determined by the headend (the electronic control center of the cable system) is considered a Multiple System Operator (MSO). The Cable Industry has had an accelerated growth rate since its beginning in 1952, when it consisted of 70 operating systems and 14,000 subscribers. In 1985, there were 6,600 operating systems and 32 million subscribers which grew to 11,351 operating systems and 56.5 million subscribers in 1995 (1995 Cable Television & Cable Factbook). Nearly 85 percent of these cable operators have membership with Cable Television Laboratories (Cable Labs), a research and development consortium established in 1988. The National Cable Television Organization (NCTA) is a membership organization that does research in CATV technologies and often makes recommendations to the Federal Communications Commission (FCC) in areas dealing with CATV.

The following is a description of the six largest MSOs.

12.3.1 TCI Group

Communications, Inc. was formed in 1968, when a small cable company, Community Television, Inc., and a small common carrier microwave company, Western Microwave, Inc., merged. The two wholly-owned TCI subsidiaries were re-named Community Tele-Communications, Inc. (CTCI) and Western Tele-Communications, Inc. (WTCI), respectively. TCI's original name was American Tele-Communications, Inc. but was changed in late 1968 to avoid conflict with other competitors. TCI became the nation's largest cable operator in 1982, when it passed the two million subscribers milestone. In 1993, a joint venture between TCI, Sega of America and Time Warner formed to develop the Sega Channel. Today TCI distributes cable TV to 14 million customers through its more than 30,000 employees. It invests in technology development, delivers digital television, telephone and Internet service through its broadband network. ATCI is a leading operator of cable television, telephone networks and programming interests in Europe, Latin America and Asia.

12.3.2 Cablevision

Cablevision Systems Corporation is the nation's sixth largest operator of cable television systems. Headquartered in Woodbury, N.Y., the company serves 2.8 million cable television customers in 19 states. Cablevision owns American Movie Classics, Bravo, SportsChannel and The Independent Film Channel. Cablevision is now turning its talent for innovation to new areas of electronic communications. In December 1996, the company launched Optimum Online, a high-speed multimedia communications service, to 15,000 homes in the northern part of the township of Oyster Bay, N.Y. During 1997, Cablevision expects to make this Internet- and online-access service available to more than 150,000 Long Island homes, as well as to Cablevision customers in Connecticut.

12.3.3 Comcast Corporation

Today, Comcast provides cable to over 3.4 million subscribers in 18 states. It also markets cellular communications to the 7.5 million population tri-state area of Delaware, New Jersey, and Pennsylvania. Comcast provides a combination of cable and telephone services in the United Kingdom. Interestingly, the telephone service Comcast provides in the U.K. is over separate telephone cables, not over its CATV coax cables.

12.3.4 Continental Cablevision

In 1964, Continental Cablevision was founded in Tiffin, Ohio. Today it is the nation's third largest MSO serving over 4 million subscribers. It has played a major role in the development of several leading programming services including helping found the Cable-Satellite Public Affairs Network (C-SPAN). Continental is also an owner of Turner Broadcasting System (CNN, TNT, Cartoon Network), Entertainment Television and Viewers Choice, the nation's leading pay-per-view service. In addition, Continental and the Hearst Corporation created and co-own New England Cable News, an award-winning 24-hour regional news network now seen in more than 1 million

New England homes. Continental is also a partner in PrimeStar, the nation's first direct broadcast satellite (DBS) service.

12.3.5 Cox Communications, Inc.

Cox Communications, Inc. is among the nation's largest multiple system operators, serving 3.2 million customers. The acquisition of Times Mirror Cable Television increased the number of customers served from 1.9 million to 3.2 million. To reflect the increasing concentration on new businesses, such as data services and telephony, the company dropped "cable" from its name, becoming Cox Communications, Inc. The company formed a telecommunications Venture with Sprint, TCI and Comcast which won licenses to deliver PCS wireless communications in 31 major metropolitan areas. As a full-service provider of telecommunications services, Cox has interests in wired telecommunications, including cable television and telephone services, wireless telecommunications including personal communications services (PCS), and direct-to-home (DTH) satellite television and programming networks.

12.3.6 Time Warner

Time Warner is the nation's second largest MSO with over 11.5 million subscribers. During 1994 and early 1995, Time Warner Cable and Time Warner Inc. entered into agreements to increase the number of subscribers under their management from 7.5 million to 11.5 million. A merger between Time Warner Inc. and Cablevision Industries added another 1.3 million subscribers. In the last few years, Time Warner continued its aggressive rollout of fiber-optic cable and multifaceted entry into the telephone business and launched the Full Service Network® (FSN). FSN debuted on December 14, 1994, in Orlando, offering true video-on-demand with full VCR functionality (the ability to fast-forward, rewind and pause), a variety of interactive home-shopping services, video games and an interactive program guide.

12.4 Internet Service Providers

Already in 1996 Oddessey's *Homefront* study indicated that 35 percent of United States households own a personal computer and 14 percent of these have on-line access. These figures have been steadily increasing. Recent market surveys of first time buyers indicated that access to the Internet was their primary reason for purchase of a home computer. The growth rate of the Internet astounded most in the telecommunications industry. In the early 1970's, ARPANET (Internet parent) was introduced. The 1980's brought the linking of public and commercial computer networks. Internet traffic has been doubling every year since 1988 (Robert Hobbes). The Internet has emerged not just as a way to send an email or download an occasional file, but as a place to visit, full of people and ideas; it became "cyberspace." The Internet frenzy is a worldwide phenomena. Europe, Asia and even the developing countries are building and expanding the Internet infrastructure.

Users gain access to the Internet via an Internet Service Provider (ISP) or a commercial online service (AOL, MSN, etc.). Internet service providers provide fast and inexpensive connection to the Web, usenet newsgroups, email, and personal home pages. Commercial online services usually provide custom content and other services, some of which typically at an extra cost. A survey done at Odyssey in San Francisco found that 48 percent of on-line households were using an ISP, while 35 percent were using commercial online services as of July 1996. This was a major change from six months earlier when the majority of households were using commercial online services.

The phenomenal growth of Internet usage was not accurately predicted by the telephone companies, ISPs, commercial online services or the MSOs who were unprepared to accommodate the volume of customers. Telephone switching systems designed for voice traffic were ill-suited for the long holding times of Internet connection. AOL's dial up ports quickly became exhausted. The result has been busy signals and periodic shutdowns creating irate customers who in turn have created lawsuits. This chaos is creating a demand for high speed Internet access that by-pass the telephone network.

Cable operators have begun to offer high-speed Internet access via cable modems. Cable modems typically operate at 40 Mb/s, but this access is shared by other customers. Some customers in Northern California have recently complained that during the evening busy hours their cable modem service is slower than a dial-up modem. The telephone companies finally found a market for their ISDN service: high-speed internet access. ISDN provides access up to 128 kb/s—almost three times faster that the fastest modem. Initial installation of an ISDN line typically runs approximately $400 plus a monthly charge of approximately $50 for unlimited access. While ISDN provides higher speed access, the calls still go through telephone switches designed for voice traffic. ISDN is a dial-up service that is not "always on." The latest addition to the telephone company arsenal is the Asymmetrical Digital Subscriber Line (ADSL). ADSL operates over conventional copper telephone loops, provides about a 10-fold increase in speed over ISDN, is "always on," and takes traffic around telephone switching systems. ADSL is only recently begun to be deployed in significant volume and, unfortunately, will not operate over all loop facilities.

Worldwide, there are over 5,500 ISPs. Some only offer access to the Internet; others provide value added online services. Below is a list of the major ISPs in the U.S.

aaaa.net	Admiral Online
Alpine Web, Inc.	America Online
AT&T Network Commerce	Blue Sky Internet
City Online	CompUnet, Inc.
Digital Entertainment Inc.	EarthLink Network
Epoch Internet	Fiberlink Communications
Franklin interNet	Golden Minutes On-line
GTE Internet (Uunet)	IBM Internet Connection
IDT Corp	Infinity Internet Services
Infopage Services	INTELI-NET.COM
Inter-Tech	Intertelavideo.Com
MacConnect	MCI
Micro-Net Online Services	Miracle Net
MSN Internet Access	Net Asset, LLC
NetAccess 800	NetSafe, Inc.
PCstarnet Internet	Premier Access
Prime Matrix Internet	Prodigy Internet
PSINet Inc.	SouthNet TeleComm
Services, Inc.	Start/Net
Surfcheap.com (Wolftalk)	TEK Interactive Group, Inc.
Trim Online	TumbleWeed Internet
U.S. Dial, Inc.	USA Street
UUNET Technologies	VPM Internet Services
WebbyNet	WEBster Computing

The following are profiles of the leading players.

12.4.1 UUNET

Worldcom recently acquired MCI and UUNET Technologies, Inc. UUNET was founded in 1987 with its headquarters in Fairfax, VA. Uunet Worldcom has a global presence and provides ISP connectivity in 76 countries servicing about 70,000 business users.

UUNET network includes 1,000 Points of Presence (POPs) around the world. The U.S. backbone network consists of DS-3 and OC-12 trunks with OC-12 (622 Mbps) ATM "metro rings" in 10 metropolitan multi-hub areas. These locations include Seattle, Los Angeles, Dallas, Houston, Chicago, Boston, Atlanta, New York City, Washington, D.C., and the San Francisco Bay area.

12.4.2 AOL

America Online, Inc., headquarter in Dullas, VA was founded in 1985. AOL operates worldwide and leads the industry in content and interactive services. AOL has a membership of over 12 million households, making it the largest interactive online community in the world. AOL Studios' original content products span popular consumer categories such as local content, entertainment, romance, women and sports.

In January 1998, America Online, Inc. acquired CompuServe's worldwide consumer online services and created CompuServe Interactive Services, a wholly owned subsidiary of America Online. CompuServe Interactive Services has approximately 2 million members worldwide.

12.4.3 BBN

BBN, now acquired by GTE, has for the past 25 years partnered with ARPANET to build the Internet by participating in the design and operation of the most advanced computer networks in the world. BBN currently operates the Defense Data Network, and is building the data network used by the U.S. Treasury. BBN also operates and maintains the access portion of the AOL nationwide.

GTE Internetworking draws upon its BBN Technologies division's expertise. BBN Technologies gains this expertise from externally funded R&D of advanced networking technologies including satellites, digital radio, multigigabit routers, security, and speech, as well as GTE's expertise in telecommunications services including local and long distance, wireless, paging, video, and Internet.

12.4.4 PSINet

PSINet, headquarter in Washington, DC, operates over 400 POPs worldwide. It is one of the independent, facilities-based, ISPs and one of the largest providers of IP-based communications services for businesses. PSINet focuses on managing IP-based VPNs (Virtual Private Networks), Internet security, electronics commerce, voice-fax over IP, and web hosting.

12.5 Competitive Local Exchange Carriers

With the enactment of the 1996 Telecommunications Act, the notion of a natural monopoly for local access passed into history. An environment was created encouraging other companies, called Competitive Local Exchange Carriers (CLECs), to compete with the existing local telephone company, called Incumbent Local Exchange Carriers. To help them get established, Congress required the ILECs to wholesale their services to the CLECs for resale. Most CLECs have complained that local service resale is not profitable and have stated their intention to become "facilities based," owning at least their own switching systems. The loop facilities are more problematic for the CLECs, and most CLECs intend to lease "unbundled" loops from the ILECs especially outside of downtown metropolitan areas. An exception are those building networks on wireless technologies such as Local Multichannel Distribution Service (LMDS) or Low Earth Orbit Satellites (LEOs).

The following are example of the major CLECs.

12.5.1 Time Warner Telecommunications

Time Warner Telecom is a good example of a cable company deciding to move into telecommunications as a facilities-based CLEC. Formed in 1993 as a partnership of Time Warner Cable with US WEST, Time Warner Telecom earned revenues of $122 million in 1998 representing a 120 percent increase over the previous year, although this represented a $73 million operating loss. Headquartered in Englewood, CO, the company had had 898 employees at the end of 1998.

Time Warner Telecom has an extensive fiber network with 272,390 fiber miles of optical cable installed. The company provides 78,000 access lines from 16 digital switches. It offers a wide range of business telephony services, primarily to medium- and large-sized business customers and to other carriers. Recently Time Warner Telecom added high-speed dedicated Internet access to its service offerings.

12.5.2 Frontier

Founded in 1899, and headquartered in Rochester, NY, Frontier Corporation is the 12th largest local exchange service provider. The company acts as both a CLEC and interexchange carrier providing data, long distance, local telephone, cellular, paging, Internet and frame relay to business customers nationwide. With more than 8,000 employees, Frontier earned revenues of $177.4 million in 1988, a 28.5 percent increase over the previous year. It currently serves more than two million customers. Frontier Communications also sells wholesale transmission capacity to carriers.

Frontier Corporation in divided into four business centers: Frontier Communications, Frontier GlobalCenter and Frontier ConferTech. Frontier Communications. Frontier's CLEC operations reach nearly 70 percent of the U.S. business population.

12.5.3 Allegiance

Founded in April 1997 by a management team led by Royce J. Holland, the former president of MFS, Allegiance Telecom, Inc. provides local calling, long distance calling, data transmission and Internet service. Allegiance is certified to provide competitive local exchange services in nine states, including California, Georgia, Illinois, New Jersey, New York, Maryland, Massachusetts, Pennsylvania, Texas and Virginia.

Although at one time Allegiance Telecom considered getting into the local exchange market via resale of ILEC services, the company has now chosen to be facilities-based and to use unbundled loop facilities of the ILEC. To that end, Allegiance Telecom is collocated in 101 ILEC central offices for unbundled loops. By the end of 1998, it had sold 43,100 lines providing an annualized recurring revenue run rate in excess of $50 million. Internet services now available from Allegiance Telecom include domain name service, business email service, web site hosting and dedicated high-speed Internet access.

Allegiance achieved an industry "first" when it completed its implementation of electronic bonding with Bell Atlantic inking their operations support systems (OSS). Electronic bonding is one of the 14 point checklist items for the RBOCs to get into the long distance business.

12.5.4 Level 3

USA Today called Level 3 "A dream team with a network dream and a killer business plan." Level 3 astonished the industry when it announced that it intended to build a full services nationwide network based on IP end to end.

The company was founded in 1998 from the Kiewit Diversified Group. This was the same company that founded MFS, later sold to Worldcom. In fact many of its key executives are from MFS including its CEO, Jim Crowe who was previously CEO.

The eyes of the telecommunications world are truly on Level 3. Many have said that one day a new PSTN could be built based entirely on IP. Level 3 is doing just that. Level 3 has set an ambitious pace for itself, targeting to have service up in 14 cities within its first year. They intend to be in 25 cities by the end of 1999 offering telephony, private lines and Internet services.

12.5.5 NextLink

NextLink was founded in 1994 by cellular phone pioneer Craig McCaw. Headquartered in Bellevue, WA., NextLink currently operates 22 facilities-based networks providing switched local and long distance services in 36 markets in 14 states. Revenues earned in 1998 was $139.7 million, a 143 percent increase from the previous year. NextLink provides local, long distance, internet services, centrex, voice messaging and private line services.

In January 1999, NextLink agreed to buy the 40 LMDS licenses owned by WNP Communications Inc. for $542 million. In addition, NextLink agreed to pay $137.7 million to buy out Nextel Communications Inc.'s 50 percent stake in Nextband, a joint venture with NextLink that had acquired 42 LMDS licenses. With these licenses, NextLink becomes the 700 lb. gorilla in the LMDS market able to reach 95 percent of customers in the top 30 U.S. markets.

The company is also building fiber-optic phone networks (InterNext) in most major U.S. cities allowing them to reach 40 percent of their customers with wireline access. It will use LMDS to connect companies in cases where it would be too expensive to use fiber.

12.5.6 McLeodUSA

Founded in June 1991, and providing fiber-optic maintenance services for the Iowa Communications Network, McLeodUSA has grown to become a major CLEC providing integrated telecommunications services to small- and medium-sized businesses, and dial tone services to residential customers. McLeodUSA's 186,000 customers are located primarily in Iowa and Illinois.

Headquartered in Cedar Rapids, Iowa, with 5,300 employees McLeodUSA earned $604 million in revenues in 1998. McLeodUSA is a facilities-based CLEC provider with 9 class 5 switching machines and 398,000 local lines. It has installed over 7,100 route miles of fiber optic cable.

McLeodUSA continues its expansion through acquisition, announcing in March 1999, a merger with the South Dakota-based CLEC, Dakota Telecommunications Group, Inc. The company will be entering the PCS market, having won 26 PCS licenses covering Iowa, Illinois, Minnesota, Nebraska and South Dakota.

TERMINOLOGY

10BaseT: 10-Mbps baseband Ethernet specification using two pairs of twisted-pair cabling (Category 3, 4, or 5): one pair for transmitting data and the other for receiving data. 10BaseT, which is part of the IEEE 802.3 specification, has a distance limit of approximately 100 meters per segment.

AAL: ATM Adaptation Layer. Service-dependent sublayer of the data link layer. The AAL accepts data from different applications and presents it to the ATM layer in the form of 48-byte ATM payload segments. AALs consist of two sublayers, CS and SAR. AALs differ on the basis of the source-destination timing used, whether they use CBR or VBR, and whether they are used for connection-oriented or connectionless mode data transfer. At present, the four types of AALs recommended by the ITU-T are AAL1, AAL2, AAL3/4, and AAL5.

ABR: Available Bit Rate. Traffic sources adjust their transmission rate in response to information they receive describing the status of the network and its capability to successfully deliver data.

Access server: Communications processor that connects asynchronous devices to a LAN or WAN through network and terminal emulation software. Performs both synchronous and asynchronous routing of supported protocols. Sometimes called a network access server. Compare with communication server.

Address: Data structure or logical convention used to identify a unique entity, such as a particular process or network device.

Aerial plant: Cable that is suspended in the air on telephone or electric utility poles.

Algorithm: Well-defined rule or process for arriving at a solution to a problem. In networking, algorithms are commonly used to determine the best route for traffic from a particular source to a particular destination.

Alternative access provider: A telecommunications provider, other than the local telephone company that provides a connection between a customer's premises (usually a large business customer) and the point of presence of the long distance carrier, or portions thereof.

Amplifier: A device that boosts the strength of an electronic signal. In a cable system, amplifiers are spaced at regular intervals throughout the system to keep signals picture-perfect no matter where you live.

ANSI: American National Standards Institute. Voluntary organization comprised of corporate, government, and other members which coordinates standards-related activities, approves U.S. national standards, and develops positions for the United States in international standards organizations. ANSI helps develop international and U.S. standards relating to, among other things, communications and networking. ANSI is a member of the IEC and the ISO.

API: Application-Programming Interface. Specification of function-call conventions that defines an interface to a service.

ARPA: Advanced Research Projects Agency. Research and development organization that is part of DOD. ARPA is responsible for numerous technological advances in communications and networking. ARPA evolved into DARPA, and then back into ARPA again (in 1994).

Asynchronous Transfer Mode (ATM): Is the transfer mode in which the information is organized into cells conveyed in fixed-length of 53-byte. It is asynchronous in the sense that the recurrence of cells containing information from an individual user is not necessarily periodic. Fixed-length cells allow cell processing to occur in hardware, thereby reducing transit delays.

Asynchronous transmission: Term describing digital signals that are transmitted without precise clocking. Such signals generally have different frequencies and phase relationships. Asynchronous transmissions usually encapsulate individual characters in control bits (called start and stop bits) that designate the beginning and end of each character. Compare with isochronous transmission, plesiochronous transmission, and synchronous transmission.

ATM layer: Service-independent sublayer of the data link layer in an ATM network. It receives the 48-byte payload segments from the AAL and attaches a 5-byte header to each, producing standard 53-byte ATM cells.

Backbone: The part of a network that acts as the primary path for traffic that is most often sourced from, and destined for, other networks.

Bandwidth:
1. A measurable characteristic defining the available resources of a device in a specific time period.
2. Frequency spectrum used to transmit pictures, sounds or data. The average television station uses a bandwidth of 6 megahertz.

Best-effort delivery: Describes a network system that does not use a sophisticated acknowledgment system to guarantee reliable delivery of information.

BISDN: Broadband ISDN. ITU-T communication standards designed to handle high-bandwidth applications such as video. BISDN currently uses

ATM technology over SONET-based transmission circuits to provide data rates from 155 to 622 Mbps and beyond.

Bridge: Device that connects and passes packets between two network segments that use the same communications protocol. Bridges operate at the data link layer (layer 2) of the OSI reference model. In general, a bridge will filter, forward, or flood an incoming frame based on the MAC address of that frame.

Broadband: A service or system requiring transmission channels capable of supporting rates greater than 1.5 Mb/s.

Cable television: Communications system that distributes broadcast and non-broadcast signals, as well as a multiplicity of satellite signals, original programming and other signals by means of a coaxial cable and/or optical fiber.

Call: An association between two or more users or between a user and a network entity that is established by the user of the network capabilities. This association may have zero or more connections.

Call admission control: Traffic management mechanism used in ATM networks that determine whether the network can offer a path with sufficient bandwidth for a requested VCC.

CATV: Cable television. Communication system where multiple channels of programming material are transmitted to homes using broadband coaxial cable. Formerly called Community Antenna Television.

Cell relay: Network technology based on the use of small, fixed-size packets, or cells. Because cells are fixed-length, they can be processed and switched in hardware at high speeds. Cell relay is the basis for many high-speed network protocols including ATM, IEEE 802.6, and SMDS.

Central office: The central location, in a traditional public network telecommunication environment, wherein access is available to signals traveling in both the forward and reverse directions.

Channel: A communication path. Multiple channels can be multiplexed over a single cable in the cable television environment.

Channel capacity: Maximum number of channels that a cable system can carry simultaneously.

Circuit: Communications path between two or more points.

Circuit switching: Switching system in which a dedicated physical circuit path must exist between sender and receiver for the duration of the "call." Used heavily in the telephone company network.

Coaxial cable: Actual line of transmission for carrying television signals. Its principal conductor is either a pure copper or copper-coated wire, surrounded by insulation and then encased in aluminum.

Competitive access provider (CAP): A telecommunications entity engaged in providing competitive access service.

Compression: The running of a data set through an algorithm that reduces the space required to store or the bandwidth required to transmit the data set. Compare with companding and expansion.

Connection: An association established by a layer (AAL) between two or more users of the layer service for the transfer of information.

Connectionless: Term used to describe data transfer without the existence of a virtual circuit. Compare with connection-oriented.

Connectionless service: A service, which allows the transfer of information among service subscribers without the need for an end-to-end call establishment procedure.

Connection-oriented: Term used to describe data transfer that requires the establishment of a virtual circuit.

Constant Bit Rate (CBR): A service class intended for real-time applications, i.e. those requiring tightly constrained delay and delay variation, as would be appropriate for voice and video applications. The consistent availability of a fixed quantity of bandwidth is considered appropriate for CBR service.

Contention: Access method in which network devices compete for permission to access the physical medium. Contrast with circuit switching and token passing.

CPE: Customer Premises Equipment. Terminating equipment, such as terminals, telephones, and modems, supplied by the Telephone Company, installed at customer sites, and connected to the telephone company network.

CRC: Cyclic Redundancy Check. Error-checking technique in which the frame recipient calculates a remainder by dividing frame contents by a prime binary divisor and compares the calculated remainder to a value stored in the frame by the sending node.

DARPA: Defense Advanced Research Projects Agency. U.S. government agency that funded research for and experimentation with the Internet. Evolved from ARPA, and then, in 1994, back to ARPA.

Data link layer: Layer 2 of the OSI reference model. This layer provides reliable transit of data across a physical link. The data link layer is concerned with physical addressing, network topology, line discipline, error notification, ordered delivery of frames, and flow control.

Datagram: Logical grouping of information sent as a network layer unit over a transmission medium without prior establishment of a virtual circuit. IP datagrams are the primary information units in the Internet.

DBS: Direct Broadcasting Satellite System in which signals are transmitted directly from a satellite to a home rooftop-receiving dish.

De facto standard: Standard that exists by nature of its widespread use. Compare with de jure standard.

Delay: The elapsed time between the instant when user information is submitted to the network and when the user at the other hand receives it.

Demographics: Breakdown of television viewers by such factors as age, sex, income levels, education and race. These figures are used in selling advertising time.

Discrambler: An electronic circuit that restores a scrambled video signal to its standard form.

Digital compression: An engineering technique for converting a cable television signal into a digital format (in which it can easily be stored).

Downstream: Flow of signals from the cable system headend through the distribution network to the customer.

Drop cable coaxial: Cable that connects a residence or service location from a tap on the nearest feeder cable.

Dynamic routing: Routing that adjusts automatically to network topology or traffic changes. Also called adaptive routing.

E.164: ITU-T recommendation for international telecommunication numbering, especially in ISDN, BISDN, and SMDS. An evolution of standard telephone numbers.

Earth station: Structure, referred to as a "dish," used for receiving

Edge device: Network entity such as a LAN segment, host, or router that connects to an ATM switch via an edge card. Edge devices send and receive the data that passes through the ATM network.

End user: A human being, organization, or telecommunications system that accesses the network in order to communicate via the services provided by the network.

Ethernet: Baseband LAN specification invented by Xerox Corporation and developed jointly by Xerox, Intel, and Digital Equipment Corporation. Ethernet networks use CSMA/CD and run over a variety of cable types at 10 Mbps. Ethernet is similar to the IEEE 802.3 series of standards.

Extended subsplit: A frequency division scheme that allows bi-directional traffic on a single cable. Reverse path signals come to the headend from 5 to 45 Mhz. Forward path signals go from the headend from 54 to the upper frequency limit.

FCC: Federal Communications Commission. U.S. government agency that supervises, licenses, and controls electronic and electromagnetic transmission standards.

Feeder cable: Cables that run down streets in the served area and distribute the signal to the individual taps.

Feeder line: Cable distribution lines that connect the main trunk line or cable to the smaller drop cable.

Fiber optics: Very thin and pliable tubes of glass or plastic used to carry wide bands of frequencies.

Firewall: Router or access server, or several routers or access servers, designated as a buffer between any connected public networks and a private network. A firewall router uses access lists and other methods to ensure the security of the private network.

Flow control: Technique for ensuring that a transmitting entity, such as a modem, does not overwhelm a receiving entity with data. When the buffers on the receiving device are full, a message is sent to the sending device to suspend the transmission until the data in the buffers has been processed. In IBM networks, this technique is called pacing.

Frame relay: Industry-standard, switched data link layer protocol that handles multiple virtual circuits using HDLC encapsulation between connected devices. Frame Relay is more efficient than X.25; the protocol for which it is generally considered a replacement.

Frequency bandwidth: A measurable characteristic defining the number of cycles that can be conveyed or transported by a device in one second, measured in Hertz (cycles per second).

FTP: File Transfer Protocol. Application protocol, part of the TCP/IP protocol stack, used for transferring files between network nodes. FTP is defined in RFC 959. FTP enables you to transfer files between your computer and a remote computer. A file can be anything—a text, or formatted document created using a word processor; a spreadsheet or software program; or a picture, movie, or sound file. A common use of FTP on the Internet is to download software programs.

Gateway: In the IP community, an older term referring to a routing device. Today, the term router is used to describe nodes that perform this function, and gateway refers to a special-purpose device that performs an application layer conversion of information from one protocol stack to another.

Guardband: Guardband provides for slot timing uncertainty due to inaccuracy of the ranging.

GUI: Graphical user interface. User environment that uses pictorial as well as textual representations of the input and output of applications and the hierarchical or other data structure in which information is stored. Conventions such as buttons, icons, and windows are typical, and many actions are performed using a pointing device (such as a mouse). Microsoft Windows and the Apple Macintosh are prominent examples of platforms utilizing a GUI.

Hardware: Equipment involved in production, storage, distribution or reception of electronic signals, such as the headend or the coaxial cable.

HDTV: A television signal with greater detail and fidelity than the current.

Headend: The central location, in an MSO environment, that has access to signals traveling in both the forward and reverse directions.

Header: Protocol control information located at the beginning of a protocol data unit.

HFC System: Hybrid Fiber Coax Cable Television System. Fiber is used for the trunks and coax is used to access the end nodes.

Hierarchical routing: Routing based on a hierarchical addressing system. For example, IP routing algorithms use IP addresses, which contain network numbers, subnet numbers, and host numbers.

High split: A frequency division scheme that allows bi-directional traffic on a single cable. Reverse path signals propagate to the headend from 5 to 174 Mhz. Forward path signals go from the headend from 234 Mhz to the upper frequency limit. A guardband is located from 174 to 234 Mhz.

Home passed: Total number of homes, which have the potential for being hooked up to the cable system.

HTML: Hypertext markup language. Simple hypertext document formatting language that uses tags to indicate how a given part of a document should be interpreted by a viewing application, such as a WWW browser.

HUBS: Local distribution centers where signals are taken from a master feed, and transmitted over cable to customers.

IETF: Internet Engineering Task Force. Task force consisting of over 80 working groups responsible for developing Internet standards. The IETF operates under the auspices of ISOC.

Interexchange carriers: A long distance carrier between serving areas of LATAs.

Internet: Term used to refer to the largest global internetwork, connecting tens of thousands of networks worldwide and having a "culture" that focuses on research and standardization based on real-life use. Many leading-edge network from ARPANET. At one time, called the DARPA Internet. Not to be confused with the general term Internet.

Interoperability: Ability of computing equipment manufactured by different vendors to communicate with one another successfully over a network.

IP: Internet Protocol. Network layer protocol in the TCP/IP stack offering a connectionless internet work service. IP provides features for addressing, type-of-service specification, fragmentation and reassembly, and security. IP is documented in RFC 791.

IP address: 32-bit address assigned to hosts using TCP/IP. An IP address belongs to one of five classes (A, B, C, D, or E) and is written as 4 octets separated with periods (dotted decimal format). Each address consists of a network number, an optional subnetwork number, and a host number. The network and subnetwork numbers together are used for routing, while the host number is used to address an individual host within the network or subnetwork. A subnet mask is used to extract network and subnetwork information from the IP address. Also called an Internet address.

IP over ATM: Specification for running IP over ATM in a manner that takes full advantage of the features of ATM.

ISDN: Integrated Services Digital Network. Communication protocol offered by telephone companies that permit telephone networks to carry data, voice, and other source traffic.

Isochronous: The time characteristics of an event or signal recurring at known, periodic time intervals.

LAN: Local-Area Network. High-speed, low-error data network covering a relatively small geographic area (up to a few thousand meters). LANs connect workstations, peripherals, terminals, and other devices in a single building or other geographically limited area. LAN standards specify cabling and signaling at the physical and data link layers of the OSI model. Ethernet, FDDI, and Token Ring are widely used LAN technologies.

LATA: Local Access and Transport Area. Geographic telephone dialing area serviced by a single local telephone company. Calls within LATAs are called "local calls." There are well over 100 LATAs in the United States.

Latency:

1. Delay between the time a device requests access to a network and the time it is granted permission to transmit.
2. Delay between the time when a device receives a frame and the time that frame is forwarded out the destination port.

Layer: A subdivision of the Open System Interconnection (OSI) architecture, constituted by subsystems of the same rank.

Line Of Sight (LOS): Characteristic of certain transmission systems such as laser, microwave, and infrared systems in which no obstructions in a direct path between transmitter and receiver can exist.

Link: Network communications channel consisting of a circuit or transmission path and all related equipment between a sender and a receiver. Most often used to refer to a WAN connection. Sometimes referred to as a line or a transmission link.

Local bridge: Bridge that directly interconnects networks in the same geographic area.

Local Exchange Carrier (LEC): A local telephone company within a serving area or LATA.

Local loop: The set of facilities used by a telephone company to transport signals between a central office, roughly similar to a cable TV headend, and a customer location. The LOCAL LOOP using twisted pair copper wire typically stretches a maximum of 18,000 feet between CO and customer premises.

MAC: Media Access Control. Lower of the two sublayers of the data link layer defined by the IEEE. The MAC sublayer handles access to shared media, such as whether token passing or contention will be used.

MAC address: An address that identifies a particular Medium Access Control (MAC) sublayer service access point.

MAN: Metropolitan-Area Network. Network that spans a metropolitan area. Generally, a MAN spans a larger geographic area than a LAN, but a smaller geographic area than a WAN. Compare with LAN and WAN.

MBONE: Multicast Backbone. The multicast backbone of the Internet. MBONE is a virtual multicast network composed of multicast LANs and the point-to-point tunnels that interconnect them.

Mesh: Network topology in which devices are organized in a manageable, segmented manner with many, often redundant, interconnections strategically placed between network nodes.

Mid Split: A frequency division scheme that allows bi-directional traffic on a single cable. Reverse   channel signals propagate to the headend from 5 to 108 Mhz. Forward path signals go from the headend from 162 Mhz to the upper frequency limit. The guardband is located from 108 to 162 Mhz.

Modem: Modulator-Demodulator. Device that converts digital and analog signals. At the source, a modem converts digital signals to a form suitable for transmission over analog communication facilities. At the destination, the analog signals are returned to their digital form. Modems allow data to be transmitted over voice-grade telephone lines.

Modulation: Process by which the characteristics of electrical signals are transformed to represent information. Types of modulation include AM, FM, and PAM.

MSO: Multiple System Operators Company that owns and operates more than one cable television system.

Multi-access network: Network that allows multiple devices to connect and communicate simultaneously.

Multicast address: Single address that refers to multiple network devices. Synonymous with group address. Compare with broadcast address and unicast address.

Multiplexing: The potential transmission of several feeds of the same cable network with the same programming available at different times of the day. This is seen as one possible use of the additional channel capacity that may be made available by digital compression. Multiplexing is also used by some cable networks to mean transmitting several slightly different versions of the network, for example several MTV channels carrying different genres of music.

Multipoint access: User access in which more than one terminal equipment is supported by a single network termination.

NAP: Network Access Point. Location for interconnection of Internet service providers in the United States for the exchange of packets.

NCTA: National Cable Television Association. The major trade association for the cable television industry.

Near Video On Demand (NVOD): An entertainment and information service that ''broadcasts'' a common set of programs to customers on a scheduled basis. At least initially, NVOD services are expected to focus on delivery of movies and other video entertainment. NVOD typically features a schedule of popular movies and events, offered on a staggered-start basis (every 15 to 30 minutes, for example). See Video on Demand.

Network: Collection of computers, printers, routers, switches, and other devices that are able to communicate with each other over some transmission medium.

Network management: Generic term used to describe systems or actions that help maintain, characterize, or troubleshoot a network.

Network operator: Person who routinely monitors and controls a network, performing such tasks as reviewing and responding to traps, monitoring throughput, configuring new circuits, and resolving problems.

NHRP: Next Hop Resolution Protocol. Protocol used by routers to dynamically discover the MAC address of other routers and hosts connected to a NBMA network. These systems can then directly communicate without requiring traffic to use an intermediate hop, increasing performance in ATM, Frame Relay, SMDS, and X.25 environments.

NIC: Network Interface Card. Board that provides network communication capabilities to and from a computer system.

N-ISDN: Narrowband ISDN. Communication standards developed by the ITU-T for baseband networks. Based on 64-kbps B channels and 16- or 64-kbps D channels. Contrast with BISDN.

NOC: Network Operations Center. Organization responsible for maintaining a network.

Node: A devise that consists of an access unit and a single point of attachment of the access unit for the purpose of transmitting and receiving data.

OAM cell: Operation, Administration, and Maintenance cell. ATM Forum specification for cells used to monitor virtual circuits. OAM cells provide a virtual circuit-level loopback in which a router responds to the cells, demonstrating that the circuit is up, and the router is operational.

Packet: Logical grouping of information that includes a header containing control information and (usually) user data. Packets are most often used to refer to network layer units of data. The term datagram, frame, message, and segment are also used to describe logical information groupings at various layers of the OSI reference model and in various technology circles.

Payload: Portion of a frame that contains upper-layer information (data).

Pay-Per-View: Cable programming for whom customers pay on a one-time basis (e.g., for prizefights, Broadway shows and movie premieres).

Peak rate: Maximum rate, in kilobits per second, at which a virtual circuit can transmit.

Physical layer: In Open System Interconnections (OSI) architecture, the layer that provides services to transmit bits or groups of bits over a transmission link between open systems.

Protocol: A set of rules and formats that determines the communication behavior of layer entities in the performance of the layer functions.

Proxy: Entity that, in the interest of efficiency, essentially stands in for another entity.

PSTN: Public Switched Telephone Network. General term referring to the variety of telephone networks and services in place worldwide. Sometimes called plain old telephone service (POTS).

QOS: Quality of Service. Measure of performance for a transmission system that reflects its transmission quality and service availability.

QOS parameters: Quality of Service parameters. Parameters that control the amount of traffic the source router in an ATM network sends over a SVC. If any switch along the path cannot accommodate the requested QOS parameters, the request is rejected, and a rejection message is forwarded back to the originator of the request.

RBOC: Regional Bell Operating Company. Local or regional telephone company that owns and operates telephone lines and switches in one of seven U.S. regions. The RBOCs were created by the divestiture of AT&T. Also called Bell Operating Company (BOC).

Redundancy:

1. In internetworking, the duplication of devices, services, or connections so that, in the event of a failure, the redundant devices, services, or connections can perform the work of those that failed. See also Redundant system.

2. In telephony, the portion of the total information contained in a message that can be eliminated without loss of essential information or meaning.

Redundant system: Computer, router, switch, or other computer system that contains two or more of each of the most important subsystems, such as two disk drives, two CPUs, or two power supplies. For example, on a fully redundant ATM switch, there are two NP cards with disks, two switch cards, and two power trays.

Reliability: Ratio of expected to receive (keepalive) from a link. If the ratio is high, the line is reliable. Used as a routing metric.

RF: Radio Frequency. Generic term referring to frequencies that correspond to radio transmissions. Cable TV and broadband networks use RF technology.

Router:

1. Network layer device that uses one or more metrics to determine the optimal path along which network traffic should be forwarded. Routers

forward packets from one network to another based on network layer information. Occasionally called a gateway (although the definition of gateway is becoming increasingly outdated). Compare with Gateway.

2. A router is an integral component of today's networks. It does just what its name implies: it routes information from one place to another or from one network to another. It is a network "traffic cop," directing packets, or digital chunks of information through the internetwork highways. It can also segment network traffic, prevent network "traffic jams" (congestion) and provide security. The router can be either a separate unit or can be a card inserted into an IBM-compatible PC.

Routing: Process of finding a path to a destination host. Routing is very complex in large networks because of the many potential intermediate destinations a packet might traverse before reaching its destination host.

Routing protocol: Protocol that accomplishes routing through the implementation of a specific routing algorithm. Examples of routing protocols include IGRP, OSPF, and RIP.

Routing table: Table stored in a router or some other internetworking device that keeps track of routes to particular network destinations and, in some cases, metrics associated with those routes.

Satellite: Domestic Communications. Device located in geostationary orbit above the earth which receives transmissions from separate points and retransmits them to cable systems, DBS and others over a wide area.

Satellite communication: Use of orbiting satellites to relay data between multiple earth-based stations. Satellite communications offer high bandwidth and a cost that is not related to distance between earth stations, long propagation delays, or broadcast capability.

Scrambling: A signal security technique for rendering a TV picture unviewable, while permitting full restoration with a properly authorized decoder or descrambler.

Set Top Box: Any of several different electronic devices that may by used in a customer's home to enable services to be on that customer's television set. If the "set top" device is for extended tuning of channels only, it is called a Converter. It restores scrambled or otherwise protected signals, it is a Descrambler.

Signaling: Process of sending a transmission signal over a physical medium for purposes of communication.

Single-mode fiber: Fiber-optic cabling with a narrow core that allows light to enter only at a single angle. Such cabling has higher bandwidth than multimode fiber, but requires a light source with a narrow spectral width (for example, a laser). Also called monomode fiber.

SNMP: Simple Network Management Protocol. Network management protocol used almost exclusively in TCP/IP networks. SNMP provides a means to monitor and control network devices, and to manage configurations, statistics collection, performance, and security.

SONET: Synchronous Optical Network. High-speed (up to 2.5 Gbps) synchronous network specification developed by Bellcore and designed to run on optical fiber. STS-1 is the basic building block of SONET. Approved as an international standard in 1988.

Spanning tree: Loop-free subset of a network topology.

Spoofing:
1. Scheme used by routers to cause a host to treat an interface as if it were up and supporting a session. The router spoofs reply to keepalive messages from the host in order to convince that host that the session still exists. Spoofing is useful in routing environments such as DDR, in which a circuit-switched link is taken down when there is no traffic to be sent across it in order to save toll charges.
2. The act of a packet illegally claiming to be from an address from which it was not actually sent. Spoofing is designed to foil network security mechanisms such as filters and access lists.

SS7: Signaling System number 7. Standard CCS system used with BISDN and ISDN. Developed by Bellcore.

Star topology: LAN topology in which end points on a network are connected to a common central switch by point-to-point links. A ring topology that is organized as a star implements a unidirectional closed-loop star, instead of point-to-point links. Compare with bus topology, ring topology, and tree topology.

Statistical multiplexing: Technique whereby information from multiple logical channels can be transmitted across a single physical channel. Statistical multiplexing dynamically allocates bandwidth only to active input channels, making better use of available bandwidth and allowing more devices to be connected than with other multiplexing techniques. Also referred to as statistical time-division multiplexing or stat mux. Compare with ATDM, FDM, and TDM.

Sublayer: A subdivision of a layer in the Open System Interconnection (OSI) reference model.

Subnetwork: Subnetworks are physically formed by connecting adjacent nodes with transmission links.

Subscriber: Customer paying a monthly fee to cable system operators for the capability of receiving diversity or programs and services.

Switch:
1. Network device that filters, forwards, and floods frames based on the destination address of each frame. The switch operates at the data link layer of the OSI model.
2. General term applied to an electronic or mechanical device that allows a connection to be established as necessary and terminated when there is no longer a session to support.

Switched connection: A connection established via signaling.

Synchronization: Establishment of common timing between sender and receiver.

T1: Digital WAN carrier facility. T1 transmits DS-1-formatted data at 1.544 Mbps through the telephone-switching network, using AMI or B8ZS coding.

TCP: Transmission Control Protocol. Connection-oriented transport layer protocol that provides reliable full-duplex data transmission. TCP is part of the TCP/IP protocol stack. See also TCP/IP.

TCP/IP: Transmission Control Protocol/Internet Protocol. Common name for the suite of protocols developed by the U.S. DOD in the 1970's to support the construction of worldwide internetwork. TCP and IP are the two best-known protocols in the suite.

TDM: Time-Division Multiplexing. Technique in which information from multiple channels can be allocated bandwidth on a single wire based on preassigned time slots. Bandwidth is allocated to each channel regardless of whether the station has data to transmit. Compare with ATDM, FDM, and statistical multiplexing.

Throughput: Rate of information arriving at, and possibly passing through, a particular point in a network system.

Token: Frame that contains control information. Possession of the token allows a network device to transmit data onto the network.

Token ring: Token-passing LAN developed and supported by IBM. Token ring runs at 4 or 16 Mbps over a ring topology. Similar to IEEE 802.5.

Topology: Physical arrangement of network nodes and media within an enterprise networking structure.

Traffic parameter: A parameter for specifying a particular traffic aspect of a connection.

Traffic policing: Process used to measure the actual traffic flow across a given connection and compare it to the total admissible traffic flow for that connection. Traffic outside of the agreed upon flow can be tagged (where the CLP bit is set to 1) and can be discarded en route if congestion develops. Traffic policing is used in ATM, Frame Relay, and other types of networks. Also known as admission control, permit processing, rate enforcement, and UPC (usage parameter control).

Traffic shaping: Use of queues to limit surges that can congest a network. Data is buffered and then sent into the network in regulated amounts to ensure that the traffic will fit within the promised traffic envelope for the particular connection. Traffic shaping is used in ATM, Frame Relay, and other types of networks. Also known as metering, shaping, and smoothing.

Transit delay: The time difference between the instant at which the first bit of a PDU crosses one designated boundary, and the instant at which the last bit of the same PDU crosses a second designated boundary.

Transmission link: The physical unit of a subnetwork that provides the transmission connection between adjacent nodes.

Transmission medium: The material on which information signals may be carried; e.g., optical fiber, coaxial cable, and twisted-wire pairs.

Transmission system: The interface and transmission medium through which peer physical layer entity transfer bits.

Trunk: Physical and logical connection between two ATM switches across which traffic in an ATM network travels. An ATM backbone is composed of a number of trunks.

Trunk cable: Cables that carry the signal from the headend to groups of subscribers.

UBR: Unspecified Bit Rate. QOS class defined by the ATM Forum for ATM networks. UBR allows any amount of data up to a specified maximum to be sent across the network, but there are no guarantees in terms of cell loss rate and delay. Compare with ABR (available bit rate), CBR, and VBR.

UDP: User Datagram Protocol. Connectionless transport layer protocol in the TCP/IP protocol stack. UDP is a simple protocol that exchanges datagrams without acknowledgments or guaranteed delivery, requiring that error processing and retransmission be handled by other protocols. UDP is defined in RFC 768.

Unbundling: The separation and discrete offering of the components of the local telephone service. UNBUNDLING of network components facilitates the provision of ''pieces'' of the local network, such as local switching and transport, by telephone company competitors.

UNI: User-Network Interface. ATM Forum specification that defines an interoperability standard for the interface between ATM-based products (a router or an ATM switch) located in a private network and the ATM switches located within the public carrier networks. Also used to describe similar connections in Frame Relay networks.

Unicast: An address specifying a single network device.

Upstream: Flow of any information from the customer, through the cable system, to the headend.

URL: Universal Resource Locator. Standardized addressing scheme for accessing hypertext documents and other services using a WWW browser. See also WWW browser.

Variable Bit Rate (VBR): A type of telecommunication service characterized by a service bit rate specified by statistically expressed parameters which allows the bit rate to vary within defined limits.

VCI: virtual channel identifier. 16-bit field in the header of an ATM cell. The VCI, together with the VPI, is used to identify the next destination of a cell as it passes through a series of ATM switches on its way to its destination. ATM switches use the VPI/VCI fields to identify the next network VCL that a cell needs to transit on its way to its final destination. The function of the VCI is similar to that of the DLCI in Frame Relay.

Video dialtone: A means by which telephone companies may provide transmission facilities and for on-telco video programming as well as certain enhanced services to third party programmers.

Video on Demand (VOD): An entertainment and information service that allows customers to order programs from library of material at any time.

Virtual Channel (VC): The communication channel that provides for the sequential unidirectional transport of ATM cells.

Virtual circuit: Logical circuit created to ensure reliable communication between two network devices. A virtual circuit is defined by a VPI/VCI pair, and can be either permanent (a PVC) or switched (a SVC). Virtual circuits are used in Frame Relay and X.25. In ATM, a virtual circuit is called a virtual channel. Sometimes abbreviated VC.

Virtual path: Logical grouping of virtual circuits that connect two sites.

WAN: Wide-Area Network. Data communications network that serves users across a broad geographic area and often uses transmission devices provided by common carriers. Frame Relay, SMDS, and X.25 are examples of WANs. Compare with LAN and MAN.

WWW: World Wide Web. Large network of Internet servers providing hypertext and other services to terminals running client applications such as a WWW browser.

X.25: ITU-T standard that defines how connections between DTE and DCE are maintained for remote terminal access and computer communications in PDNs. X.25 specifies LAPB, a data link layer protocol, and PLP, a network layer protocol. To some degree, Frame Relay has superseded X.25.

ACRONYMS

AAL ATM Adaptation Layer

ABR Available Bit Rate

ADSL Asymetric Digital Subscriber Line

ANSI American National Standards Institute

API Application Program Interface

APON ATM Passive Optical Network

ASK Amplitude Shift Keying

ATM Asynchronous Transfer Mode

BECN Backward Explicit Congestion Notification

BER Bit Error Rate

B-ICI BISDN Inter Carrier Interface

BISDN Broadband Integrated Services Digital Network

BSS Broadband Switching System

BW Bandwidth

CAC Connection Admission Control

CATV Community Antenna Television

CBR Constant Bit Rate

CCITT International Telephone and Telegraph Consultative Committee

CDMA Code Division Multiple Access

CdPN Called Party Number

CDV Cell Delay Variation

CES Circuit Emulation Service

CgPN Calling Party Number

CLP Cell Loss Priority

CPCS Common Part Convergence Sublayer

CPE Customer Premises Equipment

CRC Cyclic Redundancy Check

CSI Carrier Selection Information

CSMA/CA Carrier Sense Multiple Access/Collision Avoidance

CSMA/CD Carrier Sense Multiple Access with Collision Detection

DBS Digital Broadcasting System

DPSK   Differential Phase Shift Keying

DRA   Dynamic Rate Adaptation

DS-0   Digital Signal level -0 (DS0=64 Kb/s)

DSP   Digital Signal Processors

DTE   Dtat Terminal Equipment

E1   ETSI—Digital Signal Level 1 (E1=2.048 Mb/s)

EFCI   Explicit Forward Congestion Indication

ETSI   European Telecommunications Standards Institute

FEBE   Far End Block Error

FECN   Forward Explicit Congestion Notification

FERF   Far End Receive Failure

FRS   Frame Relay Service

FSAN   Full Services Access Network

FTTC   Fiber to the curb

FTTH   Fiber to the home

GEO   Geo-stationary Earth Orbit

HDTV   High Definition TV

HE   Head End

HEC   Header Error Control

HFC   Hybrid Fiber Coax

IAM   Initial Address Message

ICIP   Inter Carrier Interface Protocol

IEEE   Institute Of Electrical and Electronics Engineers

IETF   Internet Engineering Task Force

ILEC   Independent Local Exchange Carrier

IN   Intelligent Network

IP   Internet Protocol

ISDN   Integrated Services Digital Network

ITU   International Telecommunication Union

IXC   Inter Exchange Carrier

LAN/WAN   Local/Wide Area Network

LATA   Local Access and Transport Area

LEC   Local Exchange Carrier

LEO Low earth orbit (satellite)

LLC Logical Link Control

LMDS Local Multipoint Distribution Service

MAC Medium Access Control

MEO Medium Earth Orbiting

MIB Management Information Base

MMDS Multi-Channel/Multi-Point Distribution System

MPEGx Motion Picture Editors's Group Compression Algorithm x

NNI Network Node Interface

NPC Network Parameter Control

NT Network Terminal

ONU Optical Network Unit

OSI Open System Inerconnect

PCR Peak Cell Rate

PDH Pleisiochronous Digital Hierarchy

PLCP Physical Layer Convergence Procedure

PMD Physical Medium Dependent

POTS Plain Old Telephone Service

PV Personal Computer

PVC Permanent Virtual Connection

QAM Quadrature Amplitude Modulation

QOS Quality of Service

QPSK Quaternary Phase Shift Keying

RF Radio Frequency

SAAL Signaling ATM Adaptation Layer

SCR Sustained Cell Rate

SDH Synchronous Digital Hierarchy

SMDS Switched Multi-megabit Data Service

SNI Service Node Interface

SNMP Simple Network Management Protocol

SONET Synchronous Optical NETwork

SPE Synchronous Payload Envelope

SRTS Synchronous Residual Time Stamp

SS7   Signaling System Number 7

SSCOP   Service Specific Connection Oriented Protocol

SSCS   Service Specific Convergence Sublayer

STB   Set Top Box

STM   Syncronous Transfer Mode

SVC   Switched Virtual Connection

SWAP   Shared Wireless Access Protocol

TDMA   Time Division Multiple Access

UNI   User Network Interface

UPC   Usage Parameter Control

VBR   Variable Bit Rate

VCC   Virtual Channel Connection

VCI   Virtual Channel Identifier

VPC   Virtual Path Connection

VPCI   Virtual Path Connection Identifier

BIBLIOGRAPHY

ANSI/IEEE. 1985. "Carrier Sense Multiple Access with Collision Detection (CSMA/CD) Access Method and Physical Layer Specifications." ANSI/IEEE Std. 802.3-1985.

A. Azzam, "Access Technologies" Cross Industry Working: The Way Things Work on the NII." Revised June 10, 1998.

A. Azzam, et al. "ATM over ADSL" ATM Forum Magazine (53 Byte). 1997.

A. Azzam, et al. Brandt M. Dajer, M. Eng, J. Lin., D. Mollenauer, J. Siller, C. Grobicki, C. Sriram, K. Ulm, J. November 1993. "IEEE P 802.14 Cable-TV Functional Requirements and Evaluation Criteria."

J. Bingham and K. Jacobson, "A Proposal for a MAC Protocol to Support Both QPSK and SDMT." IEEE 802.14-95/137.

Scott Bradner, Nework World, 1998.

Eric Brown, Special to PC World, 1998.

David D. Clark, "Workshop on Internet Economics." March 1995, M.I.T.

Dave Cobbley, "The Connected Home: beyond the last mile." XIWT White Paper, August 1998.

Computers: Consumer Service and The Internet (Industry Surveys). A division of McGraw-Hill. Vol. 166, No. 42, section 2. October 15, 1998.

B. Currivan, "CATV Upstream Channel Model, Rev 1.0" IEEE 802.14-95/133, Data-Over-Cable Service Interface Specifications Radio Frequency Interface Specification.

Martin De Prycker, "Asynchronous Transfer Mode Solution for B-ISDN." 1993.

Martin De Prycker "White Paper on DMT Technology Using DMT Technology for ADSL Systems." December 11, 1996.

Sharon Fisher, "Utilicorp Alliance Draws Industry Attention."

D. Gingold, "Integrated Digital Services for Cable: Economics, Architecture, and the Role of Standards—MIT Research Program on Communications Policy." IEEE 802.14-96/230. September 1996.

Jack Glas, "The Principle of Spread Spectrum Communication: Thesis." August 1996.

N. Golmie, "Performance Evaluation of Contention Resolution Algorithms: Ternary-tree vs. p-Persistence." IEEE 802.14/96-241. September 1996.

N. Golmie, "Performance Evaluation of MAC Protocol Components for HFC Networks." Broadband Access Systems, Proc. SPIE 2917, (Boston, Massachusetts), November 1996 at 120–130.

Bill Griffin (GTE Lab), "Internet Access Performance" XIWT White Paper—August 1998.

D. Grossman (Motorola), "Security Overview." IEEE 802.14-97/030. March 1997.

Michael Hauben and Rhonda Hauben, "Netizens: On the History and Impact of Usenet and the Internet" (IEEE Computer Society Press, 1997).

Goeff Huston, "ISP Survival Guide: Strategies for Running a Competitive ISP" (John Wiley and Sons, Inc., 1999).

IEEE Project 802.14/a Draft 3 Revision 1. Cable-TV Access Method and Physical Layer Specification. March 1998.

ISO/IEC10039 ISO/IEC 10039:1991 "Information technology—Open Systems Interconnection—Local area networks—Medium Access Control (MAC) Service Definition."

ITU-T Recommendation H.222.0 (1995).

ITU-T Recommendation I.211.

ITU-T Recommendation J.83 (10/95) "Digital multi-programme systems for television sound and data services for cable distribution."

Intelogis (PLUG-IN™ Technology) "Power Line Communications White Paper" November 1998.

Larry Irving, "Rules and Policies for Local Multipoint Distribution Service" First Report and Order and Fourth Notice of Proposed Rulemaking, CC Doc. No. 92-297.

Samir Kamal (HP) IEEE/XIWT Workshop on Broadband Access. November 9–10, 1995.

J. Karaoguz, and Gottfried Ungerboeck. "Formal Proposal: Frequency Agile Multi-Mode (FAMM) Single-Carrier Modems for Upstream Transmission in HFC Systems." IEEE 802.14-95/131. November 7, 1995.

J. Tom Kolze (ComStar Commnuication), "Comparisons of Various Advanced (Upstream) PHY Modulation Techniques." IEEE 802.14/98-019a.

T. Kwok, "Communications Requirements of Multimedia Applications: A preliminary study." International Conference on Selected Topics in Wireless Communications (Vancouver, Canada. 1992).

M. Laubach, "ATM HFC Overview." IEEE 802.14-95/022. March 1995.

Tom McCabe (Bosch Telecom Inc.), "LMDS White Paper" 1998.

R. D. Mistry, et al. "Management Requirements for a Full Services Access Network," FSAN98 Third Workshop on Full Services Access Networks, Venice, Italy, March 1998.

John Montgomery, "The Orbiting Internet: Fiber in the Sky." November 1997.

Andrew Morriss (the freeman), "Does the Internet prove the need for government investment?" November 1998.

Cesare Mossotto, "Evolution towards a full service network in Italy," Third Workshop on Full Services Access Networks, Venezia, Italy, March 1998.

Michael Murphy, "High Tech Stocks and Mutual Funds, 2nd edition." 1999.

James Niccolai, "Year of the (Un)Wired House" IDG News Service January 1999.

Andrew Odlyzko (AT&T Labs—Research), "Smart and stupid networks: Why the Internet is like Microsoft." Revised version, October 6, 1998.

Richard R. Peterson, "The DBS Connection/Direct Broadcast Satellite- A New Generation of Television In America." Version 32. April 2, 1997.

R. Prodan, "Letter From CableLabs—Regarding Annex A/B)." IEEE 802.14-96/205. July 1996.

Michael Propp, "The Use of Existing Electrical Powerlines for High Speed Communications to the Home." April 7, 1998.

S. Quinn, "Requirements on HFC Access Networks—A Public Network Operator's Perspective." IEEE 802.14-95/099. September 1995.

Denny Radford (Member IEEE), "New Spread Spectrum Technologies Enable Low Cost Control Applications for Residential and Commercial Use." July 1998.

SP-RFI-I02-971008 (Interim Specification). 1997.

Shlomo Rakib (Terayon Communication Systems), "Synchronous Code Division Multiple Access (S-CDMA)." National Cable Television Association Convention. April 30, 1996.

Dennis M. Ritchie and Ken Thompson, "The UNIX Time-Sharing System," Association for Computing Machinery (ACM), Vol. 17, July–December 1974.

H. Schulzrinne, GMD Fokus, "ATM: Dangerous at Any Speed?" 1998.

Stephen Segaller, "Nerds 2.0.1: A Brief History of the Internet." (TV Books, L.L.C, 1998).

H. Sorre (Alcatel), ''SkyBridge''—White Paper, March 1997.

Spectrum magazine (February 1999).

P. Spruyt and G. Van der Plas, ''Evolution from ADSL to VDSL: The Technological Challenges,'' Proceedings of ISS'97, September 1997, at 515–521. ''Draft Physical Layer Specification for CAP/QAM-Based Rate Adaptive Digital Subscriber Line (RADSL),'' T1E1 Contribution No. 97-104R2a, October 15, 1997.

P. Spruyt, M. Mielants and S. Braet (Alcatel, Corporate Research Center), ''ADSL AND VDSL'' 1998 White Paper.

P. Sriram, ''ADAPt MAC PDU and MAC-PHY Services in Support of ATM.'' IEEE 802.14-95/168. November 1995.

Telecommunication, Code of Federal Regulations, 47 CRF 15.107.

Feeney L. Tracy (Ericsson), Partially contributed to ''Advanced Internet Application.'' Feb. 1999.

Antonov Vadim, ''ATM: Another Technological Mirage Why ATM Is Not The Solution,'' White Paper. October, 1996.

J. Vandenameele, ''How to upgrade CATV networks to provide interactive ATM-based services.'' GLOBECOM 95. (Singapore, November 1995).

G. Van der Plas, ''APON: An ATM-based FITL System.'' EFOC&N'93, 1993.

W. Verbiest, ''Integrated Broadband Access.'' Proc. Fourth IEEE Conference.

Cerf Vint (MCI), ''Evolution of the Internet.'' December 1996 XIWT Workshop White Paper.

Virginia Polytechnic Institute and State University, ''LMDS White Paper'' 1998.

Phillip J. Worthy (Nortel Networks), Partially contributed to ''Advanced Internet Applications.'' February 1999.

XIWT Symposium on Cross-Industry Activities for Information Infrastructure Robustness ''Critical Infrastructure: The Path Ahead.'' November 1998.

INDEX

6

6 MHz, 31, 93, 153, 155, 157–158, 168, 173, 187, 218, 297

10

10BaseT, 113, 170, 248, 250, 343

A

AAL1, 61, 343
AAL-2, 61
AAL5, 61, 343
Abeline, 85–86, 323
ABR, 63, 67, 69–70, 118, 171, 182, 188–189, 259, 261–264, 296, 299–300, 319, 343, 358, 361
ABR connection, 264
ABR service, 70, 263–264
AC power, 44, 159, 161, 163, 237, 245
ADSL, 3–4, 10, 18–21, 25, 41, 47, 56, 61–62, 92, 98, 103, 113, 123–127, 130–132, 134–143, 145, 147, 150, 152, 167, 187, 228, 232, 236, 249, 251–255, 261, 264, 267, 280–282, 292–305, 308–310, 330–331, 333, 336, 361
ADSL access, 10, 299–300
ADSL Forum, 309
ADSL market, 309
ADSL modem, 92, 141–142, 295, 298–299, 302, 310
ADSL service, 143, 252
ADSL subscriber, 280, 298
ADSL-lite, 141–143, 252
Ameritech, 14, 329–332
AOL, 268, 284, 287, 336, 338
APON, 5, 18, 47, 110–114, 117–121, 135, 261, 361
APON networks, 111

APON system, 113–114
Arbitrage, 25–26
ARPANET, 11, 30, 336, 338, 350
AT&T, 4, 14–17, 23–24, 28, 103, 109, 151, 186, 249, 280, 291, 305, 331, 337, 354
ATM, 5, 10–11, 30, 47–59, 61–63, 66–71, 81, 85, 94, 96–103, 105, 109, 111–115, 118, 126, 131, 135–139, 144, 152, 165–167, 170–171, 174–178, 181–186, 200, 205, 259, 261–262, 293, 296–297, 303, 308–311, 318–319, 323, 326, 338, 343–345, 347, 350, 353–354, 358–359, 361, 363
ATM cell, 61, 99, 114, 137, 139, 174, 176, 181, 185, 296
ATM Forum, 48, 53–54, 61–62, 311, 318–319
ATM header, 56, 70, 137–138
ATM interface, 138–139, 296
ATM layer, 59, 70, 343
ATM network, 10, 344, 347, 358
ATM service, 69, 261
ATM switch, 359
ATM user, 53, 68
Availability, 280

B

Backbone, 84–85, 344
BBN, 338
Bell Atlantic Corporation, 331
BellSouth, 5, 14, 45, 109, 111, 121, 331–332
Bent pipe, 199
BER, 133–134, 137, 145, 292, 295, 361
Best effort, 101
B-ISDN, 47–48, 50, 109, 296
BOCs, 8–10
Broadband access, 2, 5, 7–10, 13, 34, 46, 105, 110–111, 120–121, 206, 215, 235–236, 257, 329, 333
Business case, 90–91, 267

C

Cable modem, 10, 47, 61–62, 66, 87, 92–94, 152, 154, 156–157, 159–161, 164–179, 181–188, 232, 236, 253–255, 267, 280–282, 292, 297–302, 311–317, 336
Cable modem capacity, 297
Cable modem market, 303
Cable modem operation, 185
Cable modem technology, 296
Cable network, 30, 32, 43, 86–87, 91–92, 153–156, 158–161, 167–168, 170, 172, 174, 179, 218, 297, 299, 352
Cable plant, 95, 161–164, 173
Cable TV, 152, 155, 218–219, 316–317, 354
CableLabs, 87, 311, 317–318
Cablevision, 334–335
CAC, 67–69, 361
CAP, 140–141, 146, 149, 308–309, 346
Carrier frequency, 246
CATV, 1–2, 4, 7–9, 15, 31–32, 86, 90, 93, 105–106, 124, 144, 151–154, 157–159, 211, 225, 254, 256, 301, 303–305, 317–318, 330, 333–334, 345, 361
CATV companies, 303
CATV franchises, 330
CATV network, 16
CATV services, 331
CATV subscribers, 317
CBR, 60, 69, 118, 170–171, 182, 188–189, 261–262, 299–300, 343, 346, 358, 361
CDMA, 95, 187–189, 246, 361
CDV, 66, 68–69, 262, 361
CEBus, 243, 246–247
Central Office, 41, 135, 138, 294, 345
  see also CO
Channel, 4, 19, 31, 41–42, 50, 62, 64, 70, 93, 123–124, 128, 134, 136–140, 144, 146, 153–154, 158, 160–161, 163–164, 168, 173–174, 178, 181–182, 184–185, 187–188, 218, 246, 294–295, 297, 301, 309, 351–352, 356–357, 359
Channel Acquisition, 178
Channel allocation, 153
Channel capacity, 154, 345
Channel surfing, 163
Cisco, 24, 85, 96, 102, 271

CLASS™, 100
Class 4, 44, 196, 284
Class 5, 41, 43, 87, 94, 196, 284, 342
CLEC, 4, 15–22, 42, 47, 87, 106, 141, 214–215, 233, 329, 339–340, 342
CNRI, 286, 290
CO, 41–43, 130, 302, 310, 315, 339, 351
Coaxial cable, 345, 349, 358
Comcast Corporation, 96, 330, 334
Congress, 15, 17, 22, 80–81, 154, 219, 339
Constant bit rate, 258
  see also CBR
Constellation, 187, 194, 198, 207
Contention, 170, 177–178, 180–185, 315, 351
Contention Resolution, 183–184
Continental Cablevision, 22, 313, 315, 332, 334
Cox Communications, 313, 315, 335
CRISP, 289–290
Cryptography, 89
CSMA, 239, 241, 246–247, 249, 252, 347, 361
CSMA/CA, 239, 241, 247, 361
Cybernet Policy, 22
Cyberterrorism, 289

D

DARPA, 79, 82–83, 344, 346, 350
Data-over-powerline. See powerline
DAVIC, 149, 166, 266, 309–311, 319–321
DBS, 22, 86, 90, 93, 158, 193, 216–219, 236, 254, 330, 335, 347, 355, 361
DIFFSERV, 103
Digital Broadcasting, 193, 266–267, 361
Digital powerline, 232
Digital Television, 274
DirecTV, 9, 158, 219, 330–331
Distant learning, 266–267
Distributed architecture, 284
Distribution service, 259–261
DLC, 19–21, 41–42, 94
DLSAM, 20
DMT, 131–133, 135–141, 146–148, 187, 228, 292–295, 308–309, 316

DMT tones. See DMT
DOCSIS™, 165–166, 174, 187–190, 267, 301
DOD, 82, 344, 357
Downlink, 204–205, 209, 217–218
Downstream, 3–4, 92, 95, 110, 113–114, 116–117, 119–120, 124, 126, 129, 131–136, 138, 140, 142, 144–147, 149–150, 157, 160, 162–163, 168–169, 172–175, 177–178, 181, 184–186, 200–201, 224, 231, 292–295, 297–299, 315, 347
Downstream transmission, 132, 150, 295
DSL, 3–4, 18–21, 42, 51, 126–128, 130, 150, 252, 293, 301, 308–309
DSP, 31, 142, 362

E

E911, 284
E-commerce, 186, 268, 291
Economic model, 87, 90–91
ECTF, 326
Electronic Games, 266, 268
email, 70, 181, 238, 263, 270, 336
Encryption, 89, 110, 119–120, 179, 193, 224, 241, 300, 312
Ethernet, 97, 170, 242–244, 249–251, 302, 343, 347, 350

F

Fault Tolerant, 281
FCC, 2, 6, 10, 14, 17–22, 24–26, 28, 42, 93, 96, 153–154, 162, 197, 203, 206, 214–215, 219, 228, 249, 251, 274, 284, 301, 312, 324, 333, 348
FDM, 129, 157, 173, 251, 294–295, 356–357
Fiber node, 157, 162
Fiber ring, 88
Fiber-to-Home, 109
  see also FTTH
Frequency agile, 160
FSAN, 110–113, 117, 309, 362
FSK, 125, 228, 245

FTTC, 112, 125, 149, 279, 296, 319, 362
FTTH, 111, 296, 319, 362

G

Gemini project, 277
GEO, 193–196, 362
Geosynchronous, 192, 196, 217
GigaPoP, 84
GII, 34, 37, 46, 151
Gore, Al, 34, 80, 269
GSM, 37, 48, 208

H

HDLC, 131, 348
HDSL, 18, 21, 43, 113, 127, 145, 293
HDTV, 10, 125, 274, 317, 349, 362
Headend, 64, 88, 91, 93–95, 155–157, 162, 167–169, 171, 173–174, 176–185, 187–189, 212–213, 225, 300, 302, 314–315, 333, 347–349, 351–352, 358–359
HEC, 55, 58, 138, 176, 362
HFC, 4, 9, 62, 87–92, 94–95, 121, 153–154, 156–160, 163, 178, 183, 267, 279, 299–303, 305, 311–312, 314–315, 317, 319, 321, 331–332, 349, 362
HFC access shortfalls, 89
HFC network, 88, 91, 94, 267
HFC operation and maintenance, 89
HFC reliability, 89
HFC security, 89
HFC topology, 89, 92
High-speed wireless access, 191
Hi-PHY, 189
Home networking, 235–236
Home Office, 127, 267
Home passed, 91–93, 299
Home shopping, 266
HomePNA, 248–250, 252–254
HomeRF, 238–242, 248–249, 253

I

IECs, 15
IEEE 802.14, 61, 164–167, 171, 174, 179, 183–184, 187–190, 298, 300, 311–312, 315–317, 319
IETF, 81, 195, 310, 318, 324–325, 349, 362
ILEC, 16–22, 39, 42, 47, 87, 89–90, 94, 106, 329, 339–340, 362
IMTC, 326–327
IN, 44–45, 100, 362
Ingress noise, 160–162, 164
Intelligent Network, 37, 362
  see also IN
Interactive services, 45, 86, 90–91, 259–260
Internet, 2, 4, 7, 9–11, 13–14, 16, 22–28, 30–33, 36–37, 45, 50, 52, 79–85, 87, 90, 94, 96–103, 124–126, 134, 150, 152, 157, 159, 165, 170, 186, 192, 195–197, 200–202, 210, 223, 225–227, 231–232, 236, 239–240, 243, 245, 252–255, 257–259, 264–269, 271–273, 275, 277–278, 281–284, 288–289, 291–292, 296, 298, 301–303, 310, 314, 321–325, 327, 332–333, 334–341, 346–352, 357, 360, 362
Internet access, 9, 75, 77, 79, 124, 202, 232, 267, 302, 336, 340
Internet access charges, 90
Internet and QOS, 264
Internet network, 240, 265, 281, 302–303
Internet telephony, 22–26
Internet2 (See also UCAID)
I2, 79, 84–85, 272, 275, 322
IP, 4, 9–11, 18, 23–24, 26–28, 30, 47–48, 62, 85, 87, 94, 96–103, 118, 133–134, 152, 165–166, 170–171, 174–177, 181–186, 192, 195, 201, 205, 225, 244, 253, 255, 264, 267, 269, 272, 274, 280, 282, 284, 292, 297–298, 301–302, 305, 312, 323, 339, 341, 347–350, 356–358, 362
IP address, 28, 186, 350
IP datagram, 96
IP over ATM, 99, 165, 350
IP packet, 118, 165
IP router, 94, 302
IP traffic, 94, 170, 302
IPv6, 85

Iridium, 193, 196, 206–210
ISDN, 36, 40, 45, 47–50, 53–54, 61–62, 65, 261, 296, 362
ISOC, 324–325, 349
ISP, 10, 26–27, 87, 96, 101–102, 192, 212, 232, 239, 284, 336–339
ITC, 322
ITU, 30, 34, 45, 47–49, 53–57, 59–65, 67, 111, 125, 138, 165, 172, 174, 176, 249, 258–259, 261, 265, 289, 303, 307, 309–310, 318–319, 323, 326, 343–344, 347, 353, 360, 362

J

J.83, 165, 172

K

Ka-band, 199, 204, 207
killer app, 266, 296

L

LAN, 7–8, 28, 66, 134, 170, 233, 235–236, 238–239, 241–242, 249, 251–252, 292, 319, 343, 347, 350–351, 356–357, 359, 362
LATA, 15, 54, 331, 350–351, 362
Law Enforcement, 271
Layer 2 versus layer 3, 98
Legacy network, 87
LEO, 47, 193, 195–197, 199, 202–205, 207, 363
Line-Of-Sight, 212
  see also LOS
LMDS, 6, 193, 210–216, 219, 261, 321, 339, 341, 363
LOS, 211–213, 218, 351
LTA, 34, 37–38, 46

M

MAC, 157, 161, 165, 169–171, 173–174, 176–182, 188, 229, 243, 247, 249, 297,

300–301, 315–317, 319, 345, 351, 353, 363
MAC address, 178–179
MAC layer, 165, 170, 178, 316–317
MAC user, 55, 182, 159
MBone, 273
MCI, 17, 24, 28, 75, 77, 84, 103, 280, 291, 305, 323, 337
McLeodUSA, 342
MCNS, 164–165, 183–184, 311, 313–315
MEO, 193, 195, 363
Messaging services, 260
MFJ, 15, 330
Minislot, 173, 181
MMDS, 22, 31, 331–332, 363
Modernization, 46, 91–92
MPEG, 31, 60, 125, 135, 158, 174–176, 185–186, 218, 295, 299
MQAM, 146, 150, 309
MSO, 22, 159, 166–167, 186, 279–281, 290–291, 305, 311, 333, 336
MSOs, 87, 90–91, 311–312
Multicast, 85, 278, 352

N

NAP, 352
NASA, 79, 82–83, 194, 265, 275
Network management, 62, 82–83, 186, 207–208, 305, 309, 314
Network termination, 66, 139, 352
Network Unification, 30
NextLink, 6, 216, 341
NGDLC, 41–43
NGI, 79–84, 275, 277, 282, 321–322
NIC, 353
NIH, 79, 83, 265, 277
NII, 269–271
NIST, 80, 82–83, 277
Noise, 133, 158, 160–164, 227, 246, 295
Nomadicity, 191–192
NPC, 67–68, 363
NSF, 80, 82–84, 276–277, 323
NVOD, 186, 266–267, 291, 352

O

OAM, 58, 66, 111, 139, 353
Optical, 5, 20, 31–32, 98, 105–107, 109, 111, 114, 117–119, 125, 145, 157, 222, 233, 272, 340, 345, 356, 358
Optoelectronic, 107
OSI, 54, 317, 345–346, 350–351, 353, 356–357, 363

P

PAD, 240, 245
PBX, 28, 60, 66
PCS, 94–96, 179, 229, 306, 318, 331, 335, 342
Phone Doubler, 269
Phoneline, 248–254
Physical layer, 55, 172, 187, 228, 251–252
Policy, 22, 78, 154, 166, 286, 289, 322
POS, 85–86
POTS, 14, 18, 25, 36, 90, 94, 113, 118, 125–126, 130, 132, 134, 136, 143–144, 280, 292–293, 295, 301–302, 305, 354, 363
Power line, 162, 222–225, 227, 243–244 see also Powerline
Power Line Carrier, 221
Power outage, 301
Powerline, 6–7, 221, 223, 225–233, 242–247, 256
Privacy, 89, 110, 157, 159, 167, 178–179, 229, 267, 289
PSINet, 337, 339
PSTN, 24, 28–30, 39, 43, 45, 47, 94, 102, 223, 225, 227, 231, 239, 241, 253, 269, 282–283, 341, 354

Q

QAM, 93, 125, 132, 140–141, 146, 149, 157–158, 168, 172–173, 187, 295, 297, 308–309, 363
QBone, 85–86

QOS, 10, 51, 62, 64, 66–69, 85, 99, 101, 137, 166, 181–182, 190, 259, 261, 264–265, 319, 354, 358, 363
QPSK, 158, 168, 173, 187–188, 215, 218, 297, 363

R

RADSL, 127, 140, 309
RBOCs, 14–15, 18, 39, 87, 267, 279, 281, 290–291, 305, 329–330, 332, 341, 354
Registration, 165, 168, 178–180, 218
Regulation, 23, 25, 154
Regulatory, 13, 25, 34
Reliability, 27–28, 41, 44, 66, 89–90, 99–100, 120, 154, 157, 159, 186, 203, 211, 224, 247, 267, 281, 283–286, 290, 301, 305, 354
Retrieval services, 260
Revenue, 90, 267–268
RF, 42, 93, 129–130, 145, 147, 149, 161–162, 172, 174, 177, 201, 213, 219, 224–227, 237–238, 242, 256, 280, 300, 315, 354, 363
RF power, 162, 174, 177
RF spectrum, 93
RF wireless, 237
Routers, 26, 30, 32, 85, 94, 96–97, 99, 101–103, 249, 251, 338, 348, 353, 356
RSVP, 103, 278

S

Satellite, 2, 6, 16, 27, 37, 155, 192–209, 217–218, 220, 270–271, 277, 335, 345, 347, 363
SBC Communications, 14, 22, 329
Scalability, 97, 99, 102, 313
SCP, 44
SCTE, 311–312
SDMT, 146, 148–149, 309
Service differentiation, 93
Set Top Box, 355, 364

Signaling, 45, 62–63, 70, 319, 355–356, 363–364
SkyBridge, 6, 193, 195–202
SONET, 5, 43, 55–56, 59, 85, 108–109, 139, 233, 318, 345, 356, 363
Spectrum, 93, 96, 295, 314, 344
Splitter, 5, 109, 113–114, 118, 126, 130, 136, 141–143, 160, 302
Spread spectrum, 228, 245–246, 248
SS7, 45, 100, 284, 356, 364
Standard, 11, 43, 48, 51, 74, 125, 131–133, 136, 140, 142–143, 165, 189, 205, 214, 216, 218, 243, 246–249, 253, 258, 274, 294–295, 308–312, 316–318, 321, 344, 347–348, 356, 359–360
Surveillance, 260, 270
SWAP, 238–239, 364
SWAP LITE, 239

T

T1 line, 18, 41–43
T1E1, 56, 127, 131, 140, 142, 146, 292, 294–295, 308–310
TA '96, 15, 17–18, 25, 34, 220
TCI Group, 333
TCP/IP, 10, 73, 195, 350, 357
TDMA, 109–110, 173, 180–181, 187–189, 239, 241, 364
Telecommunication Act. See TA '96
Telecommuting, 36, 89, 134, 263, 267, 292
Teleconferencing, 24, 273, 326
Teledesic, 6, 193, 196, 202–206
Tele-immersion, 278
Telemedicine, 36, 196, 266–267, 271
Telemetry, 266, 269
Telephone wiring, 237
Time Warner, 4, 16, 151, 271, 291, 313, 315, 332–333, 335, 339, 340
Topology, 54, 87, 89
Traffic shaping, 68, 358
Transformer, 129, 226, 232, 294
Twisted pair, 7, 40, 51, 106–107, 124–128, 134, 137, 248, 252, 292–294, 296, 302, 351

U

UAWG, 310
UBR, 69–70, 118, 261–264, 358
UCAID, 84–85
UDP, 25, 358
Unbundling, 16–17, 19–21, 42, 267, 358
UNI, 53, 56–57, 62, 68, 113, 319, 359, 364
Unspecified Bit Rate, 69, 261, 263
  See also UBR
UPC, 67–68, 358, 364
Uplink, 204–205, 209, 217, 219
Upstream, 3–4, 92, 95, 109–110, 113–118,
  120, 124, 129, 131–132, 134–135, 138,
  140, 142, 144–145, 149–150, 157, 160–
  164, 168–169, 173–175, 177–178, 181–
  182, 184–185, 187–189, 224, 229, 292–
  301, 316
Upstream bandwidth, 93, 115, 300
Upstream channel, 169, 178
Upstream data, 168
Upstream direction, 117, 160, 169
Upstream PMD, 173
Upstream RF, 315
Upstream spectrum, 149
Upstream transmission, 149, 169
US West, 14, 332
User access, 200, 205, 208
UUNET, 337–338

V

Variable Bit Rate, 60, 69, 258, 261–263,
  359, 364

  See also VBR
vBNS, 82, 84–85, 103, 323
VBR, 60, 69, 118, 171, 182, 188–189, 261–
  263, 299–300, 343, 358–359, 364
VDSL, 3, 21, 41, 47, 89, 110, 112–113, 121,
  123, 125, 127, 129, 130, 135, 143–150,
  152, 293, 308–310, 332
Video telephony, 93, 266–268, 321
Viper, 274
Virtual classrooms, 271
Visible Human, 277
VocalTec, 24
VOD, 91, 133–134, 266–267, 270, 274, 292,
  296, 306, 359
Voice Over IP, 94
VON, 324, 325

W

WDM, 98, 109–110, 114
White House, 74, 79, 266, 275, 301, 322
Work at home, 36, 186, 266, 291
WTO, 26
WWW, 48, 134, 181, 292, 298, 302, 349,
  359–360

X

X.25, 48, 348, 353, 359–360
XIWT, 269–270, 286, 323–324